# Sorting Things Out

**Inside Technology**
edited by Wiebe E. Bijker, W. Bernard Carlson, and Trevor Pinch

# Sorting Things Out
## Classification and Its Consequences

Geoffrey C. Bowker
Susan Leigh Star

The MIT Press
Cambridge, Massachusetts
London, England

First MIT Press paperback edition, 2000

© 1999 Massachusetts Institute of Technology

Set in New Baskerville by Wellington Graphics.

Printed and bound in the United States of America.

Library of Congress Cataloging-in-Publication Data

Bowker, Geoffrey C.
  Sorting things out : classification and its consequences /
Geoffrey C. Bowker, Susan Leigh Star.
    p. cm. — (Inside technology)
  Includes bibliographical references (p.  ) and index.
    ISBN 978-0-262-02461-7 (hc. : alk. paper) — 978-0-262-52295-3 (pb. : alk. paper)
    1. Knowledge, Sociology of.  2. Classification.  I. Star, Susan
Leigh, 1954–  .  II. Title.  III. Series.
  BD175.B68   1999
  001′.01′2—dc21                                                99-26894
                                                                      CIP

20  19  18  17  16  15  14  13  12

For Allan Regenstreif and Adele Clarke
chosen family
hors de catégorie

# Contents

# *Acknowledgments*

This project has taken several years and spanned two countries and several institutions. We have, therefore, many people to thank, and in these few pages we will do those who we mention scant justice.

None of the work below would have been possible without the support of the National Science Foundation (NSF EVS grant no. SBR 9514744).

We would like to thank our colleagues on the Classification Project, University of Illinois: Stefan Timmermans, Niranjan Karnik, Laura Neumann, Jesper Döpping, Theresa Chi Lin, and Randi Markussen for ongoing discussions and insight. Stefan and Laura assisted with the interviews at Iowa. The Graduate School of Library and Information Science at the University of Illinois gave us continual support—we thank all faculty members and doctoral students there for their unfailing intellectual generosity. We also thank the Advanced Information Technologies Group, University of Illinois, for project support for parts of the study, especially Robert Alun Jones.

We are much indebted to the members of the Nursing Interventions Project at the University of Iowa. In particular, Joanne McCloskey and Gloria Bulechek provided many helpful suggestions during conversations. We would further like to thank the following Iowa team members who have graciously allowed us to interview them: Laurie Ackerman, Sally Blackman, Gloria Bulechek, Joan Carter, Jeanette Daly, Janice Denehy, Bill Donahue, Chris Forcucci, Orpha Glick, Mary Kanak, Vicki Kraus, Tom Kruckeberg, Meridean Maas, Joanne McCloskey, Barbara Rakel, Marita Titler, Bonnie Wakefield, and Huibin Yue. Bill Donahue was a great support in facilitating access to the Nursing Interventions Classification (NIC) list-serve archives and in helping us during our trips to Iowa.

Earlier versions of several of the chapters have been published as follows: chapter 1 "How Things (actor-net)Work: Classification, Magic, and the Ubiquity of Standards", in a special issue of the Danish philosophical journal *Philosophia* titled "Thinking in the World—Humans, Things, Nature", 25 (3–4), 1997: 195–220; chapter 3 "The History of Information Infrastructures—the Case of the International Classification of Diseases", in *Information Processing and Management*, special issue on the history of information science, 32(1), 1996, 49–61; chapter 4 "Situations vs. Standards in Long-Term, Wide-Scale Decision Making: the Case of the International Classification of Diseases", *Proceedings of the Twenty-Fourth Annual Hawaii International Conference on System Science*, 1991: 73–81; chapter 5 Susan Leigh Star and Geoffrey Bowker. 1997. "Of Lungs and Lungers: The Classified Story of Tuberculosis", *Mind, Culture and Activity* 4: 3–23. Reprinted in Anselm Strauss and Juliet Corbin, ed. 197–227, *Grounded Theory in Practice*, Thousand Oaks, CA: SAGE, 1997; chapter 7 "Infrastructure and Organizational Transformation: Classifying Nurses' Work" in Wanda Orlikowski et al. (eds.), *Information Technology and Changes in Organizational Work*, London: Chapman and Hall, 1996, 344–370; and chapter 8 "Lest We Remember: Organizational Forgetting and the Production of Knowledge", in *Accounting, Management and Information Technology*, 7 (3) 1997: 113–138.

Robert Dale Parker suggested several helpful references on the literary background of women and disease in the nineteenth century; Helen Watson Verran and Marc Berg gave us very useful comments on earlier drafts of sections of this book. Conversations with Kari Thoresen about the notion of texture in organizations were very helpful, as were ongoing conversations with the late Anselm Strauss about trajectory. The work of Mark Casey Condon on the nature of time morality in a men's homeless shelter was helpful in thinking through issues in the chapter on tuberculosis. For various parts of the argument, we received most helpful comments from Ann Bishop, Emily Ignacio, Bill Anderson, Susan Anderson, Howard S. Becker, Isabelle Baszanger, Nick Burbules, Kathy Addelson, Dick Boland, Chuck Goodwin, Chuck Bazerman, Cheris Kramarae, Donna Haraway, Linnea Dunn, Bruno Latour, JoAnne Yates, Gail Hornstein, Ina Wagner, Joan Fujimura, Alberto Cambrosio, Jürg Strübing, John Law, John Bowers, Kjeld Schmidt, Kari Thoresen, Niranjan Karnik, Karen Ruhleder, Emily Ignacio, Joseph Goguen, Mike Lynch, Charlotte Linde, Marc Berg, Adele Clarke, Alice Robbin, Ole Hanseth, John

Garrett, Mike Robinson, Tone Bratteteig, Rogers Hall, Susan Newman, Susanne Bødker, Allan Riegenstreif, and Jean Lave. This raw list cannot do justice to their help. Benoit Malin was a constant source of inspiration. Ann Fagot-Largeault was a great guide through the material philosophy of classification, both in her magnificent book on the International Classification of Diseases (ICD) and in person. John King, Dick Boland, and Bill Turner provided a great deal of help in thinking through the wider implications of our work. Kay Tomlinson gave a careful and generous reading of the final manuscript. Brad Allen provided invaluable research assistance and a gift of energy. Doug Nelson provided support and healing.

Xerox PARC provided us with office space and valuable collegial input during a sabbatical in 1997–98; we are grateful to Lucy Suchman, Julian Orr, Susan Newman, David Levy, Randy Trigg, and Jeannette Blomberg.

The United Nations and WHO archives both provided hospitality and materials; notably Liisa Fagerlund and Sahil Mandel gave of their time and energy. The South Africa collection of the Hoover Institution, Stanford University was most welcoming. The library at the University of Illinois, Urbana Champaign was a great treasure.

The authors gratefully acknowledge the following for permission to use sources:

• The South African Institute of Race Relations for permission to reproduce the information in tables 6.1 and 6.2. From Muriel Horrell. 1969. *A Survey of Race Relations.*

• Institute for Operations Research and the Management Sciences (INFORMS) for permission to reprint material for table 1.1 from Susan Leigh Star and Karen Ruhleder 1996. "Steps Toward an Ecology of Infrastructure: Design and Access for Large Information Spaces," *Information Systems Research* 17: 111–134.

• Willem Struik of Kagiso Publishers, Pretoria, for permission to reprint the photograph in figure 6.3 from Bruwer, J., J. Grobbelaar and H. van Zyl. 1958. *Race Studies (Differentiated Syllabus) for Std VI*, Voortrekkerpers, Johannesburg, and Alexander Butchart for providing a copy of the photograph.

• Times Media Picture Library, Times Media, Ltd. of Johannesburg for permission to reprint the photograph in figure 6.2.

• Bill Poole of Leco and Associates, Pittsburgh, for permission to reprint their advertisement "monitoring the classification: the recovery room," figure 3.2.

• The Hoover Institution for its help in locating and reproducing the photograph in figure 6.1. The photo was originally in a pamphlet in their collection, "The Fight for Freedom in South Africa and What It Means for Workers in the United States," produced by Red Sun Press Publications, nd.

• Mosby Publishers, St. Louis, for permission to reproduce illustrations from the *Nursing Interventions Classification,* eds. Joanne McCloskey and Gloria Bulecheck, second edition, figures 7.1 to 7.5.

• The World Health Organization, Geneva, for permission to reproduce charts from the international classification of diseases, ICD-9 and ICD-10, figures 2.1 and 2.2.

• The picture of the Wilkinson's family (father, mother and two children) appears courtesy of Terry Shean/Sunday Times, Johannesburg.

### Envoi

We would hate to have to assign a Dewey classification number to this book, which straddles sociology, anthropology, history and information systems, and design. Our modest hope is that it will not find its way onto the fantasy shelves.

# Introduction: To Classify Is Human

In an episode of *The X-Files*, a television show devoted to FBI investigations of the paranormal, federal agents Mulder and Scully investigated a spate of murders of psychics of all stamps: palm readers, astrologers, and so forth. The plot unfolded thusly: The murderer would get his fortune read or astrological chart done, and then brutaly slay the fortune-teller. It emerged during the show that the reason for these visits was that he wanted to understand what he was doing and why he was doing it, and he thought psychics could help him understand his urges to kill people. Only one psychic, an insurance salesman with the ability to scry the future, was able to prdict his murderous attacks and recognize the criminal. When finally the murderer met this psychic, he burst into his impassioned plea for an explanation of what he was doing. "Why am I compelled to kill all these people," the salesman responded in a world-weary tone such as one might take with a slow child: "Don't you get it, son? You're a homicidal maniac." The maniac was delighted with this insight. He then proceeds to try to kill again. The salesman's answer is both penetrating and banal—what it says about classification systems is the topic of this book. Why is it so funny?

Our lives are henged round with systems of classification, limned by standard formats, prescriptions, and objects. Enter a modern home and you are surrounded by standards and categories spanning the color of paint on the walls and in the fabric of the furniture, the types of wires strung to appliances, the codes in the building permits allowing the kitchen sink to be properly plumbed and the walls to be adequately fireproofed. Ignore these forms at your peril—as a building owner, be sued by irate tenants; as an inspector, risk malpractice suits denying your proper application of the ideal to the case at hand; as a parent, risk toxic paint threatening your children.

To classify is human. Not all classifications take formal shape or are standardized in commercial and bureaucratic products. We all spend large parts of our days doing classification work, often tacitly, and we make up and use a range of ad hoc classifications to do so. We sort dirty dishes from clean, white laundry from colorfast, important email to be answered from e-junk. We match the size and type of our car tires to the amount of pressure they should accept. Our desktops are a mute testimony to a kind of muddled folk classification: papers that must be read by yesterday, *but that have been there since last year;* old professional journals that really should be read and even in fact may someday be, *but that have been there since last year;* assorted grant applications, tax forms, various work-related surveys and forms waiting to be filled out for everything from parking spaces to immunizations. These surfaces may be piled with sentimental cards that are already read, *but which cannot yet be thrown out,* alongside reminder notes to send similar cards to parents, sweethearts, or friends for their birthdays, all piled on top of last year's calendar (which—who knows?—may be useful at tax time). Any part of the home, school, or workplace reveals some such system of classification: medications classed as not for children occupy a higher shelf than safer ones; books for reference are shelved close to where we do the Sunday crossword puzzle; door keys are color-coded and stored according to frequency of use.

What sorts of things order these piles, locations, and implicit labels? We have certain knowledge of these intimate spaces, classifications that appear to live partly in our hands—definitely not just in the head or in any formal algorithm. The knowledge about which thing will be useful at any given moment is embodied in a flow of mundane tasks and practices and many varied social roles (child, boss, friend, employee). When we need to put our hands on something, it is there.

Our computer desktops are no less cluttered. Here the electronic equivalent of "not yet ready to throw out" is also well represented. A quick scan of one of the author's desktops reveals eight residual categories represented in the various folders of email and papers: "fun," "take back to office," "remember to look up," "misc.," "misc. correspondence," "general web information," "teaching stuff to do," and "to do." We doubt if this is an unusual degree of disarray or an overly prolific use of the "none of the above" category so common to standardized tests and surveys.

These standards and classifications, however imbricated in our lives, are ordinarily invisible. The formal, bureaucratic ones trail behind them the entourage of permits, forms, numerals, and the sometimes-visible work of people who adjust them to make organizations run smoothly. In that sense, they may become more visible, especially when they break down or become objects of contention. But what *are* these categories? Who makes them, and who may change them? When and why do they become visible? How do they spread? What, for instance, is the relationship among locally generated categories, tailored to the particular space of a bathroom cabinet, and the commodified, elaborate, expensive ones generated by medical diagnoses, government regulatory bodies, and pharmaceutical firms?

Remarkably for such a central part of our lives, we stand for the most part in formal ignorance of the social and moral order created by these invisible, potent entities. Their impact is indisputable, and as Foucault reminds us, inescapable. Try the simple experiment of ignoring your gender classification and use instead whichever toilets are the nearest; try to locate a library book shelved under the wrong Library of Congress catalogue number; stand in the immigration queue at a busy foreign airport without the right passport or arrive without the transformer and the adaptor that translates between electrical standards. The material force of categories appears always and instantly.

At the level of public policy, classifications such as those of regions, activities, and natural resources play an equally important role. Whether or not a region is classified as ecologically important, whether another is zoned industrial or residential come to bear significantly on future economic decisions. The substrate of decision making in this area, while often hotly argued across political camps, is only intermittently visible. Changing such categories, once designated, is usually a cumbersome, bureaucratically fraught process.

For all this importance, classifications and standards occupy a peculiar place in studies of social order. Anthropologists have studied classification as a device for understanding the cultures of others— categories such as the raw and the cooked have been clues to the core organizing principles for colonial Western understandings of "primitive" culture. Some economists have looked at the effects of adopting a standard in those markets where networks and compatibility are crucial. For example, videotape recorders, refrigerators, and personal computer software embody arguably inferior technical standards, but standards that benefited from the timing of their historical entry into the marketplace. Some historians have examined the explosion of natural history and medical classifications in the late nineteenth century, both as a political force and as an organizing rubric for complex bureaucracies. A few sociologists have done detailed studies of individual categories linked with social movements, such as the diagnosis of homosexuality as an illness and its demedicalization in the wake of gay and lesbian civil rights. Information scientists work every day on the design, delegation, and choice of classification systems and standards, yet few see them as artifacts embodying moral and aesthetic choices that in turn craft people's identities, aspirations, and dignity.[1] Philosophers and statisticians have produced highly formal discussions of classification theory, but few empirical studies of use or impact.

Both within and outside the academy, single categories or classes of categories may also become objects of contention and study. The above-mentioned demedicalization of the category homosexual in the American Psychiatric Association's (APA) *Diagnostic and Statistical Manual 3* (the DSM, a handbook of psychiatric classification) followed direct and vigorous lobbying of the APA by gay and lesbian advocates (Kirk and Kutchins 1992). During this same era, feminists were split on the subject of whether the categories of premenstrual syndrome and postpartum depression would be good or bad for women as they became included in the DSM. Many feminist psychotherapists were engaged in a bitter argument about whether to include these categories. As Ann Figert (1996) relates, they even felt their own identities and professional judgments to be on the line. Allan Young (1995) makes the complicating observation that psychiatrists increasingly use the language of the DSM to communicate with each other and their accounting departments, although they frequently do not believe in the categories they are using.

More recently, as discussed in chapter 6, the option to choose multiple racial categories was introduced as part of the U.S. government's routine data-collection mission, following Statistical Directive 15 in October 1997. The Office of Management and Budget (OMB) issued the directive; conservatively, its implementation will cost several million dollars. One direct consequence is the addition of this option to the U.S. census, an addition that was fraught with political passion. A march on Washington concerning the category took the traditional ultimate avenue of mass protest for American activists. The march was conducted by people who identified themselves as multiracial, and their families and advocates. At the same time, it was vigorously opposed by many African-American and Hispanic civil rights groups (among several others), who saw the option as a "whitewash" against which important ethnic and policy-related distinctions would be lost (Robbin 1998).

Despite the contentiousness of some categories, however, none of the above-named disciplines or social movements has systematically addressed the pragmatics of the invisible forces of categories and standards in the modern built world, especially the modern information technology world. Foucault's (1970; 1982) work comes the closest to a thoroughgoing examination in his arguments that an archaeological dig is necessary to find the origins and consequences of a range of social categories and practices. He focused on the concept of order and its implementation in categorical discourse. The ubiquity described by Foucault appears as an iron cage of bureaucratic discipline against a broad historical landscape. But there is much more to be done, both empirically and theoretically. No one, including Foucault, has systematically tackled the question of how these properties inform social and moral order via the new technological and electronic infrastructures. Few have looked at the creation and maintenance of complex classifications as a kind of work practice, with its attendant financial, skill, and moral dimensions. These are the tasks of this book.

Foucault's practical archaeology is a point of departure for examining several cases of classification, some of which have become formal or standardized, and some of which have not. We have several concerns in this exploration, growing both from the consideration of classification work and its attendant moral dimensions. First, we seek to understand the role of invisibility in the work that classification does in ordering human interaction. We want to understand how these categories are made and kept invisible, and in some cases, we want to challenge the silences surrounding them. In this sense, our job here is to find tools for seeing the invisible, much as Émile Durkheim passionately sought to convince his audience of the material force of the social fact—to see that society was not just an idea—more than 100 years ago (Durkheim 1982).

The book also explores systems of classification as part of the built information environment. Much as a city planner or urban historian would leaf back through highway permits and zoning decisions to tell a city's story, we delve the dusty archives of classification design to understand better how wide-scale classification decisions have been made.

We have a moral and ethical agenda in our querying of these systems. Each standard and each category valorizes some point of view and silences another. This is not inherently a bad thing—indeed it is inescapable. But it *is* an ethical choice, and as such it is dangerous—not bad, but dangerous. For example, the decision of the U.S. Immigration and Naturalization Service to classify some races and classes as desirable for U.S. residents, and others as not, resulted in a quota system that valued affluent people from northern and western Europe over those (especially the poor) from Africa or South America. The decision to classify students by their standardized achievement and aptitude tests valorizes some kinds of knowledge skills and renders other kinds invisible. Other types of decisions with serious material force may not immediately appear as morally problematic. The collective standardization in the United States on VHS videotapes over Betamax, for instance, may seem ethically neutral. The classification and standardization of types of seed for farming is not obviously fraught with moral weight. But as Busch (1995) and Addelson (1994) argue, such long-term, collective forms of choice are also morally weighted. We[2] are used to viewing moral choices as individual, as dilemmas, and as rational choices. We have an impoverished vocabulary for collective moral passages, to use Addelson's terminology. For any individual, group or situation, classifications and standards give advantage or they give suffering. Jobs are made and lost; some regions benefit at the expense of others. How these choices are made, and how we may think about that invisible matching process, is at the core of the ethical project of this work.

### Working Infrastructures

*Sorting Things Out* stands at the crossroads of the sociology of knowledge and technology, history, and information science. The categories represented on our desktops and in our medicine cabinets are fairly ad hoc and individual, not even legitimate anthropological folk or ethno classifications. They are not often investigated by information scientists (but see Kwasnik 1988, 1991; Beghtol 1995; Star 1998). But everyone uses and creates them in some form, and they are (increasingly) important in organizing computer-based work. They often have old and deep historical roots. True, personal information managers are designed precisely to make this process transparent, but even with their aid, the problem continues: we still must design or select categories, still enter data, still struggle with things that do not fit. At the same time, we rub these ad hoc classifications against an increasingly elaborate large-scale system of formal categories and standards. Users of the Internet alone navigate, now fairly seamlessly, more than 200 formally elected Internet standards for information transmission each time they send an email message. If we are to understand larger scale classifications, we also need to understand how desktop classifications link up with those that are formal, standardized, and widespread.

Every link in hypertext creates a category. That is, it reflects some judgment about two or more objects: they are the same, or alike, or functionally linked, or linked as part of an unfolding series. The rummage sale of information on the World Wide Web is overwhelming, and we all agree that finding information is much less of a problem than assessing its quality—the nature of its categorical associations and by whom they are made (Bates, in press). The historical cultural model of social classification research in this book, from desktop to wide-scale infrastructure, is a good one through which to view problems of indexing, tracking, and even compiling bibliographies on the Web. In its cultural and workplace dimensions, it offers insights into the problematics of design of classification systems, and a lens for examining their impact. It looks at these processes as a sort of crafting of treaties. In this, a cross-disciplinary approach is crucial. Any information systems design that neglects use and user semantics is bound for trouble down the line—it will become either oppressive or irrelevant. Information systems mix up the conventional and the formal, the hard technical problems of storage and retrieval with the hard interactional problems of querying and organizing.

Information systems are undergoing rapid change. There is an explosion of information on the Web and associated technologies, and fast moving changes in how information may converge across previously disparate families of technology—for instance, using one's television to retrieve email and browse the Web, using one's Internet connections to make telephone calls. Whatever we write here about the latest electronic developments will be outdated by the time this book sees print, a medium that many would argue is itself anachronistic.

Conventions of use and understandings of the impact of these changes on social organization are slower to come. The following example illustrates the intermingling of the conventional and the local in the types of classificatory links formed by hypertext. A few years ago, our university was in the enviable position of having several job openings in library and information science. Both the authors were on the search committee. During the process of sifting through applications and finding out more about candidates, the need arose to query something on the candidate's resume. We used the Alta Vista search engine to find the candidate's email address. (Of course, the first thing one really does with Alta Vista is ego surfing—checking one's own name to see how many times it appears on the Web—but we had already done that.) His email address and formal institutional home page appeared in about fifteen seconds on our desktop, but so did his contributions to a discussion on world peace, a feminist bulletin board, and one of the more arcane alt.rec Usenet groups. We found ourselves unable to stop our eyes from roving through the quoted Usenet posts—category boundaries surely never meant to be crossed by a job search committee. Fortunately for us as committee members, we interpreted what we found on the Web as evidence that the applicant was a more well rounded person than his formal CV resume had conveyed. He became a more interesting candidate.

But of course, it might have gone badly for him. In less than a minute we had accessed information about him that crossed a social boundary of de facto privacy, access, and awareness context (Glaser and Strauss 1965). The risk of random readership had been there in some sense when he posted to a public space, but who on a search committee in the old days of a couple of years ago could possibly be bothered searching listserv archives? Who would have time? There are many ethical and etiquette-related questions here, of course, with the right to privacy not least among them. The incident also points to the fact that as a culture we have not yet developed conventions of classification for the Web that bear much moral or habitual conviction in daily practice. The label alt.rec does not yet have the reflex power that the label private does on a desk drawer or notebook cover. We would never open someone's desk drawer or diary. We are not usually known to be rude people, but we have not yet developed or absorbed routine similar politeness for things such as powerful Web search engines. We were thus somewhat embarrassed and confused about the morality of mentioning the alt.rec postings to the committee.

As we evolve the classifications of habit—grow common fingertips with respect to linkages and networks—we will be faced with some choices. How standardized will our indexes become? What forms of freedom of association (among people, texts and people, and texts) do we want to preserve and which are no longer useful? Who will decide these matters?

## Investigating Infrastructure

People do many things today that a few hundred years ago would have looked like magic. And if we don't understand a given technology today it looks like magic: for example, we are perpetually surprised by the mellifluous tones read off our favorite CDs by, we believe, a laser. Most of us have no notion of the decades of negotiation that inform agreement on, inter alia, standard disc size, speed, electronic setting, and amplification standards. It is not dissimilar to the experience of magic one enjoys at a fine restaurant or an absorbing play. Common descriptions of good waiters or butlers (one thinks of Jeeves in the Wodehouse stories) are those who clear a table and smooth the un- folding of events "as if by magic." In a compelling play, the hours of rehearsal and missteps are disappeared from center stage, behind a seamless front stage presentation. Is the magic of the CD different from the magic of the waiter or the theater ensemble? Are these two kinds of magic or one—or none?

This book is an attempt to answer these questions, which can be posed more prosaically as:

• What work do classifications and standards do? Again, we want to look at what goes into making things work like magic: making them fit together so that we can buy a radio built by someone we have never met in Japan, plug it into a wall in Champaign, Illinois, and hear the world news from the BBC.

• Who does that work? We explore the fact that all this magic involves much work: there is a lot of hard labor in effortless ease.[3] Such invisible work is often not only underpaid, it is severely underrepresented in theoretical literature (Star and Strauss 1999). We will discuss where all the "missing work" that makes things look magical goes.

• What happens to the cases that do not fit? We want to draw attention to cases that do not fit easily into our magical created world of stan- dards and classifications: the left handers in the world of right-handed magic, chronic disease sufferers in the acute world of allopathic medi- cine, the vegetarian in MacDonald's (Star 1991b), and so forth.

These are issues of great import. It is easy to get lost in Baudrillard's (1990) cool memories of simulacra. He argues that it is impossible to sort out media representations from "what really happens." We are unable to stand outside representation or separate simulations from nature. At the same time, he pays no attention to the work of constructing the simulations, or the infrastructural considerations that underwrite the images or events (and we agree that separating them ontologically is a hopeless task). The hype of our postmodern times is that we do not need to think about this sort of *work* any more. The real issues are scientific and technological, stripped of the conditions of production—in artificial life, thinking machines, nanotechnology, and genetic manipulation. . . . Clearly each of these *is* important. But there is more at stake—epistemologically, politically, and ethically—in the day-to-day work of building classification systems and producing and maintaining standards than in abstract arguments about representation. Their pyrotechnics may hold our fascinated gaze, but they cannot provide any path to answering our moral questions.

### Two Definitions: Classification and Standards

Up to this point, we have been using the terms classification and standardization without formal definition. Let us clarify the terms now.

### Classification
*A classification is a spatial, temporal, or spatio-temporal segmentation of the world.* A "classification system" is a set of boxes (metaphorical or literal) into which things can be put to then do some kind of work—bureaucratic or knowledge production. In an abstract, ideal sense, a classification system exhibits the following properties:

1. *There are consistent, unique classificatory principles in operation.*   One common sort of system here is the *genetic* principle of ordering. This refers not to DNA analysis, but to an older and simpler sense of the word: classifying things by their origin and descent (Tort 1989). A genealogical map of a family's history of marriage, birth, and death is genetic in this sense (even for adopted children and in-laws). So is a flow chart showing a hierarchy of tasks deriving from one another over time. There are many other types of classificatory principles—sorting correspondence by date received (temporal order), for example, or recipes by those most frequently used (functional order).

2. *The categories are mutually exclusive.*   In an ideal world, categories are clearly demarcated bins, into which any object addressed by the system will neatly and uniquely fit. So in the family genealogy, one mother and one father give birth to a child, forever and uniquely attributed to them as parents—there are no surrogate mothers, or

---

### *What* **Are** *You?*

I grew up in Rhode Island, a New England state largely populated by Italian-Americans and French-Canadians that is known chiefly for its small stature. When I was a kid in our neighborhood, the first thing you would ask on encountering a newcomer was "what's your name?" The second was "what are you?" "What are you" was an invitation to recite your ethnic composition in a kind of singsong voice: 90 percent of the kids would say "Italian with a little bit of French," or "half-Portuguese, one-quarter Italian and one-quarter Armenian." When I would chime in with "half-Jewish, one-quarter Scottish and one-quarter English," the range of responses went from very puzzled looks to "does that mean you're not Catholic?" Wherein, I guess, began my fascination with classification, and especially with the problem of residual categories, or, the "other," or not elsewhere classified.

—Leigh Star

---

issues of shared custody or of retrospective DNA testing. A rose is a rose, not a rose sometimes and a daisy other times.

3. *The system is complete.* With respect to the items, actions, or areas under its consideration, the ideal classification system provides total coverage of the world it describes. So, for example, a botanical classifier would not simply ignore a newly discovered plant, but would always strive to name it. A physician using a diagnostic classification must enter *something* in the patient's record where a category is called for; where unknown, the possibility exists of a medical discovery, to be absorbed into the complete system of classifying.

No real-world working classification system that we have looked at meets these "simple" requirements and we doubt that any ever could. In the case of unique classificatory systems, people disagree about their nature; they ignore or misunderstand them; or they routinely mix together different and contradictory principles. A library, for example, may have a consistent Library of Congress system in place, but supplement it in an ad hoc way. Best sellers to be rented out to patrons may be placed on a separate shelf; very rare, pornographic, or expensive books may be locked away from general viewing at the discretion of the local librarian. Thus, the books are moved, without being formally reclassified, yet carry an additional functional system in their physical placement.

For the second point, mutual exclusivity may be impossible in practice, as when there is disagreement or ambivalence about the membership of an object in a category. Medicine is replete with such examples, especially when the disease entity is controversial or socially stigmatized. On the third point, completeness, there may be good reasons to ignore data that would make a system more comprehensive. The discovery of a new species on an economically important development site may be silenced for monetary considerations. An anomaly may be acknowledged, but be too expensive—politically or bureaucratically—to introduce into a system of record keeping. In chapter 2, we demonstrate ways of reading classification systems so as to be simultaneously sensitive to these conceptual, organizational, and political dimensions.

Consider the International Classification of Diseases (ICD), which is used as a major example throughout this book. The full title of the current (tenth) edition of the ICD, is: "ICD-10—International Statistical Classification of Diseases and Related Health Problems; Tenth Revision." Note that it is designated a statistical classification: Only diseases that are statistically significant are entered here (it is not an attempt to classify all diseases).

The ICD is labeled a "classification," even though many have said that it is a "nomenclature" since it has no single classificatory principle (it has at least four, which are not mutually exclusive, a point developed in chapter 4). A nomenclature simply means an agreed-upon naming scheme, one that need not follow any classificatory principles. The nomenclature of streets in Paris, for example, includes those named after intellectual figures, plants and trees, battles, and politicians, as well as those inherited from former governments, such as Rue de Lutèce (Lutèce was the ancient Roman name for Paris). This is no classificatory system. Nomenclature and classification are frequently confused, however, since attempts are often made to model nomenclature on a single, stable system of classification principles, as for example with botany (Bowker, in press) or anatomy. In the case of the ICD, diagnostic nomenclature and the terms in the ICD itself were conflated in the American system of diagnosis-related groups (DRGs), much to the dismay of some medical researchers. In many cases the ICD represents a compromise between conflicting schemes." The terms used in categories C82–C85 for non-Hodgkin's lymphomas are those of the Working Formulation, which attempted to find common ground among several major classification systems. The terms used in these schemes are not given in the Tabular List but appear in the Alphabetical Index; exact equivalence with the terms appearing in the Tabular List is not always possible" (ICD-10, 1: 215).

The ICD, however, presents itself clearly as a classification scheme and not a nomenclature. Since 1970, there has been an effort under- way by the WHO to build a distinct International Nomenclature of Diseases (IND), whose main purpose will be to provide: "a single recommended name for every disease entity" (ICD-10, 1: 25).

For the purposes of this book, we take a broad enough definition so that anything consistently called a classification system *and treated as such* can be included in the term. This is a classic Pragmatist turn— things perceived as real are real in their consequences (Thomas and Thomas 1917). If we took a purist or formalist view, the ICD would be a (somewhat confused) nomenclature and who knows what the IND would represent. With a broad, Pragmatic definition we can look at the work that is involved in building and maintaining a family of entities that people call classification systems rather than attempt the Herculean, Sisyphian task of purifying the (un)stable systems in place. Howard Becker makes a cognate point here:

Epistemology has been a . . . negative discipline, mostly devoted to saying what you shouldn't do if you want your activity to merit the title of science, and to keeping unworthy pretenders from successfully appropriating it. The sociol- ogy of science, the empirical descendant of epistemology, gives up trying to decide what should and shouldn't count as science, and tells what people who claim to be doing science do. (Becker 1996, 54–55)

The work of making, maintaining, and analyzing classification systems is richly textured. It is one of the central kinds of work of modernity, including science and medicine. It is, we argue, central to social life.

### Standards

Classifications and standards are closely related, but not identical. While this book focuses on classification, standards are crucial compo- nents of the larger argument. The systems we discuss often do become standardized; in addition, a standard is in part a way of classifying the world. What then are standards? The term as we use it in the book has several dimensions:

1. A "standard" is any set of agreed-upon rules for the production of (textual or material) objects.

2. A standard spans more than one community of practice (or site of activity). It has temporal reach as well in that it persists over time.

3. Standards are deployed in making things work together over distance and heterogeneous metrics. For example, computer protocols for Internet communication involve a cascade of standards (Abbate and Kahin 1995) that need to work together well for the average user to gain seamless access to the web of information. There are standards for the components to link from your computer to the phone network, for coding and decoding binary streams as sound, for sending messages from one network to another, for attaching documents to messages, and so forth.

4. Legal bodies often enforce standards, be these mandated by professional organizations, manufacturers' organizations, or the state. We might say tomorrow that volapük, a universal language that boasted some twenty-three journals in 1889 (Proust 1989, 580), or its successor Esperanto shall henceforth be the standard language for international diplomacy. Without a mechanism of enforcement, however, or a grassroots movement, we shall fail.

5. There is no natural law that the best standard shall win—QWERTY, Lotus 123, DOS, and VHS are often cited as examples in this context. The standards that do win may do so for a variety of other reasons: they build on an installed base, they had better marketing at the outset, or they were used by a community of gatekeepers who favored their use. Sometimes standards win due to an outright conspiracy, as in the case of the gas refrigerator documented by Cowan (1985).

6. Standards have significant inertia and can be very difficult and expensive to change.

It was possible to build a cathedral like Chartres without standard representations (blueprints) and standard building materials such as regular sizes for stones, tools, and so forth (Turnbull 1993). People invented an amazing array of analog measuring devices (such as string lengths). Each cathedral town posted the local analog metric (a length of metal) at its gates, so that peripatetic master builders could calibrate their work to it when they arrived in the town. They did not have a wide-scale measurement system such as our modern metric or decimal systems. (Whether as a result of this local improvisation or not, Turnbull notes, many cathedrals did fall down!)

It is no longer possible to build a complex collective project without standardized measurements. Consider a modern housing development where so much needs to come together from distant and proximate sources—electricity, gas, sewer, timber sizes, screws, nails and so on. The control of standards is a central, often underanalyzed feature of economic life (see the work of Paul David—for example David and Rothwell 1994—for a rich treatment). It is key to knowledge production as well. Latour (1987) speculates that far more economic resources are spent creating and maintaining standards than in producing "pure" science. There are a number of histories of standards that point to the development and maintenance of standards as being critical to industrial production.

*labor*

At the same time, just as with classifications, these dimensions of standards are in some sense idealized. They embody goals of practice and production that are never perfectly realized, like Plato's triangles. The process of building to a standardized code, for example, usually includes a face-to-face negotiation between builder(s) and inspector(s), which itself includes a history of relations between those people. Small deviations are routinely overlooked, unless the inspector is making a political point. The idiom "good enough for government use" embodies the common-sense accommodations of the slip between the ideal standard and the contingencies of practice.

In this and in many other ways, then, classifications and standards are two sides of the same coin. Classifications may or may not become standardized. If they do not, they are ad hoc, limited to an individual or a local community, and/or of limited duration. At the same time, every successful standard imposes a classification system, at the very least between good and bad ways of organizing actions or things. And the work-arounds involved in the practical use of standards frequently entail the use of ad hoc nonstandard categories. For example, a patient may respond to a standardized protocol for the management of chronic back pain by approximating the directions and supplementing them with an idiosyncratic or alternative medical classification scheme. If the protocol requires a number of exercises done three times a day, patients may distinguish good days from bad days, vacation days from working days, and only do the exercises when they deem them necessary.

Classifications and standards are related in another sense, which concerns the use of a classification by more than one social world or community of practice, and the impact that use has on questions of membership and the taken-for-grantedness of objects (Cambrosio and Keating 1995). Throughout this book, we speak of classifications as objects for cooperation across social worlds, or as boundary objects (Star and Griesemer 1989). Drawing from earlier studies of interdisciplinary scientific cooperation, we define boundary objects as those objects that both inhabit several communities of practice *and* satisfy the informational requirements of each of them. In working practice, they are objects that are able both to travel across borders and maintain some sort of constant identity. They can be tailored to meet the needs of any one community (they are plastic in this sense, or customizable). At the same time, they have common identities across settings. This is achieved by allowing the objects to be weakly structured in common use, imposing stronger structures in the individual-site tailored use. They are thus both ambiguous and constant; they may be abstract or concrete. In chapter 9, we explore in detail the abstract ramifications of the use of classifications by more than one community and the connection with the emergence of standards.

### The Structure of This Book

To explore these questions, we have written a first chapter detailing some key themes of the work to follow. We have then divided the middle of the book into three parts, which look at several classification systems. We have structured these studies around three issues in turn: classification and large-scale infrastructures (part I), classification and biography (part II), and classification and work practice (part III). Weaving these three themes together, we can explore the texture of the space within which infrastructures work and classification systems from different worlds meet, adjust, fracture, or merge. In two concluding chapters, we elaborate some theoretical conclusions from these studies.

### Part I: Classification and Large-Scale Infrastructures

Classification systems are integral to any working infrastructure. In part I (chapters 2 to 4) we examine how a global medical classification system was developed to serve the conflicting needs of multiple local, national, and international information systems.

Our investigation here begins in the late nineteenth century with another kind of information explosion—the development of myriad systems of classification and standardization of modern industrial and scientific institutions.

In the nineteenth century people learned to look at themselves as surrounded by tiny, invisible things that have the power of life or death: microbes and bacteria. They learned to teach their children to wash their hands of germs before eating, and later, to apply antiseptic salve to a cat scratch or an inflamed fingernail. Company washrooms sprouted signs admonishing employees to wash hands before returning to work, especially if they worked with food served to others. In this period, people also learned how to perform surgery that would not usually be fatal and how to link gum disease with bacteria between the teeth.

At the same time they learned these practices about germs, another ubiquitous set of tiny, invisible things were being negotiated and sewn into the social fabric. These were formal, commodified classifications and standards, both scientific and commercial. People classified, measured, and standardized just about everything—animals, human races, books, pharmaceutical products, taxes, jobs, and diseases. The categories so produced lived in industry, medicine, science, education, and government. They ranged from the measurement of machine tools to the measurement of people's forearms and foreheads. The standards were sometimes physically tiny measures: how big should a standard size second of time be, an eyeglass screw, or an electrical pulse rate?[4] At other times, they were larger: what size should a railroad car be, a city street, or a corporation? Government agencies, industrial consortia, and scientific committees created the standards and category systems. So did mail-order firms, machine-tool manufacturers, animal breeders, and thousands of other actors. Most of these activities became silently embodied in the built environment and in notions of good practice. The decisions taken in the course of their construction are forever lost to the historical record. In fact, their history is considered by most to be boring, trivial, and unworthy of investigation.

There are some striking similarities to our own late twentieth-century historical moment in that faced by Europeans at the end of the nineteenth century. A new international information-sharing and gathering movement was starting, thanks to the advent of wide-scale international travel, international quasigovernmental governance structures, and a growing awareness that many phenomena (like epidemics and markets) would not be confined to one country. In the nineteenth century, for the first time people faced large numbers of bodies and their microbes moving rapidly across national borders and between large bureaucracies—and at an unprecedented rate. Especially in the case of epidemics, international public health became an urgent necessity. Attempts to control these passengers represent one of the first large-scale western medical classification schemes: ships that

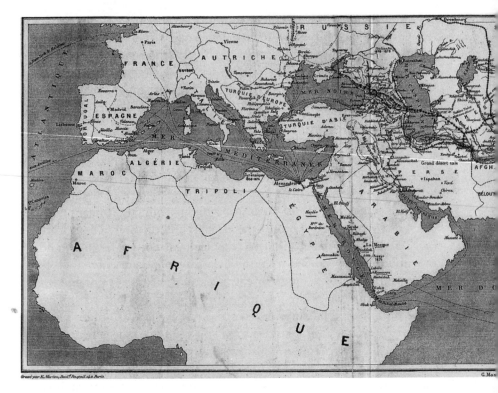

**Figure I.1**
Map indicating the geographical distribution of the sources of cholera and
"the progress of cholera epidemics" by land and sea routes. The progression
by land is shown by the line with small vertical marks (1823–1847), by sea in
1865 via ship, and new progressions overland from 1892. Note the sea routes
marked between Mecca and Marseilles.
Source: A. Proust 1892.

called at ports on the way back from Mecca had to follow a period of
quarantine during which anyone infected would become sympto-
matic—thus emulating the slower timeline of horse or camel travel (see
figure I.1).

After quarantine, one was given a "clean bill of health" and allowed
freedom of transport. This was a costly delay for the ships, and so a
black market in clean bills of health appeared shortly thereafter . . . .
The problem of tracking who was dying of what and where on earth
became a permanent feature of international bureaucracy (see figure
I.2).

Constructing such a list may appear to be to us a comparatively
straightforward task, once the mechanisms for reporting were in place.

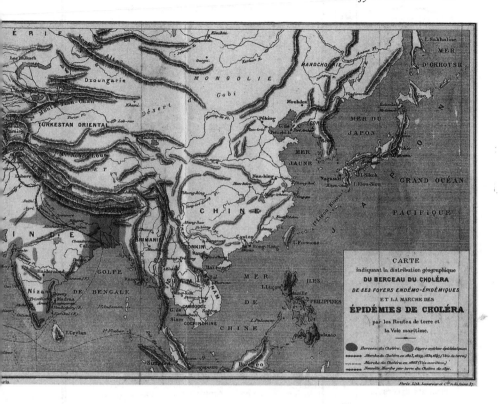

CARTE
indiquant la distribution géographique
**DU BERCEAU DU CHOLÉRA**
*DE SES FOYERS ENDÉMO-ÉPIDÉMIQUES*
ET LA MARCHE DES
**ÉPIDÉMIES DE CHOLÉRA**
par les Routes de terre et
la Voie maritime.

For over 100 years, however, there has never been consensus about disease categories or about the process of collecting data. So one culture sees spirit possession as a valid cause of death, another ridicules this as superstition; one medical specialty sees cancer as a localized phenomenon to be cut out and stopped from spreading, another sees it as a disorder of the whole immune system that merely manifests in one location or another. The implications for both treatment and classification differ. Trying to encode both causes results in serious information retrieval problems.

In addition, classifications shift historically. In Britain in 1650 we find that 696 people died of being "aged"; 31 succumbed to wolves, 9 to grief, and 19 to "King's Evil." "Mother" claimed 2 in 1647 but none in 1650, but in that year 2 were "smothered and stifled" (see figure I.3). Seven starved in 1650 (Graunt 1662), but by 1930 the WHO would make a distinction: if an adult starved to death it was a misfortune; if a child starved, it was homicide. Death by wolf alone becomes impossible by 1948, where death from animals is divided between venomous and nonvenomous, and only dogs and rats are singled out for categories of their own (ICD-5 1948, 267).

N°

## PATENTE DE SANTÉ

Nom du bâtiment... .
Nature du bâtiment..
Pavillon.............
Tonneaux...........
Canons.............
Appartenant au port d
Destination .........
Nom du capitaine.....
Nom du médecin......
Équipage (tout compris)...............
Passagers...........
Cargaison........ ...
État hygiénique du navire...............
État hygiénique de l'équipage (couchage, vêtements, etc.)....
État hygiénique des passagers ........ .
Vivres et approvisionnements divers.....
Eau...............

Malades à bord {

État { du port.....
sanitaire { des environs

Il a été constaté dans le port ou ses environs pendant la dernière semaine écoulée :

...... cas de choléra.
...... cas de fièvre jaune.
...... cas de peste.

Délivrée le du mois
d 189 ,
à heure du .

*ADMINISTRATION SANITAIRE DE FRANCE*

---

N°     RÉPUBLIQUE FRANÇAISE     PORT

### ADMINISTRATION SANITAIRE

## PATENTE DE SANTÉ

Nous, de la santé à
certifions que le bâtiment ci-après désigné part de ce port dans les conditions suivantes, dûment constatées :

Nom du bâtiment.....
Nature du bâtiment...
Pavillon .............
Tonneaux............
Canons...............
Appartenant au port d
Destination ..........
Nom du capitaine.....
Nom du médecin......
Équipage (tout compris)
Passagers ...........
Cargaison.... .......

Malades à bord {

État hygiénique du navire................
État hygiénique de l'équipage (couchage, vêtements, etc.).....
État hygiénique des passagers ..........
Vivres et approvisionnements divers......
Eau..................

Conformément aux articles 30, 31, 32 et 33 du règlement, l'état sanitaire du navire a été vérifié, la visite médicale a été passée au moment de l'embarquement des passagers et il a été constaté qu'il n'existait à bord, *au moment du départ*, aucun malade atteint d'affection pestilentielle (choléra, fièvre jaune, peste), ni linge sale, ni substance susceptible de nuire à la santé du bord.

Nous certifions, en outre, { du port est..........
que l'état sanitaire { des environs est.....

et qu'il a été constaté dans le { .....cas de choléra
port (ou ses environs) pendant { .....cas de fièvre jaune
la dernière semaine écoulée { .....cas de peste

En foi de quoi, nous avons délivré la présente patente, à , le du mois d
189 , à heure du

L'Expéditionnaire          *Sceau de l'Administration,*
de la Patente,

LE     DE LA SANTÉ,

---

### PRESCRIPTIONS EXTRAITES DU RÈGLEMENT GÉNÉRAL
#### DE POLICE SANITAIRE MARITIME

VOIR AU VERSO.

*Figure I.2*
French bill of health. An original "clean bill of health."
Source: A. Proust 1892.

The first part of this book is dedicated to understandig the construction of the International Classification of Diseases (ICD): a classification scheme with its origins in the late nineteenth century but still present today—indeed, it is ubiquitous in medical bureaucracy and medical information systems. The ICD constitutes an impressive attempt to coordinate information and resources about mortality and morbidity globally. For the background research for understanding international processes of classification, we went to Geneva and studied the archives of the WHO and its predecessors such as the League of Nations and the Office Internationale d'Hygiène Publique. Roughly every ten years since the 1890s, the ICD has been revised. The UN and the WHO have kept some records of the process of revision; others are to be found in the file cabinets of individuals involved in the revision process.

What we found was not a record of gradually increasing consensus, but a panoply of tangled and crisscrossing classification schemes held together by an increasingly harassed and sprawling international public health bureaucracy. Spirit possession and superstition never do reconcile, but for some data to be entered on the western-oriented death certificate, it becomes possible from the WHO point of view for a death to be assigned the category "nonexistent disease."

One of the other major influences on keeping medical records has been insurance companies, as we discuss in chapter 4. As the working lives of individuals became more closely tied up with the state and its occupational health concerns, the classification of work-related diseases (including industrial accidents) became very important. Life expectancy measures were equally important, both for estimating the available labor force and for basic planning measures. Of course, occupational and nonwork related medical classifications did not always line up: companies might have been reluctant to take responsibility for unsafe working conditions, latency in conditions such as asbestosis makes data hard to come by; thus there may have been moral conflicts about the cause of such illnesses.

In similar fashion, any classification that touched on religious or ethical questions (and surprisingly many do so) would be disputed. If life begins at the moment of conception, abortion is murder and a fetus dead at three months is a stillbirth, encoded as a live infant death. Contemporary abortion wars in the United States and western Europe attest to the enduring and irreconcilable ontologies involved in these codifications.

| The Years of our Lord | 1647 | 1648 | 1649 | 1650 | 1651 | 1652 | 1653 | 1654 | 1655 | 1656 | 1657 | 1658 |
|---|---|---|---|---|---|---|---|---|---|---|---|---|
| Abortive and Stil-born | 335 | 329 | 327 | 351 | 389 | 381 | 384 | 433 | 483 | 419 | 463 | 467 |
| Aged | 916 | 835 | 889 | 696 | 780 | 834 | 864 | 974 | 743 | 892 | 869 | 1176 |
| Ague and Fever | 1260 | 884 | 751 | 970 | 1038 | 1212 | 282 | 1371 | 689 | 875 | 999 | 1800 |
| Apoplex and Suddenly | 68 | 74 | 64 | 74 | 106 | 111 | 118 | 86 | 92 | 102 | 113 | 138 |
| Bleach | | | 1 | 3 | 7 | 2 | | | 1 | | | |
| Blasted | | | | | 6 | 6 | | | 4 | | 5 | 5 |
| Bleeding | 4 | 1 | | | | | | | 7 | 3 | 5 | 4 |
| Bloody Flux, Scouring and Flux | 3 | 2 | 5 | 1 | 3 | 4 | 3 | 2 | | | 5 | |
| | 155 | 176 | 802 | 289 | 833 | 762 | 200 | 386 | 168 | 368 | 362 | 233 |
| Burnt and Scalded | 3 | 6 | 10 | 5 | 11 | 8 | 5 | 7 | 10 | 5 | 7 | 4 |
| Calenture | 1 | | | 1 | | 2 | 1 | 1 | | | 3 | |
| Cancer, Gangrene and Fistula | 26 | 29 | 31 | 19 | 31 | 53 | 36 | 37 | 73 | 31 | 24 | 35 |
| Wolf | | | | 8 | | | | | | | | |
| Canker, Sore-mouth and Thrush | 66 | 28 | 54 | 42 | 68 | 51 | 53 | 72 | 44 | 81 | 19 | 27 |
| Child-bed | 161 | 106 | 114 | 117 | 206 | 213 | 158 | 192 | 177 | 201 | 236 | 225 |
| Chrisoms and Infants | 1369 | 1254 | 1065 | 990 | 1237 | 1280 | 1050 | 1343 | 1089 | 1393 | 1162 | 1144 |
| Colick and Wind | 103 | 71 | 85 | 82 | 76 | 102 | 80 | 101 | 85 | 120 | 113 | 179 |
| Cold and Cough | | | | | | | 41 | 36 | 21 | 58 | 30 | 31 |
| Consumption and Cough | 2423 | 2200 | 2388 | 1988 | 2350 | 2410 | 2286 | 2868 | 2606 | 3184 | 2757 | 3610 |
| Convulsion | 684 | 491 | 530 | 493 | 569 | 653 | 606 | 828 | 702 | 1027 | 807 | 841 |
| Cramp | | | 1 | | | | | | | | | |
| Cut of the Stone | | 2 | 1 | 3 | | 1 | 1 | 2 | 4 | 1 | 3 | 5 |
| Dropsie and Tympany | 185 | 434 | 421 | 508 | 444 | 556 | 617 | 704 | 660 | 706 | 631 | 931 |
| Drowned | 47 | 40 | 30 | 27 | 49 | 50 | 53 | 30 | 43 | 49 | 63 | 60 |
| Excessive drinking | | | 2 | | | | | | | | | |
| Executed | 8 | 17 | 29 | 43 | 24 | 12 | 19 | 21 | 19 | 22 | 20 | 18 |
| Fainted in a Bath | | | | | 1 | | | | | | | |
| Falling-Sickness | 3 | 2 | 2 | 3 | | 3 | 4 | 1 | 4 | 3 | 1 | |
| Flox[1] and small Pox | 139 | 400 | 1190 | 184 | 525 | 1279 | 139 | 812 | 1294 | 823 | 835 | 409 |
| Found dead in the Streets | 6 | 6 | 9 | 8 | 7 | 9 | 14 | 4 | 3 | 4 | 9 | 11 |
| French-Pox | 18 | 29 | 15 | 18 | 21 | 20 | 20 | 20 | 29 | 23 | 25 | 53 |
| Frighted | 4 | 4 | 1 | | 3 | | 2 | | 1 | 1 | | |
| Gout | 9 | 5 | 12 | 9 | 7 | 7 | 5 | 6 | 8 | 7 | 8 | 13 |
| Grief | 12 | 13 | 16 | 7 | 17 | 14 | 11 | 17 | 10 | 13 | 10 | 12 |
| Hanged, and made-away themselves | 11 | 10 | 13 | 14 | 9 | 14 | 15 | 9 | 14 | 16 | 24 | 18 |
| Head-Ach | | 1 | 11 | 2 | | 2 | 6 | 6 | 5 | 3 | 4 | 5 |
| Jaundice | 57 | 35 | 39 | 49 | 41 | 43 | 57 | 71 | 61 | 41 | 46 | 77 |
| Jaw-faln | 1 | 1 | | | 3 | | | | 2 | 2 | 3 | 1 |
| Impostume | 75 | 61 | 65 | 59 | 80 | 105 | 79 | 90 | 92 | 122 | 80 | 134 |
| Itch | | 1 | | | | | | | | | | |
| Killed by several Accidents | 27 | 57 | 39 | 94 | 47 | 45 | 57 | 58 | 52 | 43 | 52 | 47 |
| King's Evil | 27 | 26 | 22 | 19 | 22 | 20 | 26 | 26 | 27 | 24 | 23 | 28 |
| Lethargy | 3 | 4 | 2 | 4 | 4 | 4 | 3 | 10 | 9 | 4 | 6 | 2 |
| Leprosie | | | 1 | | | | | | | | | 1 |
| Liver-grown, Spleen and Rickets | 53 | 46 | 56 | 59 | 65 | 72 | 67 | 65 | 52 | 50 | 38 | 51 |
| Lunatick | 12 | 18 | 6 | 11 | 7 | 11 | 9 | 12 | 6 | 7 | 13 | 5 |
| Meagrom | 12 | 13 | | 5 | 8 | 6 | 6 | 14 | 3 | 6 | 7 | 6 |
| Measles | 5 | 92 | 3 | 33 | 33 | 62 | 8 | 52 | 11 | 153 | 15 | 80 |
| Mother | 2 | | | | | 1 | 1 | 2 | 2 | 3 | | 3 |
| Murdered | 3 | 2 | 7 | 5 | 4 | 3 | 3 | 3 | 9 | 6 | 5 | 7 |
| Overlaid and Starved at Nurse | 25 | 22 | 36 | 28 | 28 | 29 | 30 | 36 | 58 | 53 | 44 | 50 |
| Palsie | 27 | 21 | 19 | 20 | 23 | 20 | 29 | 18 | 22 | 23 | 20 | 22 |
| ▸Plague | 3597 | 611 | 67 | 15 | 23 | 16 | 6 | 16 | 9 | 6 | 4 | 14 |
| Plague in the Guts | | | | 1 | | | 110 | 32 | | 87 | 315 | 446 |
| Pleurisie | 30 | 26 | 13 | 20 | 23 | 19 | 17 | 23 | 10 | 9 | 17 | 16 |
| Poisoned | | 3 | | 7 | | | | | | | | |
| Purples and Spotted Fever | 145 | 47 | 43 | 65 | 54 | 60 | 75 | 89 | 56 | 52 | 56 | 126 |
| Quinsie and Sore-throat | 14 | 11 | 12 | 17 | 24 | 20 | 18 | 9 | 15 | 13 | 7 | 10 |
| Rickets | 150 | 224 | 216 | 190 | 260 | 329 | 229 | 372 | 347 | 458 | 317 | 476 |
| Mother, rising of the Lights | 150 | 92 | 115 | 120 | 134 | 138 | 135 | 178 | 166 | 212 | 203 | 228 |
| Rupture | 16 | 7 | 7 | · 6 | 7 | 16 | 7 | 15 | 11 | 20 | 19 | 18 |
| Scal'd head | 2 | | | | 1 | | | | 2 | | | |
| Scurvy | 32 | 20 | 21 | 21 | 29 | 43 | 41 | 44 | 103 | 71 | 82 | 82 |
| Smothered and stifled | | | 2 | | | | | | | | | |
| Sores, ulcers, broken and bruised | 15 | 17 | 17 | 16 | 26 | 32 | 25 | 32 | 23 | 34 | 40 | 47 |
| Shot (Limbs | | | | | | | | | | | | |
| Spleen | 12 | 17 | | | | | 13 | 13 | | 6 | 2 | 5 |
| Shingles | | | | | | | | | | | | 1 |
| Starved | | 4 | 8 | 7 | 1 | 2 | 1 | 1 | 3 | 1 | 3 | 6 |
| Stitch | | | | | | | | | | | | |
| Stone and Strangury | 45 | 42 | 29 | 28 | 50 | 41 | 44 | 38 | 49 | 57 | 72 | 69 |
| Sciatica | | | | | | | | | | | | |
| Stopping of the Stomach | 29 | 29 | 30 | 33 | 55 | 67 | 66 | 107 | 94 | 145 | 129 | 277 |
| Surfet | 217 | 137 | 136 | 123 | 104 | 177 | 178 | 212 | 161 | 137 | 218 | |
| Swine-Pox | 4 | 4 | 3 | | | | 1 | 4 | 2 | 1 | 1 | 1 |
| Teeth and Worms | 767 | 597 | 540 | 598 | 709 | 905 | 691 | 1131 | 803 | 1198 | 878 | 1036 |
| Tissick | 62 | 47 | | | | | | | | | 57 | 66 |
| Thrush | | | | | | | | | | | 66 | |
| Vomiting | 1 | 6 | 3 | 7 | 4 | 6 | 3 | 14 | 7 | 27 | 16 | 19 |
| Worms | 147 | 107 | 105 | 65 | 85 | 86 | 53 | | | | | |
| Wen | 1 | | 1 | | 2 | 2 | | | 1 | | 1 | 2 |
| Suddenly | | | | | | | | | | | | |

[1] Probably a name for confluent small p

## Figure I.3
The table of casualties, England in the seventeenth century.
Source: J. Graunt 1662.

| 1659 | 1660 | 1629 | 1630 | 1631 | 1632 | 1635 | 1634 | 1635 | 1636 | 1629 1630 1631 1632 | 1633 1634 1635 1636 | 1647 1648 1649 1650 | 1651 1652 1653 1654 | 1655 1656 1657 1658 | 1629 1649 1659 | In 20 Years. |
|---|---|---|---|---|---|---|---|---|---|---|---|---|---|---|---|---|
| 421 | 544 | 499 | 439 | 410 | 445 | 500 | 475 | 507 | 523 | 1793 | 2005 | 1342 | 1587 | 1832 | 1247 | 8559 |
| 909 | 1095 | 579 | 712 | 661 | 671 | 704 | 623 | 794 | 714 | 2475 | 2814 | 3336 | 3452 | 3680 | 2377 | 15759 |
| 2303 | 2148 | 956 | 1091 | 1115 | 1108 | 953 | 1279 | 1622 | 2360 | 4418 | 6235 | 3865 | 4903 | 4363 | 4010 | 23784 |
| 91 | 67 | 22 | 36 | | 17 | 24 | 35 | 26 | | 75 | 85 | 280 | 421 | 445 | 177 | 1306 |
| | | | | | | | | | | | | 4 | 9 | 1 | 1 | 15 |
| 3 | 8 | 13 | 8 | 10 | 13 | 6 | 4 | | 4 | 54 | 14 | 5 | 12 | 14 | 16 | 99 |
| 7 | 2 | 5 | 2 | 5 | 4 | 4 | 3 | | | 16 | 7 | 11 | 12 | 19 | 17 | 65 |
| 346 | 251 | 449 | 438 | 352 | 348 | 278 | 512 | 346 | 330 | 1587 | -1466 | 1422 | 2181 | 1161 | 1597 | 7818 |
| 6 | 6 | 3 | 10 | 7 | 5 | 1 | 3 | 12 | 3 | 25 | 19 | 24 | 31 | 26 | 19 | 125 |
| | | | | | | | | 1 | 3 | | 4 | 2 | 4 | 3 | | 13 |
| 63 | 52 | 20 | 14 | 23 | 28 | 27 | 30 | 24 | 30 | 85 | 112 | 105 | 157 | 150 | 114 | 609 |
| 73 | 68 | 6 | 4 | 4 | 1 | | | 5 | 74 | 15 | 79 | 190 | 244 | 161 | 133 | 689 |
| | | | | | | | | | | | 8 | | | | | 8 |
| 226 | 194 | 150 | 157 | 112 | 171 | 132 | 143 | 163 | 230 | 590 | 668 | 498 | 769 | 839 | 490 | 3364 |
| 858 | 1123 | 2596 | 2378 | 2035 | 2268 | 2130 | 2315 | 2113 | 1895 | 9277 | 8453 | 4678 | 4910 | 4788 | 4519 | 32106 |
| 116 | 167 | 48 | 57 | | | | | 37 | 50 | 105 | 87 | 341 | 359 | 497 | 247 | 1389 |
| 33 | 24 | 10 | 58 | 51 | 55 | 45 | 54 | 50 | 57 | 174 | 207 | 00 | 77 | 140 | 43 | 598 |
| 2982 | 3414 | 1827 | 1910 | 1713 | 1797 | 1754 | 1955 | 2080 | 2477 | 5157 | 8266 | 8999 | 9914 | 12157 | 7197 | 44487 |
| 742 | 1031 | 52 | 87 | 18 | 241 | 221 | 386 | 418 | 709 | 498 | 1734 | 2198 | 2656 | 3377 | 1324 | 9073 |
| | | | | 1 | 0 | 0 | 0 | 0 | 0 | 01 | 00 | 01 | 0 | 0 | 1 | 2 |
| 6 | 4 | | | | 5 | 1 | 5 | 2 | 2 | 5 | 10 | 6 | 4 | 13 | 47 | 38 |
| 646 | 872 | 235 | 252 | 279 | 280 | 266 | 250 | 329 | 389 | 1048 | 1734 | 1538 | 2321 | 2982 | 1302 | 9623 |
| 57 | 48 | 43 | 33 | 29 | 34 | 37 | 32 | 32 | 45 | 139 | 147 | 144 | 182 | 215 | 130 | 827 |
| | | | | | | | | | | | | 2 | | | 2 | 2 |
| 7 | 18 | 19 | 13 | 12 | 18 | 13 | 13 | 13 | 13 | 62 | 52 | 97 | 76 | 79 | 55 | 384 |
| | | | | | | | | | | | | 1 | | | | 1 |
| 4 | 5 | 3 | 10 | 7 | 7 | 2 | 5 | 6 | 8 | 27 | 21 | 10 | 8 | 8 | 9 | 74 |
| 1523 | 354 | 72 | 40 | 58 | 531 | 72 | 1354 | 293 | 127 | 701 | 1846 | 1913 | 2755 | 3361 | 2785 | 10576 |
| 2 | 6 | 18 | 33 | 20 | 6 | 13 | 8 | 24 | 24 | 83 | 69 | 29 | 34 | 27 | 29 | 243 |
| 51 | 31 | 17 | 12 | 12 | 12 | 7 | 17 | 12 | 22 | 53 | 48 | 80 | 81 | 130 | 83 | 392 |
| | 9 | 1 | | | 1 | | | | 3 | 2 | 3 | 9 | 5 | 2 | 2 | 21 |
| 14 | 2 | 2 | 5 | 3 | 4 | 4 | 5 | 7 | 8 | 14 | 24 | 35 | 25 | 36 | 28 | 134 |
| 13 | 4 | 18 | 20 | 22 | 11 | 14 | 17 | 5 | 20 | 71 | 56 | 48 | 59 | 45 | 47 | 279 |
| 11 | 36 | 8 | 8 | 6 | 15 | | 3 | 8 | 7 | 37 | 18 | 48 | 47 | 72 | 32 | 222 |
| 35 | 26 | | | | | | | 4 | 2 | 0 | 6 | 14 | 14 | 17 | 46 | 051 |
| 102 | 76 | 47 | 59 | 35 | 43 | 35 | 45 | 54 | 63 | 184 | 197 | 180 | 212 | 225 | 188 | 998 |
| | | 10 | 16 | 13 | 8 | 10 | 10 | 4 | 11 | 47 | 35 | 02 | 5 | 6 | 10 | 95 |
| 105 | 96 | 58 | 76 | 73 | 74 | 50 | 62 | 73 | 130 | 282 | 315 | 260 | 35 | 428 | 228 | 1639 |
| | | | | | | 10 | | | | 00 | 10 | 01 | | | | 11 |
| 55 | 47 | 54 | 55 | 47 | 46 | 49 | 41 | 51 | 60 | 202 | 201 | 217 | 207 | 194 | 148 | 1021 |
| 28 | 54 | 16 | 25 | 18 | 38 | 35 | 20 | 20 | 69 | 97 | 150 | 94 | 94 | 102 | 66 | 537 |
| 6 | 4 | 1 | | 2 | 2 | 3 | | 2 | 2 | 5 | 7 | 13 | 21 | 21 | 9 | 67 |
| | 2 | 2 | | | | | | 2 | | 2 | 2 | 1 | | 1 | 3 | 06 |
| 8 | 15 | 94 | 12 | 99 | 87 | 82 | 77 | 98 | 99 | 392 | 356 | 213 | 269 | 191 | 158 | 1421 |
| 14 | 14 | 6 | 11 | 6 | 5 | 4 | 2 | 2 | 5 | 28 | 13 | 47 | 39 | 31 | 26 | 158 |
| 5 | 4 | | | 24 | | | | | 22 | 24 | 22 | 30 | 34 | 22 | 05 | 132 |
| 6 | 74 | 42 | 2 | 3 | 80 | 21 | 33 | 27 | 12 | 127 | 83 | 133 | 155 | 259 | 51 | 757 |
| 1 | 8 | 1 | | | | | | | 3 | 01 | 3 | 2 | 4 | 8 | 02 | 18 |
| 70 | 20 | | | 3 | 7 | | 6 | 5 | 8 | 10 | 1. | 17 | 13 | 27 | 77 | 86 |
| 46 | 43 | 4 | 10 | 13 | 7 | 8 | 14 | 10 | 14 | 34 | 46 | 111 | 123 | 215 | 86 | 529 |
| 17 | 21 | 17 | 23 | 17 | 25 | 14 | 21 | 17 | 25 | 82 | 77 | 87 | 90 | 87 | 53 | 423 |
| 36 | 14 | | 1317 | 274 | 8 | | 1 | | 10400 | 1599 | 10401 | 4290 | 61 | 33 | 103 | 16384 |
| 253 | 402 | | | | | | | | | 00 | 00 | 61 | 142 | 844 | 253 | 991 |
| 12 | 10 | 26 | 24 | 26 | 36 | 21 | | 45 | 24 | 112 | 90 | 89 | 72 | 52 | 51 | 415 |
| | | | | | | 2 | | | 2 | 00 | 4 | 10 | 00 | 00 | 00 | 14 |
| 368 | 146 | 32 | 58 | 58 | 38 | 24 | 125 | 245 | 397 | 186 | 791 | 300 | 278 | 290 | 243 | 1845 |
| 21 | 14 | 01 | 8 | 6 | 7 | 24 | 04 | 5 | 22 | 22 | 55 | 54 | 71 | 45 | 34 | 247 |
| 441 | 521 | | | | | 14 | 49 | 50 | 00 | 113 | 780 | 1190 | 1598 | 657 | 3681 |
| 210 | 249 | 44 | 72 | 99 | 98 | 60 | 84 | 72 | 104 | 309 | 220 | 777 | 585 | 809 | 369 | 2700 |
| 12 | 28 | 2 | 6 | 4 | 9 | 4 | 3 | 10 | 13 | 21 | 30 | 36 | 45 | 68 | 21 | 201 |
| | | | | | | | | | | | | 2 | 1 | 2 | | 05 |
| 95 | 12 | 5 | 7 | 9 | | 9 | | 00 | 25 | 33 | 34 | 94 | 132 | 300 | 115 | 593 |
| | | | 24 | | | | | | | 24 | | 2 | | | 2 | 26 |
| 61 | 48 | 23 | | 20 | 48 | 19 | 19 | 22 | 29 | 91 | 89 | 65 | 115 | 144 | 141 | 504 |
| 7 | 20 | | | | | | | | | | | | | | 07 | 27 |
| 7 | 7 | | | | | | | | | | | 29 | 26 | 13 | 07 | 68 |
| 1 | | | | | | 1 | | | | | | | | | 1 | 2 |
| 7 | 14 | | | | | | | | | 14 | | 19 | 5 | 13 | 29 | 51 |
| | | | | | | | | | | | | | | | | 1 |
| 22 | 30 | 35 | 39 | 58 | 50 | 58 | 49 | 33 | 45 | 114 | 185 | 144 | 173 | 247 | 51 | 937 |
| | 2 | | | | 1 | 3 | | 1 | 6 | 1 | 4 | | | | | 13 |
| 186 | 214 | | | | | | | | 6 | | 6 | 121 | 295 | 247 | 216 | 669 |
| 202 | 192 | 63 | 157 | 149 | 86 | 104 | 114 | 132 | 371 | 445 | 721 | 613 | 671 | 644 | 401 | 3094 |
| 2 | | 5 | 8 | 4 | 6 | 3 | | 10 | | 23 | 13 | 11 | | 5 | 10 | 57 |
| 839 | 1008 | 440 | 506 | 335 | 470 | 432 | 454 | 539 | 1207 | 1751 | 2632 | 2502 | 3436 | 3915 | 1819 | 14236 |
| | | 8 | 12 | 14 | 34 | 23 | 15 | 27 | | 68 | 65 | 109 | | | 8 | 242 |
| 8 | 10 | 15 | 23 | 17 | 40 | 28 | 31 | 34 | | 95 | 93 | | | 123 | 15 | 211 |
| | | 1 | 4 | 1 | 1 | 2 | 5 | 6 | 3 | 7 | 16 | 17 | 27 | 69 | 12 | 136 |
| 1 | 1 | 19 | 31 | 28 | 27 | 19 | 28 | 27 | | 105 | 74 | 424 | 224 | | 124 | 830 |
| | | | | 1 | | 4 | | | | 1 | 4 | 2 | 4 | 4 | 2 | 15 |
| | | 63 | 59 | 37 | 62 | 58 | 62 | 78 | 34 | 221 | 233 | | | | 63 | 454 |
| | | | | | | | | | | | | | | | 34190 | 229250 |

ox. See Creighton, i., 462—463.

This Table to face page 406.

For a bureaucracy to establish a smooth data collection effort, a means must be found to detour around such higher order issues. The statistical committee discussed in chapter 4, assigned with determining the exact moment of the beginning of life by number of attempted breaths and weight of fetus or infant, cuts a Solomon-like figure against such a disputed landscape. At the same time, there is an element of reductionist absurdity here—how many breaths equals "life"? If not specified, another source of quality control for data is lost; if specified, it appears to make common sense ironic. This is an issue we will revisit as well in the discussion of nursing interventions, in chapter 7.

Algorithms for codification do not resolve the moral questions involved, although they may obscure them. For decades, priests, feminists, and medical ethicists on both sides have debated the question of when a human life begins. The moral questions involved in encoding such information—and the politics of certainty and of voice involved—are much more obscure.

Forms like the death certificate, when aggregated, form a case of what Kirk and Kutchins (1992) call "the substitution of precision for validity" (see also Star 1989b). That is, when a seemingly neutral data collection mechanism is substituted for ethical conflict about the contents of the forms, the moral debate is partially erased. One may get ever more precise knowledge, without having resolved deeper questions, and indeed, by burying those questions.

There is no simple pluralistic answer to how such questions may be resolved democratically or with due process. Making all knowledge retrievable, and thus re-debatable, is an appealing solution in a sense from a purely information scientific point of view. From a practical organizational viewpoint, however, this approach fails. For example, in 1927, a manual describing simultaneous causes of death listed some 8,300 terms, which represented 34 million possible combinations that might appear on the face of a death certificate. A complete user manual for filling out the certificate would involve sixty-one volumes of 1,000 pages each. This is clearly not a pragmatic choice for conducting a task that most physicians also find boring, low-status, and clinically unimportant.

As we know from studies of work of all sorts, people do not do the ideal job, but the doable job. When faced with too many alternatives and too much information, they satisfice (March and Simon 1958). As an indicator of this, studies of the validity of codes on death certificates repeatedly show that doctors have favorite categories; these are regionally biased; and autopsies (which are rarely done) have a low rate of agreement with the code on the form (Fagot-Largeault 1989).

Even when there is relatively simple consensus about the cause of death, the act of assigning a classification can be socially or ethically charged. Thus, in some countries the death certificate has two faces: a public certificate handed to the funeral director so that arrangements can be made quickly and discreetly, and a statistical cause filed anonymously with the public health department. In this case, the doctor is not faced with telling the family of a socially unacceptable form of death: syphilis can become heart failure, or suicide can become a stroke. For example, as we discuss in chapter 4, the process of moving to an anonymous statistical record may reveal hidden biases in the reporting of death. Where the death certificate is public, stigma and the desire to protect the feelings of the family may reign over scientific accuracy.

Over the years, those designing the list of causes of death and disease have struggled with all of these problems. One of the simple but important rules of thumb to try to control for this degree of uncertainty is to distribute the residual categories. "Not elsewhere classified" appears throughout the entire ICD, but nowhere as a top-level category. So since uncertainty is inevitable, and its scope and scale essentially unknowable, at least its impact will not hit a single disease or location disproportionately. Its effects will remain as local as possible; the quest for certainty is not lost, but postponed, diluted, and abridged.

With the rise of very-large-scale information systems, the Internet, the Web, and digital libraries, we find that the sorts of uncertainties faced by the WHO are themselves endemic in our own lives. When we use email filters, for example, we risk losing the information that does not fit the sender's category: junk email is very hard to sort out automatically in a reliable way. If we have too many detailed filters, we lose the efficiency sought from the filter in the first place. As we move into desktop use of hyperlinked digital libraries, we fracture the traditional bibliographic categories across media, versions, genres, and author. The freedom entailed is that we can customize our own library spaces; but as Jo Freeman (1972) pointed out in her classic article, "The Tyranny of Structurelessness," this is also so much more work that we may fall into a lowest level convenience classification rather than a high-level semantic one. In one of our digital library projects at Illinois, for example, several undergraduates we interviewed in focus groups stated that they would just get five references for a term paper—any five—since that is what the professor wanted, and references had better be ones that are listed electronically and available without walking across campus.

The ICD classification is in many ways an ideal mirror of how people designing global information schemes struggle with uncertainty, ambiguity, standardization, and the practicalities of data quality. Digging into the archives, and reading the ICD closely through its changes, reveals some of the upstream, design-oriented decisions informing the negotiated order achieved by the vast system of forms, boxes, software, and death certificates. At the same time, we have been constantly aware of the human suffering often occasioned by the apparently bloodless apparatus of paperwork through which these data are collected.

## Part II: Classification and Biography

The second part of this book looks at two cases where the lives of individuals are broken, twisted, and torqued by their encounters with classification systems. This often invisible anguish informs another level of ethical inquiry. Once having been made, the classification systems are applied to individual cases—sometimes resulting in a kind of surreal bureaucratic landscape. Sociologist Max Weber spoke of the "iron cage of bureaucracy" hemming in the lives of modern workers and families. The cage formed by classification systems can be constraining in just this way, although cage might be too impoverished a metaphor to describe its variations and occasional stretches. In chapters 5 and 6 we look at biography and classification. We chose two examples where classification has become a direct tool mediating human suffering. Our first case concerns tuberculosis patients and the impact of disease classification on their lives. We use historical data to discuss the experience of the disease within the tuberculosis asylum.

Tuberculosis patients, like many with chronic illness, live under a confusing regime of categories and metrics (see also Ziporyn 1992). Many people were incarcerated for years—some for decades—waiting for the disease to run its course, to achieve a cure at high altitudes, or to die there. They were subjected to a constant battery of measurements: lung capacity, auscultation, body temperature and pulse rate, x-rays, and, as they were developed, laboratory tests of blood and other bodily fluids. The results of the tests determined the degree of freedom from the sanatorium regime as well as, ultimately, the date of release.

Of no surprise to medical sociologists, the interpretation and negotiations of the tests between doctor and patient were fraught with questions of the social value of the patient (middle-class patients being thought more compliant and reliable when on furlough from the asylum than those from lower classes), with gender stereotypes, and with the gradual adaptation of the patient's biographical expectations to the period of incarceration. Thomas Mann's *The Magic Mountain* and Julius Roth's *Timetables* are full of stories of classification and metrication. We examine how different time lines, and expectations about those time lines, unfold in these two remarkable volumes. Biography, career, the state of the medical art with respect to the disease, and the public health adjudication of tuberculosis are all intertwined against the landscape of the sanatorium.

Life in the sanatorium has a surreal, almost nightmarish quality, as detailed by Mann, Roth, and many other writers throughout the twentieth century. This sense comes precisely from the misalignment of a patient's life expectations, the uncertainties of the disease and of the treatment, and the negotiations laden with other sorts of interactional burdens. It is one thing to be ill and in the hospital with an indefinite release date. It is quite another when the date of release includes one's ability to negotiate well with the physicians, their interpretation of the latest research, and the exigencies of public health forms and red tape. We call this agglomeration *torque,* a twisting of time lines that pull at each other, and bend or twist both patient biography and the process of metrication. When all are aligned, there is no sense of torque or stress; when they pull against each other over a long period, a nightmare texture emerges.

A similar torque is found in the second case in this section, that of race classification and reclassification under apartheid in South Africa. Between 1950 and the fall of apartheid forty years later, South Africans were ruled under an extremely rigid, comprehensive system of race classification. Divided into four main racial groups— white/European, Bantu (black), Asian and coloured (mixed race)— people's lives were rigidly segregated. The segregation extended from so-called petty apartheid (separate bus stops, water fountains, and toilets) to rights of work, residency, education, and freedom of movement. This system became the target of worldwide protest and eventually came to a formal end. These facts are common knowledge. What has been less well documented or publicized are the actual techniques used to classify people by race. In chapter 6, we examine in detail some cases of mixed-race people who applied to be reclassified after their initial racial designation by the state. These borderline cases serve to illuminate the underlying architecture of apartheid. This was a mixture of brute power, confused eugenics, and appropriations of anthropological theories of race. The scientific reason given for apartheid by the white supremacist Nationalist party was "separate development"—the idea that to develop naturally, the races must develop separately.

In pursuing this ideology, of course, people and families that crossed the color barrier were problematic. If a natural scientific explanation was given for apartheid, systematic means should be available to winnow white from black, coloured from black and so on. As the chapter delineates, this attempt was fraught with inconsistencies and local work-arounds, as people never easily fit any categories. Over 100,000 people made formal appeals concerning their race classification; most were denied.

Although it lies at a political extreme, these cases form a continuum with the classification of people at different stages of tuberculosis. In both cases, biographies and categories fall along often conflicting trajectories. Lives are twisted, even torn, in the attempt to force the one into the other. These torques may be petty or grand, but they are a way of understanding the coconstruction of lives and their categories.

### Part III: Classification and Work Practice

In part III, chapters 7 and 8, look at how classification systems organize and are organized by work practice. We examine the effort of a group of nursing scientists based at the University of Iowa, led by Joanne McCloskey and Gloria Bulechek, to produce a classification of nursing interventions. Their Nursing Intervention Classification (NIC) aims at depicting the range of activities that nurses carry out in their daily routines. Their original system consisted of a list of some 336 interventions; each comprised of a label, a definition, a set of activities, and a short list of background readings. Each of those interventions is in turn classified within a taxonomy of six domains and twenty-six classes. For example, one of the tasks nurses commonly perform is preparing and monitoring intravenous medication. The nursing intervention "epidural analgesia administration" is defined as:

"preparation and delivery of narcotic analgesics into the epidural space;" another common one, "cough enhancement," groups activities designed to help respiration.

The Iowa NIC researchers built up their system of nursing interventions inductively. They created a preliminary list that distinguished between nursing interventions and activities, then nurtured a large grassroots network of nursing researchers.[5] This group narrowed the preliminary list of interventions to the original 336 published in NIC and further validated them via surveys and focus groups. Different interventions were reviewed for clinical relevance, and a coding scheme was developed. The classification system grew through a cooperative process, with nurses in field sites trying out categories, and suggesting new ones in a series of regional and specialist meetings. Since 1992 the nurses have added over 50 interventions to their original list. We attended a number of these meetings, and interviewed many of the nurses involved.

Caring work such as calming and educating patients, usually done by nurses, often cuts across specific medical diagnostic categories. The NIC investigators use their list of interventions to make visible and legitimate the work that nurses do. The idea is that it will be used to compare work across hospitals, specialties, and geographical areas, and to build objective research measures for the outcomes. NIC, although still relatively young, promises to be a major rallying point for nurses in the decades to come. Before NIC, much nursing work was invisible to the medical record. As one nurse poignantly said, "we were just thrown in with the cost of the room." Another said, "I am not a bed!" The traditional, quintessential nurse would be ever present, caregiving, and helpful—but not a part of the formal patient-doctor information structure. Of course, this invisibility is bound up with traditional gender roles, as with librarians, social workers, and primary school teachers.

But as with the ICD, classifying events is difficult. In the case of NIC, the politics move from a politics of certainty to a politics of ambiguity. The essence of this politics is walking a tightrope between increased visibility and increased surveillance; between overspecifying what a nurse *should* do and taking away discretion from the individual practitioner.

When discretion and the tacit knowledge that is part of every occupation meet the medical bureaucracy, which would account for every pill and every moment of health care workers' time, contradictions ensue. This is especially true in the "softer" areas of care. Social-psychological care giving is one of the areas where this dilemma is prominent. For example, NIC lists as nursing interventions "anticipatory guidance" and "mood management"—preparation for grief or surgery. Difficult though these are to capture in a classification scheme, one much more difficult is "humor." How can one capture humor as a deliberate nursing intervention? Does sarcasm, irony, or laughter count as a nursing intervention? When do you stop? How to reimburse humor, how to measure this kind of care? No one would dispute the importance of humor, but it is by its nature a situated and subjective action. A grey area of common sense remains for the individual staff nurse to define whether some of the nursing interventions are worth classifying.

There are continuing tensions within NIC between just this kind of common sense and abstracting away from the local to standardize and compare, while at the same time rendering invisible work visible. Nurses' work is often invisible for a combination of good and bad reasons. Nurses have to ask mundane questions, rearrange bedcovers, move a patient's hand so that it is closer to a button, and sympathize about the suffering involved in illness. Bringing this work out into the open and differentiating its components can mean belaboring the obvious or risking being too vague.

One of the battlefields where comparability and control appear as opposing factors is in linking NIC to costs. NIC researchers assert that the classification of nursing interventions will allow a determination of the costs of services provided by nurses and planning for resources needed in nursing practice. As the nurse above says, nursing treatments are usually bundled in with the room price. NIC is used in the development of nursing health care systems and may provide a planning vehicle for previously untracked costs. As we shall see, NIC can also be problematic for nurses. Like any other classification scheme that renders work visible, it can also render surveillance easier—and it could in the end lead to a Tayloristic dissection of the tasks of nursing (as the NIC designers are well aware). So-called unskilled tasks may be taken out of their hands and the profession as a whole may suffer a loss of autonomy and the substitution of rigid procedure for common sense.

As in the case of the ICD, there are many layers of meaning involved in developing and implementing nursing classification. NIC might look like a straightforward organizational tool: it is in fact much more than that. It merges science, practice, bureaucracy, and information systems. NIC coordinates bodies, impairments, charts, reimbursement systems, vocabularies, patients, and health care professionals. Ultimately, it provides a manifesto for nursing as an organized occupation, a basis for a scientific domain, and a tool for organizing work practices.

### Why It Is Important to Study Classification Systems

The sheer density of the collisions of classification schemes in our lives calls for a new kind of science, a new set of metaphors, linking traditional social science and computer and information science. We need a topography of things such as the distribution of ambiguity; the fluid dynamics of how classification systems meet up—a plate tectonics rather than a static geology. This new science will draw on the best empirical studies of work-arounds, information use, and mundane tools such as desktop folders and file cabinets (perhaps peering backwards out from the Web and into the practices). It will also use the best of object-oriented programming and other areas of computer science to describe this territory. It will build on years of valuable research on classification in library and information science.

We end this introduction with a future scenario that symbolizes this abstract endeavor. Imagine that you are walking through a forest of interarticulated branches. Some are covered with ice or snow, and the sun melts their touching tips to reveal space between. Some are so thickly brambled they seem solid; others are oddly angular in nature, like esplanaded trees.

Some of the trees are wild, some have been cultivated. Some are old and gnarled, and some are tiny shoots; some of the old ones are nearly dead, others show green leaves. The forest is still wild, but there are some parks, and some protocols for finding one's way along, at least on the known paths. Helicopters flying overhead can quickly tell you how many types of each tree, even each leaf, there are in the world, but they cannot yet give you a guidebook for bird-watching or forestry management. There is a lot of underbrush and a complex ecology of soil bacteria, flora, and fauna.

Now imagine that the forest is a huge information space and each of the trees and bushes are classification systems. Those who make them up and use them are the animals and plants, and the soil is a mix of the Internet, the paper world, and other communication infrastructures.

Your job is to describe this forest. You may write a basic manual of forestry, or paint a landscape, compose an opera, or improve the maps used throughout. What will your product look like? Who will use it?

In this book, we show from our studies of medical, scientific, and race classification that, like a good forest, some areas will be left wild, or in darkness, or even unmapped (that is, some ambiguity will remain). We will show that abstract schema that do not take use into account—say, maps that leave out landmarks or altitude or how readers use maps—will simply fail. (That is, common sense will be seen as the precious resource that it is.) We intuit that a mixture of scientific, poetic, and artistic talents, such as that represented in the hypertextual world, will be crucial to this task. We will demonstrate the value of a mixture of formal and folk classifications that are used sensibly in the context of people's lives.

# 1

## Some Tricks of the Trade in Analyzing Classification

My guess is that we have a folk theory of categorization itself. It says that things come in well-defined kinds, that the kinds are characterized by shared properties, and that there is one right taxonomy of the kinds.

It is easier to show what is wrong with a scientific theory than with a folk theory. A folk theory defines common sense itself. When the folk theory and the technical theory converge, it gets even tougher to see where that theory gets in the way—or even that it is a theory at all.

*(Lakoff 1987, 121)*

### Introduction: A Good Infrastructure Is Hard to Find

Information infrastructure is a tricky thing to analyze.[6] Good, usable systems disappear almost by definition. The easier they are to use, the harder they are to see. As well, most of the time, the bigger they are, the harder they are to see. Unless we are electricians or building inspectors, we rarely think about the myriad of databases, standards, and instruction manuals subtending our reading lamps, much less about the politics of the electric grid that they tap into. And so on, as many layers of technology accrue and expand over space and time. Systems of classification (and of standardization) form a juncture of social organization, moral order, and layers of technical integration. Each subsystem inherits, increasingly as it scales up, the inertia of the installed base of systems that have come before.

Infrastructures are never transparent for everyone, and their workability as they scale up becomes increasingly complex. Through due methodological attention to the architecture and use of these systems, we can achieve a deeper understanding of how it is that individuals and communities meet infrastructure. We know that this means, at the least, an understanding of infrastructure that includes these points:

- A historical process of development of many tools, arranged for a wide variety of users, and made to work in concert.

- A practical match among routines of work practice, technology, and wider scale organizational and technical resources.

- A rich set of negotiated compromises ranging from epistemology to data entry that are both available and transparent to communities of users.

- A negotiated order in which all of the above, recursively, can function together.

Table 1.1 shows a more elaborate definition of infrastructure, using Star and Ruhleder (1996), who emphasize that one person's infrastructure may be another's barrier.

This chapter offers four themes, methodological points of departure for the analysis of these complex relationships. Each theme operates as a gestalt switch—it comes in the form of an *infrastructural inversion* (Bowker 1994). This inversion is a struggle against the tendency of infrastructure to disappear (except when breaking down). It means learning to look closely at technologies and arrangements that, by design and by habit, tend to fade into the woodwork (sometimes literally!).

Infrastructural inversion means recognizing the depths of interdependence of technical networks and standards, on the one hand, and the real work of politics and knowledge production[8] on the other. It foregrounds these normally invisible Lilliputian threads and furthermore gives them causal prominence in many areas usually attributed to heroic actors, social movements, or cultural mores. The inversion is similar to the argument made by Becker (1982) in his book *Art Worlds*. Most history and social analysis of art has neglected the details of infrastructure within which communities of artistic practice emerge. Becker's inversion examines the conventions and constraints of the material artistic infrastructure and its ramifications. For example, the convention of musical concerts lasting about three hours ramifies throughout the producing organization. Parking attendants, unions, ticket takers, and theater rentals are arranged in cascading dependence on this interval of time. An eight-hour musical piece, which is occasionally written, means rearranging all of these expectations, which in turn is so expensive that such productions are rare. Or paintings are about the size, usually, that will hang comfortably on a wall. They are also the size that fits rolls of canvas, the skills of framers,

**Table 1.1**
A definition of infrastructure

- *Embeddedness.* Infrastructure is sunk into, inside of, other structures, social arrangements, and technologies,

- *Transparency.* Infrastructure is transparent to use in the sense that it does not have to be reinvented each time or assembled for each task, but invisibly supports those tasks.

- *Reach or scope.* This may be either spatial or temporal—infrastructure has reach beyond a single event or one-site practice;

- *Learned as part of membership.* The taken-for-grantedness of artifacts and organizational arrangements is a sine qua non of membership in a community of practice (Lave and Wenger 1991, Star 1996). Strangers and outsiders encounter infrastructure as a target object to be learned about. New participants acquire a naturalized familiarity with its objects as they become members.

- *Links with conventions of practice.* Infrastructure both shapes and is shaped by the conventions of a community of practice; for example, the ways that cycles of day-night work are affected by and affect electrical power rates and needs. Generations of typists have learned the QWERTY keyboard; its limitations are inherited by the computer keyboard and thence by the design of today's computer furniture (Becker 1982).

- *Embodiment of standards.* Modified by scope and often by conflicting conventions, infrastructure takes on transparency by plugging into other infrastructures and tools in a standardized fashion.

- *Built on an installed base.* Infrastructure does not grow de novo; it wrestles with the inertia of the installed base and inherits strengths and limitations from that base. Optical fibers run along old railroad lines, new systems are designed for backward compatibility; and failing to account for these constraints may be fatal or distorting to new development processes (Monteiro and Hanseth 1996).

- *Becomes visible upon breakdown.* The normally invisible quality of working infrastructure becomes visible when it breaks: the server is down, the bridge washes out, there is a power blackout. Even when there are backup mechanisms or procedures, their existence further highlights the now visible infrastructure.

- *Is fixed in modular increments, not all at once or globally.* Because infrastructure is big, layered, and complex, and because it means different things locally, it is never changed from above. Changes take time and negotiation, and adjustment with other aspects of the systems involved.[7]

Source: Star and Rohleder 1996. –

and the very doorways of museums and galleries. These constraints are mutable only at great cost, and artists must always consider them before violating them.

Scientific inversions of infrastructure were the theme of a path-breaking edited volume, *The Right Tools for the Job: At Work in Twenti-eth-Century Life Sciences* (Clarke and Fujimura 1992). The purpose of this volume was to tell the history of biology in a new way—from the point of view of the materials that constrain and enable biological researchers. Rats, petri dishes, taxidermy, planaria, drosophila, and test tubes take center stage in this narrative. The standardization of genetic research on a few specially bred organisms (notably drosophila) has constrained the pacing of research and the ways the questions may be framed, and it has given biological supply houses an important, invisible role in research horizons. While elephants or whales might answer different kinds of biological questions, they are obviously un-wieldy lab animals. While pregnant cow's urine played a critical role in the discovery and isolation of reproductive hormones, no historian of biology had thought it important to describe the task of obtaining gallons of it on a regular basis. Adele Clarke (1998) puckishly relates her discovery, found in the memoirs of a biologist, of the technique required to do so: tickle the cow's labia to make her urinate. A starkly different view of the tasks of laboratory biology emerges from this image. It must be added to the processes of stabling, feeding, impreg-nating, and caring for the cows involved. The supply chain, tech-niques, and animal handling methods had to be invented along with biology's conceptual frame; they are not accidental, but constitutive.

Our infrastructural inversion with respect to information technolo-gies and their attendant classification systems follows this line of analy-sis. Like the cow's urine or the eight-hour concert, we have found many examples of counterintuitive, often humorous struggles with constraints and conventions in the crafting of classifications. For in-stance, as we shall see in chapter 5, in analyzing the experience of tuberculosis patients in Mann's *The Magic Mountain,* we found the story of one woman who had been incarcerated so long in the sanatorium that leaving it became unthinkable. She recovered from the disease, but tried to subvert the diagnosis of wellness. When the doctors took her temperature, she would surreptitiously dip the thermometer in hot water to make it seem that she still had a fever. On discovering this, the doctors created a thermometer without markings, so that she could not tell what the mercury column indicated. They called this "the silent sister." The silent sister immediately becomes itself a telling indicator of the entangled infrastructure, medical politics, and the use of metrics in classifying tubercular patients. It tells a rich metaphorical story, and may become a concept useful beyond the rarified walls of the fictional Swiss asylum. What other silent sisters will we encounter in our infrastructural inversion—what surveillance, deception, caring, struggling, or negotiating?

In the sections below, four themes are presented that require the special double vision implied in the anecdotes above. They frame the new way of seeing that brings to life large-scale, bureaucratic classifications and standards. Without this map, excursions into this aspect of information infrastructure can be stiflingly boring. Many classifications appear as nothing more than lists of numbers with labels attached, buried in software menus, users' manuals, or other references. As discussed in chapter 2, new eyes are needed for reading classification systems, for restoring the deleted and dessicated narratives to these peculiar cultural, technical, and scientific artifacts.

## *Methodological Themes for Infrastructural Inversion*

### *Ubiquity*

The first major theme is the *ubiquity* of classifying and standardizing. Classification schemes and standards literally saturate our environment. In the built world we inhabit, thousands and thousands of standards are used everywhere, from setting up the plumbing in a house to assembling a car engine to transferring a file from one computer to another. Consider the canonically simple act of writing a letter longhand, putting it in an envelope, and mailing it. There are standards for paper size, the distance between lines in lined paper, envelope size, the glue on the envelope, the size of stamps, their glue, the ink in a pen, the sharpness of its nib, the composition of the paper (which in turn can be broken down to the nature of the watermark, if any; the degree of recycled material used in its production, the definition of what counts as recycling), and so forth.

Similarly, in any bureaucracy, classifications abound—consider the simple but increasingly common classifications that are used when you dial an airline for information ("if you are traveling domestically, press 1"; "if you want information about flight arrivals and departures. . . ."). And once the airline has you on the line, you are classified by them as a frequent flyer (normal, gold or platinum); corporate or

***Becoming an Irate***

Howard Becker relates a delightful anecdote concerning his classification by an airline. A relative working for one of the airlines told him how desk clerks handle customer complaints. The strategy is first to try to solve the problem. If the customer remains unsatisfied and becomes very angry in the process, the clerk dubs him or her "an irate." The clerk then calls the supervisor, "I have an irate on the line," shorthand for the category of an irritated passenger.

One day Becker was having a difficult interaction with the same airline. He called the airline desk, and in a calm tone of voice, said, "Hello, my name is Howard Becker and I'm an irate. Can you help me with this ticket?" The clerk began to sputter, "How did you know that word?" Becker had succeeded in unearthing a little of the hidden classificatory apparatus behind the scenes at the airline. He notes that the interaction after this speeded up and went particularly smoothly.

individual; tourist or business class; short haul or long haul (different fare rates and scheduling apply).

This categorical saturation furthermore forms a complex web. Although it is possible to pull out a single classification scheme or standard for reference purposes, in reality none of them stand alone. So a subproperty of ubiquity is interdependence, and frequently, integration. A systems approach might see the proliferation of both standards and classifications as purely a matter of integration—almost like a gigantic web of interoperability. Yet the sheer density of these phenomena go beyond questions of interoperability. They are layered, tangled, textured; they interact to form an ecology as well as a flat set of compatibilities. That is to say, they facilitate the coordination of heterogeneous "dispositifs techniques" (Foucault 1975). They are lodged in different communities of practice such as laboratories, records offices, insurance companies, and so forth.[9] There *are* spaces between (unclassified, nonstandard areas), of course, and these are equally important to the analysis. It seems that increasingly these spaces are marked as unclassified and nonstandard.

It is a struggle to step back from this complexity and think about the issue of ubiquity rather than try to trace the myriad connections in any one case. The ubiquity of classifications and standards is curiously difficult to see, as we are quite schooled in ignoring both, for a variety of interesting reasons. We also need concepts for understanding movements, textures, and shifts that will grasp patterns within the ubiquitous larger phenomenon. The distribution of residual categories ("not elsewhere classified" or "other") is one such concept. "Others" are everywhere, structuring social order. Another such concept might be what Strauss et al. (1985) call a "cumulative mess trajectory." In medicine, this occurs when one has an illness, is given a medicine to cure the illness, but incurs a serious side effect, which then needs to be treated with another medicine, and so forth. If the trajectory becomes so tangled that you cannot turn back and the interactions multiply, "cumulative mess" results. We see this phenomenon in the interaction of categories and standards all the time—ecological examples are particularly rich places to look.

### Materiality and Texture

The second methodological departure point is that classifications and standards are *material*, as well as symbolic. How do we perceive this densely saturated classified and textured world? Under the sway of cognitive idealism, it is easy to see classifications as properties of mind and standards as ideal numbers or floating cultural inheritances. But they have material force in the world. They are built into and embedded in every feature of the built environment (and in many of the nature-culture borderlands, such as with engineered genetic organisms).

All classification and standardization schemes are a mixture of physical entities, such as paper forms, plugs, or software instructions encoded in silicon, and conventional arrangements such as speed and rhythm, dimension, and how specifications are implemented. Perhaps because of this mixture, the web of intertwined schemes can be difficult to see. In general, the trick is to question every apparently natural easiness in the world around us and look for the work involved in making it easy. Within a project or on a desktop, the seeing consists in seamlessly moving between the physical and the conventional. So when computer programmers write some lines of Java code, they move within conventional constraints and make innovations based on them; at the same time, they strike plastic keys, shift notes around on a desktop, and consult manuals for various standards and other information. If we were to try to list all the classifications and standards involved in writing a program, the list could run to pages. Classifications include types of objects, types of hardware, matches between requirements categories and code categories, and metacategories such as the goodness of fit of the piece of code with the larger system under development. Standards range from the precise integration of the underlying hardware to the 60Hz power coming out of the wall through a standard size plug.

Merely reducing the description to the physical aspect such as the plug does not get us anywhere interesting about the actual mixture of physical and conventional or symbolic. A good operations researcher could describe how and whether things would work together, often purposefully blurring the physical and conventional boundaries in making the analysis. But what is missing is a sense of the landscape of work as experienced by those within it. It gives no sense of something as important as the texture of an organization: Is it smooth or rough? Bare or knotty? What is needed is a sense of the topography of all of the arrangements: Are they colliding, coextensive, gappy, or orthogonal? One way to get at these questions is to take quite literally the kinds of metaphors that people use when describing their experience of organizations, bureaucracies, and information systems, which are discussed in more detail in chapter 9.

When we think of classifications and standards as both material and symbolic, we adapt a set of tools not usually applied to them. There are tools for analyzing built structures, such as structural integrity, enclosures and confinements, permeability, and durability, among many others. Structures have texture and depth. The textural way of speaking of classifications and standards is common in organizations and groups. Metaphors of tautness, knots, fabrics, and networks pervade modern language (Lakoff and Johnson 1980).

### The Indeterminacy of the Past: Multiple Times, Multiple Voices

The third methodological theme concerns *the past as indeterminate*.[10] We are constantly revising our knowledge of the past in light of new developments in the present. This is not a new idea to historiography or to biography. We change our resumes as we acquire new skills to appear like smooth, planned paths of development, even if the change had been unexpected or undesired. When we become members of new social worlds, we often retell our life stories in new terminology. A common example of this is a religious conversion where the past is retold as exemplifying errors, sinning, and repentance (Strauss 1959). Or when one comes out as gay or lesbian, childhood behaviors and teenage crushes become indicators of early inklings of sexual choice (Wolfe and Stanley 1980).

At wider levels of scale, these revisions also mean the introduction of new voices—many possible kinds of interpretations of categories, texts, and artifacts. Multiple voices and silences are represented in any scheme that attempts to sort out the world. No one classification organizes reality for everyone—for example, the red light, yellow light, green light traffic light distinctions do not work for blind people (who need sound coding). In looking to classification schemes as ways of ordering the past, it is easy to forget those who have been overlooked in this way. Thus, the indeterminacy of the past implies recovering multivocality; it also means understanding how standard narratives that appear universal have been constructed (Star 1991a).

There is no way of ever getting access to the past except through classification systems of one sort or another—formal or informal, hierarchical or not. Take the apparently unproblematic statement: "In 1640, the English revolution occurred; this led to a twenty-year period in which the English had no monarchy." The classifications involved here, all problematic, include the following:

• The current segmentation of time into days, months, and years. Accounts of the English revolution generally use the Gregorian calendar, which was adopted some 100 years later, so causing translation problems with contemporary documents.

• The classification of peoples into English, Irish, Scots, French, and so on. These designations were by no means so clear at the time; the whole discourse of "national genius" or character only arose in the nineteenth century.

• The classification of events into revolutions, reforms, revolts, rebellions, and so forth (see Furet 1978 on thinking the French revolution). There was no concept of "revolution" at the time; our current conception is marked by the historiographical work of Karl Marx.

• What do we classify as being a "monarchy?" There is a strong historiographical tradition that says that Oliver Cromwell was a monarch—he walked, talked, and acted like one after all. Under this view, there is no hiatus at all in this English institution; rather a usurper took the throne.

There are two major historiographic schools of thought about using classification systems on the past. One maintains that we should only use classifications available to actors at the time, much as an ethnographer tries faithfully to mirror the categories of their respondents.

Authors in this tradition warn against the dangers of anachronism. Hacking (1995) on child abuse is a sophisticated version that we discuss in chapter 7. If a category did not exist contemporaneously, it should not be retroactively applied.

The other school of thought holds that we should use the real classifications that progress in the arts and sciences has uncovered. Often history informed by current sociology will take this path. For example, Tort's (1989) work on "genetic" classification systems (which were not so called at the time, but which are of vital interest to the Foucaldian problematic) imposes a post hoc order on nineteenth-century classification schemes in a variety of sciences. Even though those schemes were perceived by their creators as responding solely to the specific needs of the discipline they were dealing with (etymology, say, or mineralogy), Tort demonstrates that there was a link between many different schemes (both direct in people shifting disciplines and conceptual in their organization) that allows us to perceive an order nowhere apparent to contemporaries.

From a pragmatist point of view, both aspects are important in analyzing the consequences of modern systems of classification and standardization. We seek to understand classification systems according to the work that they are doing and the networks within which they are embedded. That entails both an understanding of the categories of those designing and using the systems, and a set of analytic questions derived from our own concerns as analysts.

When we ask historical questions about the deeply and heterogeneously structured space of classification systems and standards, we are dealing with a four-dimensional archaeology. The systems move in space, time, and process. Some of the archaeological structures we uncover are stable, some in motion, some evolving, some decaying. They are not consistent. An institutional memory about an epidemic, for example, can be held simultaneously and with internal contradictions (sometimes piecemeal or distributed and sometimes with entirely different stories at different locations) across a given institutional space.

In the case of AIDS, classifications have shifted significantly over the last twenty years, including the invention of the category in the 1980s—from gay-related immune disorder (GRID) through a chain of other monikers to the now accepted acquired immune deficiency syndrome (AIDS). It is now to some extent possible to look back at cases that might previously have been AIDS (Grmek 1990) before we had

### When Is It a Harley?

One of the ways the past becomes indeterminate is through gradual shifts in what it means to "really be" something—the essence of it.

Sitting in a tattoo parlor, surrounded by people I do not usually hang out with. Young men in black leather vests and sun-bleached hair. I turn to the waiting room reading material, which in this case is the monthly *Thunder Press,* a newsletter for motorbike aficionados. The lead article asks the question: "Is It Still a Harley" if you have customized your bike yourself? The Oregon Department of Motor Vehicles makes the definitive call: "Anything that is not totally factory built will make it a reconstructed motorcycle, and it will be called 'assembled' on the title" (69).

A major activity in the Harley social world is customizing features of one's motorcycle, and there are important symbolic and affiliative signs attached to the customizing process. Deleting the name Harley from the registration form is perceived as an insult to the owner, and this insult is stitched together in the article with others that come from the government toward bikers (restricting meeting places, insisting on helmet-wearing, being overly enthusiastic in enforcing traffic violations by bikers).

This is a pure example of the politics of essence, of identity politics. It is echoed in many areas of life, for example, in James Davis' (1991) classic study *Who Is Black?* where the question of the one-drop rule in the United States, and the rejection of mixed-race people as a legitimate category is an old and a cruel story. The central process here is the distillation of the sine qua non out from the messy and crenellated surrounds—the rejection of marginality in favor of purity.

When this occurs, the suffering of the marginal becomes privatized and distributed, creating the conditions for pluralistic ignorance ("I'm the only one"). Meeting the purity criteria of the essentialized category also becomes bureaucratized and again the onus is shifted to the individual alone. Only when the category is joined with a social movement can the black box of essence be reopened, as for example with the recent uprisings and demonstrations of mixed race Hispanic people toward the U.S. census and its rigid categories. The problem becomes clear if one is both black and Hispanic, a common combination in the Caribbean. Through which master trait will the government perceive you?

—Leigh Star

Source: Anonymous, "Is It Still a Harley," *Thunder Press* 5:4 (July 1996, 1 and 69).

the category (a problematic gaze to be sure, as Bruno Latour (forth-coming) has written about tuberculosis). There are epidemiological stories about trying to collect information about a shameful disease; there is a wealth of personal and public narratives about living with it. There is a public health story and a virology story, which use different category systems. There are the standardized forms of insurance com-panies and the categories and standards of the Census Bureau. When an attempt was made to combine these data in the 1980s to disenfran-chise young men living in San Francisco, from health insurance, the resultant political challenge stopped the combination of these data from being so used. At the same time, the San Francisco blood banks refused for years to employ HIV screening, thus denying the admis-sion of another category to their blood labeling, as Shilts (1987) tells us, with many casualties as a result. Whose story has categorical ascen-dancy here? That question is forever morally moot—all of the stories are important and all of the categories tell a different one.

### Practical Politics

The fourth major theme is uncovering *the practical politics of classifying and standardizing*. This is the design end of the spectrum of investigat-ing categories and standards as technologies. There are two processes associated with these politics: arriving at categories and standards, and, along the way, deciding what will be visible or invisible within the system.

It follows from the indeterminacy discussed above that the spread or enforcement of categories and standards involves negotiation or force. Whatever appears as universal or indeed standard, is the result of negotiations, organizational processes, and conflict. How do these negotiations take place? Who determines the final outcome in prepar-ing a formal classification? Visibility issues arise as one decides where to make cuts in the system, for example, down to what level of detail one specifies a description of work, of an illness, of a setting. Because there are always advantages and disadvantages to being visible, this becomes crucial in the workability of the schema. As well, ordinary biases of what should be visible, or legitimated, within a particular scheme are always in action. The trade-offs involved in this sort of politics are discussed in chapters 5 on tuberculosis and 7 on nursing work.

Someone, somewhere, must decide and argue over the minutiae of classifying and standardizing. The negotiations themselves form the

---

### There's No Such Thing as a Rodent

An article in the *San Jose Mercury News* by Rick Weiss declares: "Researchers say there's no such thing as a rodent." He quotes an article from *Nature,* which argues that the 2,000 species of animals ordinarily considered rodents—including rats, mice, and guinea pigs—did not evolve from a common ancestor. The finding is deeply controversial. Weiss says, "On one side are researchers who have spent their careers hunched over fossils or skeletal remains to determine which animals evolved from which." On the other, the article continues, are those who would use DNA analysis to make the determination. The fossil studiers say that DNA is not yet accurate enough. The classification of species has always been deeply controversial. Biologists speak of a rough cut among their ranks: lumpers (those who see fewer categories and more commonalties) versus splitters (those who would name a new species with fewer kinds of difference cited). There are always practical consequences for these names. Splitters, for example, often included people who wanted a new species named after them, and the more species there are, the more likely is an eponymous label. The deliberately provocative headline of this article demands a response: "well, don't tell that to my cat." We often refer implicitly in this fashion to the power of naming—blurring the name of the category with its members. (*San Jose Mercury News,* June 13, 1996: 5A by Rick Weiss)

---

basis for a fascinating practical ontology—our favorite example is when is someone really alive? Is it breathing, attempts at breathing, or movement? And how long must each of those last? Whose voice will determine the outcome is sometimes an exercise of pure power: We, the holders of western medicine and scions of colonial regimes, will decide what a disease is and simply obviate systems such as acupuncture or Aryuvedic medicine. Sometimes the negotiations are more subtle, involving questions such as the disparate viewpoints of an immunologist and a surgeon, or a public health official (interested in even *one* case of the plague) and a statistician (for whom one case is not relevant).

Once a system is in place, the practical politics of these decisions are often forgotten, literally buried in archives (when records are kept at all) or built into software or the sizes and compositions of things. In addition to our archaeological expeditions into the records of such negotiations, this book provides some observations of the negotiations in action.

Finally, even where everyone agrees on how classifications or standards should be established, there are often practical difficulties about how to craft them. For example, a classification system with 20,000 bins on every form is practically unusable for data-entry purposes. The constraints of technological record keeping come into play at every turn. For example, the original ICD had some 200 diseases not because of the nature of the human body and its problems but because this was the maximum number that would fit the large census sheets then in use.

Sometimes the decision simply about how fine-grained to make the system has political consequences as well. For instance, describing and recording someone's tasks, as in the case of nursing work, may mean controlling or surveilling their work as well, and may imply an attempt to take away discretion. After all, the loosest classification of work is accorded to those with the most power and discretion who are able to set their own terms. There are financial stakes as well. In a study of a health insurance company's system of classifying for doctor and patient reimbursement, Gerson and Star (1986) found that doctors wanted the most fine-grained of category systems, so that each procedure could be reimbursed separately and thus most profitably. Data-entry personnel and hospital administrators, among others, wanted broader, simpler, and coarser-grained categories for reasons of efficiency. These conflicts were, however, invisible to the outside world, which received only the forms for reimbursement purposes and a copy of the codebook for reference. Both the content of the categories and the structure of the overall scheme are concerns for due process within organizations—whose voice will be heard and when will enough data, of the right granularity, have been collected?

### Infrastructure and Method: Convergence

These ubiquitous, textured classifications and standards help frame our representation of the past and the sequencing of events in the present. They can best be understood as doing the ever local, ever partial work of making it appear that science describes nature (and nature alone) and that politics is about social power (and social power alone). Consider the case of psychoanalysts discussed at length in Young (1995), Kirk and Kutchins (1992), and Kutchins and Kirk (1997). To receive reimbursement for their procedures, psychoanalysts now need to couch them in a biomedical language (using the DSM).

### Fitting Categories to Circumstances

An academic friend on the East Coast tells an anecdote of negotiation with her long-term psychoanalyst about how to fill out her insurance forms. She was able to receive several free sessions of therapy a year under her health insurance plan. Each year, she and her therapist would discuss how best to categorize her. It was important to represent the illness as serious and long-term. At the same time, they were worried that the information about the diagnosis might not always remain confidential. What could they label her that would be both serious and nonstigmatizing? Finally, they settled on the diagnosis of obsessive-compulsive. No academic would ever be penalized for being obsessive-compulsive, our friend concluded with a wry laugh! (Kirk and Kutchins (1992) document similar negotiations between psychiatrists and patients.)

Theoretically, this rubric is anathema to them, systematically replacing the categories of psychoanalysis with the language of the pharmacopoeia and of the biochemistry of the brain. The DSM, however, is the lingua franca of the medical insurance companies. Thus, psychoanalysts use the categories not only to obtain reimbursement but as a shorthand to communicate with each other. There are local translation mechanisms that allow the DSM to continue to operate in this fashion and, at the same time, to become the sole legal, recognized representation of mental disorder. A "reverse engineering" of the DSM or the ICD reveals the multitude of local political and social struggles and compromises that go into the constitution of a "universal" classification.

Standards, categories, technologies, and phenomenology are increasingly converging in large-scale information infrastructure. As we have indicated in this chapter, this convergence poses both political and ethical questions. These questions are by no means obvious in ordinary moral discourse. For all the reasons given above, large-scale classification systems are often invisible, erased by their naturalization into the routines of life. Conflict and multiplicity are often buried beneath layers of obscure representation.

Methodologically, we do not stand outside these systems, nor pronounce on their mapping to some otherworldly "real" or "constructed" nature. Rather, we are concerned with what they do, pragmatically speaking, as scaffolding in the conduct of modern life. Part of that analysis means understanding the coconstruction of classification systems with the means for data collection and validation.

To clarify our position here, let us take an analogy. In the early nineteenth century in England there were a huge number of capital crimes, starting from stealing a loaf of bread and going on up. Precisely because the penalties were so draconian, however, few juries would ever impose the maximum sentence; and indeed there was a drastic reduction in the number of executions even as the penal code was progressively strengthened. There are two ways of writing this history: one can either concentrate on the creation of the law; or one can concentrate on the way things worked out in practice. This is very similar to the position taken in Latour's *We Have Never Been Modern* (1993). He argues that we can either look at what scientists say they are doing (working within a purified realm of knowledge) or at what they actually are doing (manufacturing hybrids of nature-culture). We think both are important. We advocate here a pragmatic methodological development—pay more attention to the classification and standardization work that allows for hybrids to be manufactured and so more deeply explore the terrain of the politics of science in action.

The point is that both words and deeds are valid kinds of account. Early sociology of science in the actor-network tradition concentrated on the ways in which it comes to appear that science gives an objective account of natural order: trials of strength, enrolling of allies, cascades of inscriptions, and the operation of immutable mobiles (Latour 1987, 1988). Actor network theory drew attention to the importance of the development of standards (though not to the linked development of classification systems), but did not look at these in detail. Sociologists of science invited us to look at the process of producing something that looked like what the positivists alleged science to be. We got to see the Janus face of science as both constructed and realist. In so doing we followed the actors, often ethnographically. We shared their insights. Allies must be enrolled, translation mechanisms must be set in train so that, in the canonical case, Pasteur's laboratory work can be seen as a direct translation of the quest for French honor after defeat in the battlefield (Latour 1988).

By the very nature of the method, However, we also shared the actors' blindness. The actors being followed did not themselves *see* what was excluded: they constructed a world in which that exclusion could occur. Thus if we just follow the doctors who create the ICD at the WHO in Geneva, we will not see the variety of representation systems that other cultures have for classifying diseases of the body and spirit; and we will not see the fragile networks these classification systems subtend. Rather, we will see only those who are strong enough and shaped in such a fashion as to impact allopathic medicine. We will see the blind leading the blind.

This blindness occurs by changing the world such that the system's description of reality becomes true. Thus, for example, consider the case where all diseases are classified purely physiologically. Systems of medical observation and treatment are set up such that physical manifestations are the only manifestations recorded. Physical treatments are the only treatments available. Under these conditions, then, logically schizophrenia may only result purely and simply from a chemical imbalance in the brain. It will be impossible to think or act otherwise. We have called this the principle of *convergence* (Star, Bowker and Neumann in press).

### Resistance

Reality is 'that which resists,' according to Latour's (1987) Pragmatist-inspired definition. The resistances that designers and users encounter will change the ubiquitous networks of classifications and standards. Although convergence may appear at times to create an inescapable cycle of feedback and verification, the very multiplicity of people, things and processes involved mean that they are never locked in for all time.

The methods in this chapter offer an approach to resistance as a reading of where and how political work is done in the world of classifications and standards, and how such artifacts can be problematized and challenged. Donald MacKenzie's (1990) wonderful study of "missile accuracy" furnishes the best example of this approach. In a concluding chapter to his book, he discusses the possibility of "uninventing the bomb," by which he means changing society and technology in such a way that the atomic bomb becomes an impossibility. Such change, he suggests, can be carried out in part at the overt level of political organizations. Crucially for our purposes, however, he also sensitizes the reader to the site of the development and maintenance of technical standards as a site of political decisions and struggle. Standards and classifications, however dry and formal on the surfaces, are suffused with traces of political and social work. Whether we wish to uninvent any particular aspect of complex information infrastructure is properly a political and a public issue. Because it has rarely been cast in that light, tyrannies of various sorts flourish. Some are the tyrannies of inertia—red tape—rather than explicit public policies. Others are the quiet victories of infrastructure builders inscribing their politics into the systems. Still other are almost accidental—systems that become so complex that no one person and no organization can predict or administer good policy.

The magic of modern technoscience is a lot of hard work involving smoke-filled rooms, and boring lists of numbers and settings. Tyranny or democracy, its import on our lives cannot be denied. This chapter has offered a number of points of departure for evaluation, resistance, and better analysis of one of its least understood aspects.

# I

## *Classification and Large-Scale Infrastructures*

In the following three chapters, which analyze the international classification of diseases (ICD) we look at the operation of classification systems in supporting large-scale infrastructural arrangements. Chapter 2 concentrates on the text of the ICD itself, producing a reading of this classification which has over the past century ingrained itself in a multiplicity of forms, work arrangements, and laws worldwide. We examine how its internal structure affords the prosecution of multiple agendas. Chapter 3 discusses the history of the ICD, showing how it has changed over time in step with changing information technology and changing organizational needs. Chapter 4 draws general design implications from the study of this highly effective, long-term, and wide-scale classification scheme.

# 2

## The Kindness of Strangers: Kinds and Politics in Classification Systems

Most Enlightenment naturalists joined the chorus of praise for system in the abstract; but their responses to particular systems were apt to be less cohesive. The very icons of classification—the tables and diagrams prefixed and appended to works of Enlightenment zoology to distinguish them from the unstructured productions of previous ages—could simultaneously evidence this lack of unity.

*(Ritvo 1997, 21)*

### Introduction: Formal and Informal Aspects of Classification

How people classify things, and what relationship those categories have with social organization, has long been a central topic within anthropology, especially cognitive anthropology and cognitive science. In this chapter, we touch on some of the issues raised in those disciplines, such as the relationship between what is singled out as different and what is considered normal. Our primary project is a pragmatic one, not a logical or cognitive one. We want to know empirically how people have designed and used classification systems. We want to understand how political and semantic conflicts are managed over long periods of time and at large levels of scale.

Equally, as good pragmatists, we know that things *perceived* as real are real in their consequences (Thomas and Thomas 1970 [1917]). So even when people take classifications to be purely mental, or purely formal, they *also* mold their behavior to fit those conceptions. When formal characteristics are built into wide-scale bureaucracies such as the WHO, or inscribed in hospital software standards, then the compelling power of those beliefs is strengthened considerably. They often come to be considered as natural, and no one is able completely to disregard or escape them. People constantly fiddle with them, however, and work around the formal restrictions (Hunn 1982). When we look over a long enough period of time, the formal and the informal are completely mingled in infrastructure.

There has been a recent trend in social informatics and science studies to move away from dichotomizing the formal and the informal.[11] In the early 1980s, the original éclat of discovering the failures of formalisms led to a kind of enthusiastic debunking. People do not really follow formal rules; they make up their own. They tailor rigid computer systems to their everyday working needs. Expert systems do not formally model people's thoughts as they fail to capture tacit knowledge. People do not devise formal, abstract plans and goals and then execute them, as the old cognitive model of Miller, Galanter, and Pribram (1960) would have it. Rather, they use a dynamic and situated improvisation (Suchman 1987) where plans are resources and are renegotiated as circumstances warrant. Suchman's situated action perspective constituted a powerful critique of artificial intelligence's claim that the mind could be formally specified.

Building on this initial set of findings and especially Suchman's notion that plans are also material resources for action—whether or not people follow them exactly—a more sophisticated model has emerged in recent years. Although it is true that maps do not fully capture terrains, they are powerful technologies (Becker 1986). They help to find one's way, as originally formally intended. And they serve as resources to structure all sorts of collective action—dreams of vacations, crossword puzzle solutions, explanations of social distance (Schmidt 1997, Zorbaugh 1929). Marc Berg analyzes the formalisms of medical decision making in use as powerful both formally and as spurs to informal action (1997b 1998). Just because people do not do exactly what they say will, does not mean they are doing nothing. Nor does it mean that they do not believe in the stated formal purpose and tailor their behavior to it. Obvious as this point may appear from a common sense perspective, it has not been obvious in scientific writing about cognition and classification.

In this book we offer a balanced reconsideration of classifications as formal and informal resources, often annealed together. People juggle vernacular (or folk) classifications together with the most formal category schemes (as detailed in Atran 1990). They subvert the formal schemes with informal work-arounds. Indeed, the various approaches are often so seamlessly pasted together they become impossible to distinguish in the historical record. For instance, a physician decides to diagnose a patient using the categories that the insurance company will accept. The patient then self-describes, using that label to get consistent help from the next practitioner seen. The next practitioner accepts this as part of the patient's history of illness. As many of the examples in this book will show, this convergence may then be converted into data and at the aggregate level, seemingly disappear to leave the record as a collection of natural facts (Star, Bowker, and Neumann, in press).

Any classification system embodies a dynamic compromise. Harriet Ritvo writes of zoological classification in the nineteenth century:

> But if the experts resisted granting recognition to competing claimants of the zoological territory they had staked out, they tacitly acknowledged the objections of various laymen in many ways. They even quietly incorporated vernacular categories into their classificatory schemes, especially with regard to mammals, the creatures most important to people and most like them. This consistently inconsistent practice illuminates both the nature of scientific enterprise during the period and the relation of science to the larger culture. (1997, xii)

As Ritvo shows us, these tracks do not disappear completely. Traces of bureaucratic struggles, differences in world-view, and systematic erasures do remain in the written classification system, however indirectly. The trick is to read the classification itself, restoring the narratives of conflict and compromise as we do so. This reading requires that we juggle the formal and informal aspects of classification while reading. Our reading teases out the cognitive, bureaucratic, and formal aspects of the work of designing and using classification systems. We are not here treating the generation or the detailed implementation of these categories—both topics well worthy of attention. (Young's (1995) description of posttraumatic stress disorder is a model here). For the purposes of this chapter, our emphasis is on reading the system, our argument being that one can read a surprising amount of social, political, and philosophical context from a set of categories—and that in many cases the classification system in practice is all that we have to go on.

Sitting down and reading a document like the ICD is a curiously perverse activity. The three volumes of ICD-10, more than 2,000 pages long, have very little in the way of overt narrative. There is a short history of the enterprise of producing international classifications of disease at the beginning of volume two, which contains explanatory or prescriptive notes. It provides most notably a set of rules for using the classifications of the ICD with directions on what to do in ambiguous

### I50    Heart failure

***Excludes:*** complicating:
- abortion or ectopic or molar pregnancy (O00–O07, O08.8)
- obstetric surgery and procedures (O75.4)

due to hypertension (I11.0)
- with renal disease (I13.–)

following cardiac surgery or due to presence of cardiac prosthesis (I97.1)
neonatal cardiac failure (P29.0)

**I50.0    Congestive heart failure**
Congestive heart disease
Right ventricular failure (secondary to left heart failure)

**I50.1    Left ventricular failure**
Acute oedema of lung ⎫ with mention of heart disease
Acute pulmonary oedema ⎬ NOS or heart failure
Cardiac asthma
Left heart failure

**I50.9    Heart failure, unspecified**
Biventricular failure
Cardiac, heart or myocardial failure NOS

*Figure 2.1*
Heart failure as specified in the ICD-10.
Source: ICD 10, 1: 494.

situations. Volume three is an index, a vital tool since diseases have multiple designators and so many paths into the classification system must be provided. The first volume is the largest. It is primarily a long list of numbers with names of diseases or modifying conditions. (An example is given in figure 2.1)

Reading the ICD is a lot like reading the telephone book. In fact, it is worse. The telephone book, especially the yellow pages, contains a more obvious degree of narrative structure. It tells how local businesses see themselves, how many restaurants of a given ethnicity there are in the locale, whether or not hot tubs or plastic surgeons are to be found there. (Yet most people don't curl up with a good telephone book of a Saturday night.) Aside from this direct information to be retrieved, an indirect reading can be instructive. A slim volume indicates a rural area. Those with only husband's names listed for married couples indicate a sexist society. The names of services may change over time, indicating changed community values. In the Santa Cruz, California, phone book, for example, Alcoholics Anonymous and Narcotics Anonymous are listed in emergency services; years ago they would have been listed under "rehabilitation," if at all. The changed status reflects the widespread recognition of the organizations' reliability in crisis situations, as well as acceptance of their theory of addiction as a medical condition. Under the community events section in the beginning, next to the Garlic Festival and the celebration of the anniversary of the city's founding, the Gay and Lesbian Pride Parade is listed as an annual event. Behind this simple telephone book listing lie decades of activism and conflict—for gays and lesbians, becoming part of the civic infrastructure in this way betokens a kind of public acceptance almost unthinkable thiry years ago.

ICD-10 is an equally rich text. In the example of common "heart failure," given above, several primary divisions of heart failure are spelled out: congestive, left ventricular, and so forth. Yet those failures caused by mechanical failure of a prosthesis—pacemaker breakdown— are explicitly excluded at this point. We read this and wonder: if the breakdown is due to a manufacturing defect, would that constitute criminal negligence, and so is this the reason the category is kept separate? If the person passed through an area posted as proscribed to pacemakers, could it be suicide as well as heart failure? Or an accident? Or if it were due to a contributing cause like illiteracy, and if so, is there room in the ICD to make this kind of connection? The narrative questions begin to appear. When we look in the cross-referenced section under pacemaker (cardiac prosthesis), there are two factors influencing the category of heart failure—presence of a pacemaker (Z95.0) and the activity of its maintenance and management (Z45.0). A failed pacemaker as proximal cause of death must be pieced together as a narrative by the physician, but then reencoded and reembedded in the statistical list. In final form the death certificate would read as sudden death, with pacemaker in place.

We did sit down and read the ICD, and another detailed classification system, the International Classification and Nomenclature of Viruses (INV). This chapter analyses their embedded narrative structures, formal and informal, and the narrative structures in which they are embedded. Our work here is an exercise both in restoring the stories of practical classifying, conflict, and consensus therein and in understanding the design of the list itself.

### The Classification of Acupuncture

In 1991 the World Health Organization came out with a "proposed standard international acupuncture nomenclature" (WHO 1991). The initiative began in 1960. The report notes that the whole system of acupuncture of 361 points was complete in about A.D. 300 but the past 20 years had seen an explosion, with 48 extra points being added (WHO 1991, 1–2). The WHO report gives a typical sequence of reasons for developing an international classification: "Even when the practice of acupuncture was largely restricted to China, Japan, and neighboring Asian countries, the lack of a uniform nomenclature caused serious difficulties in teaching, research, and clinical practice" (WHO 1991, 1). The bottom line was scientific development: "Putting acupuncture on a firm scientific basis requires rigorous investigation of the claims made for its efficacy. Many institutions and modern medical colleges are carrying out useful investigations to this end. Some are looking into the physiology and mode of action of acupuncture treatment, others are studying its efficacy in certain pathological conditions. These workers need to exchange information with one another regularly so as to facilitate their clinical and basic research. Such international communication is possible only if a common language is used by all concerned" (WHO 1991, 5).

So now the "triple energizer meridian," the "conception vessel" and the "governor vessel" are internationally known and accepted terms. Ironically, the classification system retains an interesting (literal!) indexicality: the 48 points were only recognized if they were at least 0.5 cun from a classic acupuncture point, where a cun is: "the distance between the interphalangeal creases of the patient's middle finger" (WHO 1991, 14).

### Formal Classification

The structural aspects of classification are themselves a technical specialty in information science, biology, and statistics, among other places. Information scientists design thesauri for information retrieval, valuing parsimony and accuracy of terms, and the overall stability of the system over long periods of time. For biologists the choice of structure reflects how one sees species and the evolutionary process. For transformed cladists and numerical taxonomists, no useful statement about the past can be read out of their classifications; for evolutionary taxonomists that is the very basis of their system. These beliefs are reflected in radically different classification styles and practices, for example, whether or not to include the fossil record in the classification system; fossils being a problem since they perpetually threaten to create another level of taxa, and so cause an expensive and painstaking reordering of the whole system (Scott-Ram 1990).

There is even a metadiscipline of classifying that examines the architectonic features of classification systems in general. Using a variety of statistical techniques, these specialists analyze data structures, overall shape and structure of taxonomies and categories, and assess the elegance and durability of a classification system much in the way an architect would assess the structural and aesthetic features of a building. The International Classification Society regularly meets to discuss these issues, as does the Special Interest Group in Classification Research of the American Society for Information Science (SIG-CR of ASIS). The kinds of readings and assessments brought to bear by these specialists has not traditionally dealt explicitly with political or cultural issues at the metalevel (although those debates are the stuff of classification design and revision for any applied group, such as the WHO).

### Practical Classifying, Folk, Vernacular, and Ethno-classifications

Practical classifying is the stuff of cultural anthropology—how people classify their everyday worlds, including everything from color to kinship. Traditionally much ethno- or folk-classification research has examined tribal categories in nonindustrial societies. How people in industrial societies categorize on an everyday basis is less well known, especially in natural workaday settings. Most of the extant research, in linguistics or cognitive psychology, has been in experimental settings highly constrained in focus.

Here, we use the term practical classifying to mean how people categorize the objects they encounter in everyday situations, including formal classification schemes. Part of reading classifications is understanding the nature of these encounters, and the interplay between vernacular and formal systems.

The kind of reading we do here emphasizes the range of ways classification systems may be fuzzy or logical, reflective at once of bureaucratic concerns, scientific grounds, formal considerations, and cognitive theories. Our reading will not resolve the divergent perspectives created by these different needs, but will hopefully restore some of the stories to the dry lists that shape so much of our lives. In the section below, we frame our reading by briefly describing some of the theories about classification that have informed cognitive and social science discussions of classification.

### Kinds of Classification in Theories about Classification

Within the field of the sociology of science, the Edinburgh School has developed a rich analysis of scientific classification. In many ways its analysis of classification goes back to Durkheim and Mauss's classic "De quelques formes primitives de classification: contribution à l'étude des représentations collectives." Durkheim and Mauss had made the strong claim that classifications of the natural world in "primitive" societies directly reflected kinship structure in the sense that they projected the microcosm of social organization onto the macrocosm of the world—social tools were used for describing the natural world. They concluded that "the history of scientific classification is one by which the element of social affect has become progressively weaker, leaving more place for the reflective thought of individuals" (Durkheim and Mauss 1969, 88). David Bloor (1982) produced a rereading of Durkheim and Mauss that both defended them against the attacks on the validity of their analysis and extended their work to scientific classifications in seventeenth-century physics. He claimed that Boyle and Newton were producing classifications of entities in the world that reproduced their theological and political beliefs; in his words, both sides in the debate "were arranging the fundamental laws and classifications of their natural knowledge in a way that artfully aligned them with their social goals" (Bloor 1982, 290). This position prefigures the mechanism Latour (1993) gives for the projection of social categories out into nature and then their reimportation in the process of political debates ("if they are out there in the world then they must be real and so we must model our society accordingly"). Bloor used firstly Hesse's network model of classifications and more recently (Barnes, Bloor, and Henry 1996) he and colleagues have offered a finitist model. Common to both philosophical descriptions is the position that no category stands alone—when a new member is added to a class, this has ramifications for the class and the system of which it is part. Just as Lakotos (1976) argued about mathematical objects, the new exemplar can change the whole nature of the system. Specific classification choices are "underdetermined and indeterminate. It will emerge as we *decide* how to develop the analogy between the finite number of our exiting examples of things and the indefinite number of things we shall encounter in the future" (Barnes, Bloor, and Henry 1996, 55). While this is a useful general model, it does not have the power to trace exactly how changes are made, this has been the great breakthrough of Rosch's prototype theory discussed below.

Starting as well from a reading of Durkheim and Mauss, Mary Douglas observed a similar kind of mechanism for the reification of social categories: "How a system of knowledge gets off the ground is the same as the problem of how any collective good is created. . . . Communities do not grow up into little institutions and these do not grow into big ones by any continuous process. For a convention to turn into a legitimate social institution it needs a parallel cognitive convention to sustain it" (Douglas 1986, 46). For her, classification systems of all types are at base social institutions that reflect and describe the way things are in the social world. Again prefiguring Latour, she argues:

Before it can perform its entropy-reducing work, the incipient institution needs some stabilizing principle to stop its premature demise. That stabilizing principle is the naturalization of social classifications. There needs to be an analogy by which the formal structure of a crucial set of social relations is found in the physical world, or in the supernatural world, or in eternity, anywhere, so long as it is not seen as a socially contrived arrangement. When the analogy is applied back and forth from one set of social relations to another, and from these back to nature, its recurring formal structure becomes easily recognized and endowed with self-validating truth. (Douglas 1986, 48)

Douglas and Bloor here draw attention to a key feature of classification systems, that they grow out of and are maintained by social institutions. Building on this broad generalization, our approach in this book is to offer fine-grained analyses of the nature of information infrastructures such as classification systems and thus to demonstrate how they simultaneously represent the world "out there," the organizational context of their application (an issue discussed in Dean 1979) and the political and social roots of that context. We suggest that at this finer grain we detect rather a coconstruction of nature and society than a projection of the social onto the natural.

A classic divide among kinds of classification systems—and one that can lead us to this kind of coconstruction—is that drawn by Taylor, who distinguishes between Aristotelian classification and prototype classifications. Experimental psychologist Eleanor Rosch (1978) defined the prototype classification. This distinction is going to be an important one through this chapter, so let us explore it in some detail. An Aristotelian classification works according to a set of binary characteristics that the object being classified either presents or does not present. At each level of classification, enough binary features are adduced to place any member of a given population into one and only one class. So we might say that a pen is an object for writing within a population consisting of pens, balls, and bottles (Taylor 1995). We would have to add in one more feature to distinguish it adequately, for example, from pens, pencils, balls, or bottles. A technical classification system operating by binary characteristics is called monothetic if a single set of necessary and sufficient conditions is adduced ("in the universe of polygons, the class of triangles consists of figures that have three sides"); polythetic if a number of shared characteristics are used. In our example, we might say a pen is thin, cylindrical, used for writing, has a ball point, and so forth (Blois 1984). Desrosières (1993) points to a typical breakdown between monothetic and polythetic classifications in the work of statisticians. He associates the former with Linnaeus and the latter with Buffon (who engaged in local classification practices, just using the set of traits needed to make a determination in a specific instance); and writes, "These local practices are often carried out by those working in statistical centers, according to a division of labor whereby the chiefs are inspired by Linnaean precepts but the working statisticians apply, without realizing it, Buffon's method" (Desrosières 1993, 296, authors' translation). Aristotelian models—monothetic or polythetic—have traditionally informed formal classification theory in a broad range of sciences, including biological systematics, geology, and physics.

According to Rosch's prototype theory, our classifications tend to be much fuzzier than we might at first think. We do not deal with a set of binary characteristics when we decide that this thing we are sitting on is a chair. Indeed it is possible to name a population of objects that people would in general agree to call chairs which have no two binary features in common.

Prototype theory proposes that we have a broad picture in our minds of what a chair is; and we extend this picture by metaphor and analogy when trying to decide if any given thing that we are sitting on counts. We call up a best example, and then see if there is a reasonable direct or metaphorical thread that takes us from the example to the object under consideration. George Lakoff (1987) and John Taylor

(1995) have powerfully developed prototype theory within the field of sociolinguistics. One finding of the theory is that different social groups tend to have quite different prototypes in mind when classifying something as, say, a piece of furniture. Thus when surveyed, a group of Germans came up consistently with a different set of best examples than did a group of Americans (Taylor 1995, 44–57). For the Americans, chair and sofa are best fits for furniture, for the Germans, asked about *möbel*, it was bed and table.

An important implication of the theory is that there are levels at which we most easily and naturally distinguish between objects in the world, and that supervenient or subvenient levels tend to be more technically defined. Looking at a picture of a Maine coon cat, a non-expert will say that this is a picture of a cat, while an expert might call it either a Maine coon cat or a vertebrate.

This distinction between two main types of classification is a very useful one. There are a number of reasons, however, for saying that it is not an absolute distinction. Indeed, one could say that we all probably have our own prototype of the ideal Aristotelian classification system, but that no one system in practice fully meets a single set of Aristotelian requirements (Sweetser 1987). As Coleman and Kay note, while blackboxing the notion of "knowledge of the occasions":

It seems that the use of some words, like *lie*, may depend on two sorts of considerations. One is the traditional question of what count as criteria for classifying a real-world thing in the category: perhaps we would like to reserve the term *semantic prototype* for this constellation of things. But a second consideration is knowledge of the occasions, reasons, etc., for deciding whether or not to classify something in a particular way. A frequent reason for reporting something as a lie is that we want to blame or criticize the person who said it. (1981, 37)

Our analysis here stresses precisely this latter criterion of "in practice." Turning to an example from the workplace, it is possible to begin to see how practice and location mediate such divisions. In the medical arena, the criterion emerged from a survey of physicians in 1979 in the United Kingdom that general practitioners, "had a constant tendency to regard a wider range of phenomena as disease" than the hospital physicians, who in turn were more inclusive than the lay public. The perceived need for medical intervention was the determining axis (Prins 1981, 176; Campbell, Scadding, and Roberts 1979). An influential factor, Prins notes, appears to have been whether or not medical intervention was required—for the lay public measles and mumps might be prototypical diseases; but arthritis, a card-carrying ICD-10 disease, might be seen rather as a condition.

So why do we sometimes appear in practice prototypical in our classifications, even if in principal we are Aristotelian? For two main reasons: because each classification system is tied to a particular set of coding practices; and because classification systems in general (we are not making this as an ex cathedra pronouncement) reflect the conflicting, contradictory motives of the sociotechnical situations that gave rise to them. Ritvo notes a similar phenomenon in eighteenth-century zoological classification, and for the same reasons; she states that:

Eighteenth-century systems reflected competing, if unacknowledged, principles of organization that undermined both their schematic novelty and their claim to be based on objective analysis of the natural world. These competing principles usually divided animals into groups based not on their physical characteristics but on subjective perceptions of them. . . . Rather than analyzing nature exclusively on its own terms—the claim embodied in their formal systems—naturalists often implicitly presented it in terms of its relationship to people, even constructing formal categories that echoed the anthropocentric and sentimental projection characteristic of both the bestiary tradition they had so emphatically discarded and (then as now) of much vernacular discourse about animals. (1997, 38–39)

Goldstein (1987, 379) also notes that prototypical categories are themselves manufactured, accented, and dramaturgically presented. In her discussion of the development of neurological categorization in the nineteenth century, she notes the,

. . . theatricality of Charcot's Friday lessons, where patients in nervous crisis and hypnotic trance were exhibited before an avid audience, including artists and litterateurs as well as physicians. When Charcot lectured on tremors, for example, the afflicted patients appeared wearing headdresses decorated with long plumes, whose distinctive, feathery vibrations illustrated the different varieties of the pathology. (Goldstein 1987, 169–171)

At any given moment, she points out, a particular category may become famous or politicized, or seize the popular imagination. This is of course the case throughout the worlds of classification.

### Practices

Consider the ICD. When originally drawn up, it had a maximum of 200 categories. As we note above, this was not the number of diseases in the world, but the number of lines on Austrian census forms. If too many diseases got identified then there would be no way of maintaining and analyzing registers of causes of death, as the technology would not hold more information.

In addition to this inheritance, there is a practical Occam's razor. When doctors come to code causes of death they are frequently faced with a set of difficult judgments (which may require an autopsy and further diagnostic work). They can simply go for the easiest way, by using a generalized 'other' category. They can then get back to dealing with their live patients (Fagot-Largeault 1989, chapter 3). So the classical beauty of the Aristotelian classification gives way to a fuzzier classification system that shares in practice key features with common sense prototype classifications—heterogeneous objects linked by metaphor or analogy.

The powerful habits of practice with respect to the humble tasks of filling out forms are often neglected in studies of classifying. Goodwin (1996) provides an elegant description of working student archaeologists matching patches of earth against a standard set of color patches in the Munsell color charts. He argues that earlier cognitive anthropological work on color assumed a universal genetic origin for color recognition, but failed to examine the kinds of practices that informed the ways in which color tests were designed and carried out in the course of this research. He notes:

Rather than standing alone as self-explicating textual objects, forms are embedded within webs of socially organized, situated practices. In order to make an entry in the slot provided for color an archaeologist must make use of another tool, the set of standard color samples provided by a Munsell chart. This chart incorporates into a portable physical object the results of a long history of scientific investigation of the properties of color. The version of this chart that archaeologists bring into the field has been tailored to the distinctive requirements of their work situation. (1996, 66)

The archaeologists constantly compare the pieces of earth against the chart, negotiate with each other, and transform their everyday terms for the earth into the formal numbered categories on the chart. The uncertainties they face along the way are removed once the numbers are selected and reported: "The definitiveness provided by a coding scheme typically erases from subsequent documentation the cognitive and perceptual uncertainties that these students are grappling with, as well as the work practices within which they are embedded"

(Goodwin 1996, 78; see also Star 1983). In general, classificatory work practices involve politics, kinds of both prototypical and Aristotelian classifications, and deletion of the practices in the production of the final formal record.

### Contradictory Requirements of Classification Systems in General

Classification systems in general inherit contradictory motives in the circumstances of their creation. This is very clearly illustrated by items in the ICD covering such charged ethical or religious issues as abortion or stillbirth. Over the years, as we will discuss in the next chapter, defining the moment of birth differed radically from Protestant to Catholic countries and with technological changes. The final definitions given in the ICD directly reflect the charged political and ethical atmosphere of the subject, distinguishing, for example, legal and illegal abortion as separate categories. In this sense, the ICD can also be read as a kind of treaty, a bloodless set of numbers obscuring the behind-the-scenes battles informing its creation. This dryness itself contains an implicit authority, appearing to rise above uncertainty, power struggles, and the impermanence of the compromises.

Indeed, one might observe that technical classification schemes are constructed in such a way as to fit our common-sense prototypical picture of what a technical classification is. Thus when the International Committee for the Nomenclature of Viruses, to which we shall return, floated the idea of using "siglas"—a series of code letters attached to the virus name to indicate its characteristics—Matthews describes the response: "Leading virology journals were only lukewarm to try out cryptogram ideas. Among comments from this period: 'Why should they be given *funny names?* Are we not exposing ourselves to the laughter of the general public? Do we want to join the ranks of old-fashioned botanists and zoologists so soon?'" (Matthews 1983, 13–14). A good technical classification should not only be correct in Aristotelian terms, it should, in good prototypical fashion, look and feel scientific. This is not an isolated case—the developers of the Nursing Interventions Classification (NIC) have made similar observations for example (as we shall see in chapter 7, they initially did not classify "leech therapy" not because it was not a scientific intervention but because it did not look and feel like one). With respect to the ICD, there has been a long debate within its patient community about naming chronic fatigue syndrome (CFS) for example (as there was for

AIDS). Consider this discussion among patients suffering from chronic fatigue syndrome. "Many patients feel that one of the greatest burdens of having chronic fatigue syndrome is the name of the illness. The word fatigue (which many patients refer to as the 'F' word) indicates everyday tiredness. It reinforces negative perceptions that remain with the public and most medical doctors, despite a decade of steady, gradual research advances" (Chronic Fatigue Syndrome Electronic Newsletter, 20 February 1997). One option was to name it after Darwin, but it was felt that although he had the scientific cachet, he didn't necessarily have the disease. Inversely, Florence Nightingale's diagnosis is somewhat more certain but less prestigious:

Nightingale's. (A general note: no historical figure has been definitively diagnosed with CFS/M.E. Purists may object to choosing any person in history, who may not have actually had the disease, as the basis for an eponym.) Florence Nightingale is a widely respected and world-renowned figure who founded the International Red Cross and the first formal school for nursing. For decades she had an undiagnosed, severely debilitating illness, whose symptoms were similar to CFS. Despite Nightingale's considerable talents and her personal character, many doubted that she had a physical illness. Her illness was quite controversial. A 1996 paper by D.A.B. Young which appeared in the British Medical Journal indicates that Nightingale's illness was likely to have been chronic brucellosis (a disease with symptoms similar but not identical to CFS). Patient groups have promoted Nightingale's birthday, May 12, as International CFIDS/M.E. Awareness Day, and Nightingale is a familiar symbol to those who know this disease. However, some argue that women's diseases often have difficulty in getting recognized and accepted. Choosing Nightingale's name as an eponym might add to, rather than offer relief from, current name-associated problems. (Chronic Fatigue Syndrome Electronic Newsletter, 20 February 1997)

More generally, Taylor from a linguistic perspective and Durkheim and Mauss—for whom primitive, social classifications, "seem to link, without any discontinuity, with the first scientific classifications" (Durkheim and Mauss 1969, 82)—from an anthropological one have observed that technical classifications grow out of and have to answer to our common sense, socially comfortable classifications. It just would not be socially feasible to call a donkey a fish, no matter how good your scientific grounds.

There is no great divide between folk and scientific classifications. Below, we discuss one particular fault line between the two: a fracture that is constantly being redefined and changing its nature as the plate of lived experience is subducted under the crust of scientific knowledge. This fault line is the ways in which temporal experiences—history, events, development, memory, evolution—are registered in and expressed by two formal classification systems, the ICD and the INV. The crack comes when the messy flow of bodily and natural experience must be ordered against a formal, neat set of categories. We will trace this particular fault line across the two classification schemes. It is the case that all complex classification schemes have multiple sets of faults and fractures arising from similar tensions. Chapter 5 sets forth a model of the fault process as it occurs for clinical and bodily trajectories in tuberculosis. On a meta level, the system of faults and tensions forms a kind of texture of any given organizational terrain; mapping this texture is a major research challenge for the field of social informatics.

### The International Classification of Diseases Is a Pragmatic Classification

To communicate information in the aggregate, we must first classify. At any time over the past 100 years one can find complaints about the Tower of Babel that afflicts the storage and communication of medical knowledge.[12] David Rothwell notes that "More than two hundred statistical systems are being used by the United States government to monitor health, occupational and environmental conditions through the country. Despite the incredible amount of information accumulated, there is no method of coordinating these data into a single coherent database, a national health information system" (1985, 169). Mark Musen complains:

The medical-informatics community suffers from a failure to communicate. The terms that WMR uses to describe patient findings generally are not recognized by Medline. The manner in which Iliad stores descriptions of diseases is different from that of Dxplain. Therapy plans generated by ONCOCIN are meaningless to the HELP system. . . . Each time another developer describes yet another formalism for encoding medical knowledge, the number of incompatibilities among these different systems increases exponentially. (Musen 1992, 435)

He points out that there is no clear relationship between "the Unified Medical Language System [UMLS] advanced by the National Library of Medicine and the Arden syntax proposed by the American Society for Testing and Materials as a standard for representing medical knowledge" (Musen 1992, 436). The ICD, he points out, originated as a means for describing causes of death; a trace of its heritage is its continued difficulty with describing chronic as opposed to acute forms of disease. This is one basis for the temporal fault lines that emerge in its usage. The UMLS originated as a means of information retrieval (the MeSH scheme) and is not as sensitive to clinical conditions as it might be (Musen 1992, 440).

The two basic problems for any overarching classification scheme in a rapidly changing and complex field can be described as follows. First, any classificatory decision made now might by its nature block off valuable future developments. If we decide that all instances of sudden infant death syndrome (SIDS) are to be placed into a single box (R95 in ICD-10), then we are not recording information that might be used by future researchers to distinguish possible multiple social or environmental causes of SIDS. We are not making it impossible to carry out such studies; but we are making it difficult to retrieve information. Inversely, if every possibly relevant piece of information were stored in the scheme, it would be entirely unwieldy.

For these reasons, the decision not to collect is the most difficult to take for people maintaining any sort of collection based on a classification system, whether it be the acquisition department of a library, the curator of an art museum, or the collector of information for vital statistics. There are always practical budget and storage issues. These are balanced against two other factors, the need for a well-ordered and in some sense parsimonious repository that can be used, and the side bets that are made about what material will be useful in the future. This latter is particularly difficult. Collectors and curators of all sorts must become future forecasters and decide the boundaries of what will be useful for the future. There is no perfect answer, only a set of practical tradeoffs. This is a problem that has plagued museums of natural history. Fossils found in the nineteenth century might come along with general information about the depth at which they were discovered and the surrounding geological features (though they often did not). Even if this information was included, it was never as precisely noted as would be useful for geologists and paleontologists today: since there was no conception at that stage of the kinds of dating techniques that are used nowadays. The museum is then faced with the choice between recording as much as possible now (which is very expensive and possibly not useful anyway) and having the collection perhaps last longer into the future, or recording a judicious amount now (which will keep the administrative costs down) and having the collection possibly be not so useful in the future. The latter has generally been the de facto choice and is generally a reasonable one to have made since new criteria of relevance cannot be predicted.

Second, different designers of the classification system have different needs, and the shifting ecology of relationships among the disciplines using the classification will necessarily be reflected in the scheme itself. As with the insurance company example above, these relationships must be resolved to make a usable form, often obscuring power relationships in the process. As Goodwin notes, "A quite different kind of multivocality, one organized by the craft requirements of a work task rather than the genres of the literary academy, can be found in mundane, bureaucratic forms" (1996, 66). But one must dig to find the voices. The process of filling out the forms may further obscure them. For example, the designers of the ICD recommend that its classification scheme be interpreted economically:

The condition to be used for single-condition morbidity analysis is the main condition treated or investigated during the relevant episode of health care. The main condition is defined as the condition, diagnosed at the end of the episode of health care, primarily responsible for the patient's need for treatment or investigation. If there is more than one such condition, the one held most responsible for the greatest use of resources should be selected. (ICD-10, 2: 96)

This reflects a constant condition of the use of the ICD: it has been recommended throughout its history that priority should be given to coding diseases that represent a threat to public health. This goal is clearly a good one; equally clearly it can discriminate selectively against the reporting of rare noncontagious conditions. In chapter 4 we discuss an ongoing battle between statisticians (who are not generally interested in the very rare occurrences of disease) and public health officials (who want to know about even one case of bubonic plague or Ebola!).

Faced with these problems, the WHO has been consistently pragmatic in its aims and clear in its explanations of the ICD. From the time of the ninth revision on, it has been recognized explicitly that "the ICD alone could not cover all the information required and that only a 'family' of disease and health related classifications would meet the different requirements in public health" (IDC-10, 2: 20). This "family" is pictured in ICD-10 as shown in figure 2.2.

The family itself is a diverse one: there are various standard modifications of the ICD. The most significant in the United States is the ICD-9-CM, where CM stands for "clinical modification." This

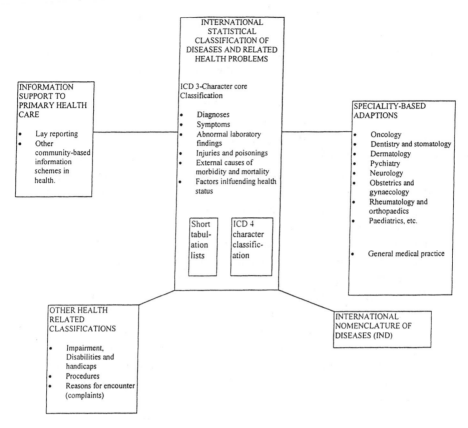

**Figure 2.2**
Pigeonholing the classification—the ICD family.
Source: Adapted from ICD-10, 1.

document has a complex history, tracing back to the development of modifications of the ICD for use in hospital information systems. It is now the classification of record in a wide variety of medical settings, and it is used for billing, insurance, and administration as well as in patient medical records. This institutional entrenchment of ICD-9-CM has made it very difficult for ICD-10 to be fully adopted in the United States: the clinical modification necessarily lags behind the production of the classification itself.

When we come to look at the ways in which culture and practice interweave in the text of the ICD, we are not unmasking a false pretender to the crown of science; we are drawing attention to an explicit, positive and practical feature of ICD design: "The ICD has developed as a practical, rather than a purely theoretical classification.

. . . There have . . . been adjustments to meet the variety of statistical applications for which the ICD is designed, such as mortality, morbidity, social security and other types of health statistics and surveys" (ICD-10, 2: 12). The preamble to the classification defines a classification of diseases as: 'a system of categories to which morbid entities are assigned according to established criteria' (ICD-10, 2, vol. 1). A *statistical* classification, such as the ICD "must encompass the entire range of morbid conditions within a manageable number of categories" (ICD-10, 2: 1). It is not meant to be a net to capture all knowledge, but rather a workable epidemiological tool. This practical goal does not make it less scientific. All classification systems are developed within a context of organizational practice. The goal of the ICD's designers is to create what Latour (1988) has called immutable mobiles, inscriptions that may travel unchanged and be combinable and comparable. Indeed the term immutable mobile might almost have been in the designers' minds when they wrote, "The purpose of the ICD is to permit the systematic recording, analysis, interpretation, and comparison of mortality and morbidity data collected in different countries or areas and at different times. The ICD is used to translate diagnoses of diseases and other health problems from words into an alphanumeric code, which permits easy storage, retrieval and analysis of the data" (ICD-10, 2: 2). The ICD has become the international tool for "standard diagnostic classification for all general epidemiological and many health management purposes" (ICD-10, 2: 2).

The world has changed since the ICD was first introduced, and the classification scheme has evolved to try to encompass these changes. The ICD is thus both highly responsive and tightly constrained. A large-scale change in the way that people die (Israel et al. 1986, 161) has led to the addition of one line in the internationally recommended death certificate (see figure 2.3). One of the main bureaucratic uses of the ICD is the recording and compiling of causes of death from bureaus of vital statistics via coroners, hospitals, doctors, or priests:

In considering the international form of medical certificate of cause of death, the Expert Committee had recognized that the situation of an aging population with a greater proportion of deaths involving multiple disease processes, and the effects of associated therapeutic interventions, tended to increase the number of possible statements between the underlying cause and the direct cause of death: this meant that an increasing number of conditions were being entered on death certificates in many countries. This led the committee to recommend the inclusion of an additional line (d) in Part 1 of the certificate. (ICD-10, 1: 18)

| Cause of Death | | Approximate interval between onset and death |
|---|---|---|
| I | | |
| Disease or condition directly leading to death* | (a) ............................. | ................................. |
| | due to (or as a consequence of) | |
| *Antecedent causes* | (b) ............................. | ................................. |
| Morbid conditions, if any, giving rise to the above cause, stating the underlying condition last | due to (or as a consequence of) | |
| | (c) ............................. | ................................. |
| | (d) ............................. | ............................. |
| II Other significant conditions contributing to the death, but not related to the disease or condition causing it | ................................. | ................................. |
| | ................................. | ................................. |

\*      This does not mean the mode of dying, e.g. heart failure, respiratory failure. It means the disease. injury, or complication that caused death.

Source: Adapted from Fagot-Largeault, 1986.

**Figure 2.3**
A standard international death certificate.
Source: Adapted from Fagot-Largeault 1986.

Thus there is now one more blank line on the form to indicate multiple causation.

A major change incorporated in the classification scheme in the last two revisions has been the so-called "dagger and asterisk" system, this is a means of cross-referencing manifestations and underlying causes for a particular disease. The ICD and its instruments have thus, through a pair of small-scale formal changes (a line here and an asterisk there), loosened up their implicit causality and thus their picture of the past. Histories now can be more fluid than they once could. More complex narratives are possible.

The classification scheme is responsive to changes in medicine and medical technology in many ways; there are constant changes in the allopathic understanding and description of diseases reflected in the classification scheme itself. The development of new diagnostic technology in the 1940s, for example, led to the reclassification of tuberculosis (otherwise there would have been too many cases). The 1955 edition of *Diagnostic Standards and Classification of Tuberculosis* notes that new laboratory tests had made it more difficult to decide whether a particular case of tuberculosis was active or inactive, since activity could now be seen at sites previously considered inactive. At the same time, one would not necessarily want to call the "new" active sites cases of tuberculosis, since they very well may not progress to the point of needing treatment. The committee cites the 1955 version of the book. "The Committee, however, recognizes the fact that all classifications are ephemeral. They are useful only as long as they serve their purpose. The purpose of a clinical classification of tuberculosis is, however, a most important one. On it depend such matters as legal requirements for isolation, medico-legal considerations with respect to compensation for disability, standards for the return of patients to work, and similar matters" (Diagnostic Standards and Classification of Tuberculosis, 1955: 6). We will discuss the classification of tuberculosis in more detail in chapter 5. For another similar example, the discovery of the lentiviruses led to the description of a new set of disease entities: slow-acting viruses from which one could suffer asymptotically for extended periods.

In the interests of creating a working infrastructure, Aristotelian principles are here deliberately violated:

C15       Malignant neoplasms of oesophagus
Note:     Two alternative subclassifications are given:
.0–.2 by anatomical description
.3–.5 by thirds

This departure from the principle that categories should be mutually exclusive is deliberate, since both forms of terminology are in use but the resulting anatomical divisions are not analogous. (ICD-10, 1: 190)

Where the state of the art is unclear, so is the scheme itself,

Note: The terms used in categories C82-C85 for non-Hodgkin's lymphomas are those of the Working Formulation, which attempt to find common ground among several major classification schemes. The terms used in these schemes are not given in the Tabular List but appear in the Alphabetical Index; exact equivalence with the terms appearing in the Tabular List is not always possible.
Includes:     morphology codes M959–M994 with behaviour code /3.

Excludes:    secondary and unspecified neoplasm of lymph nodes (C77.-).
(ICD-10, 1: 215)

There are several specialty-based adaptations of the ICD originating
in different national or international bodies (dermatology, stemming
from the British Association of Dermatologists; and, under development, rheumatology and orthopaedics from the International League
against Rheumatism). (ICD 10, 2, vol. 5–6).

The ICD is also directly responsive to other types of changes in the
world. Diseases themselves die (smallpox), are superseded (Gay-
Related Immune Disorder (GRID) becomes AIDS), are newly born
(radiation sickness with the discovery of radium), or fall into disrepute
(hysteria or neurasthenia). Since the ICD is a statistical classification,
a disease with no incidence is of no interest. Thus smallpox was still
well defined within ICD-9,

050   Smallpox
Excludes:        arthropod-borne viral diseases (060.0–066.9)
                 Boston exanthem (048)
        50.1   Variola major
               hemorrhagic (pustular) smallpox Malignant smallpox
    Purpura variolosa
        50.1   Alastrim
               Variola minor
        50.2   Modified smallpox
               Varioloid
        050.9          Smallpox, unspecified
(ICD 9CM: 11)

By ICD-10 this had collapsed into: "BO3 Smallpox," with a footnote:
"In 1980 the thirty-third World Health Assembly declared that smallpox had been eradicated. The classification is maintained for surveillance purposes" (ICD-10, 1, 150). Or again, malnutrition is defined in
relativistic fashion—as the population changes so does the definition:

The degree of malnutrition is usually measured in terms of weight, expressed
in standard deviations from the mean of the relevant reference population.
When one or more previous measurements are available, lack of weight gain
in children, or evidence of weight loss in children or adults, is usually indica-
tive of malnutrition. When only one measurement is available, the diagnosis
is based on probabilities and is not definitive without other clinical or laboratory tests. In the exceptional circumstances that no measurement of weight is
available, reliance should be placed on clinical evidence. (ICD-10, 1: 290)

In these cases, then, the fact that the world is changing is reflected
directly in the classification scheme. Another source for this recogni-

tion is the development of accident categories, which also in their explanations display a historical cultural specificity. For example, this set of accident categories describes a series of tumbles more common in the industrial world than for a nomadic tribe:

E884    Other fall from one level to another
E884.0   Fall from playground equipment
                Excludes: recreational machinery (E919.8)
E884.1   Fall from cliff
E884.2   Fall from chair
E884.3   Fall from wheelchair
E884.4   Fall from bed
E884.5   Fall from other furniture
E884.6   Fall from commode
                Toilet
E884.9   Other fall from one level to another
                Fall from:        Fall from:
                embankment      stationary vehicle
                haystack         tree

(ICD-9-CM, 289)

There is a relatively impoverished vocabulary for talking about natural accidents. The ICD is richest in its description of ways of dying in developed countries at this moment in history; it is not that other accidents and diseases cannot be described, but they cannot be described in as much detail. Differentiating insect bites and snake bites, for example, is very important for those living in the rural tropics. While arthropods, centipedes, and chiggers are singled out under "bites" in the ICD index, however, snakes are only divided into venomous and nonvenomous, as are spiders.[13] Clearly this makes sense to some extent, given that this is a pragmatic classification. There is only a point in making fine distinctions between types of accident if those distinctions might make a difference in practice to some agency— medical or other. At the same time, those agencies have traditionally been more accountable to Western allopathic medicine and to the industrial world than to traditional indigenous or alternative systems.

So the ICD bears traces of its historical situation as a tool used by public health officials in developed countries. It also reflects changes in the world at large, either the eradication of diseases or culturally charged changing understandings of certain conditions. Further, it is very much an entrenched scheme. There is a natural reluctance to operate changes, since each change renders a previous set of statistics incomparable and thence less useful.

The first and last entries in the ICD describe a sociotechnical trajectory. The first disease in the ICD over the years has been cholera; unsurprising, since cholera was the issue that in the 1850s brought participants to the table in an attempt to deal with it as an international threat. As we noted in the introduction, this threat was exacerbated by the development of steamship technology, which allowed cholera sufferers to carry the disease back to Europe before dying. The last condition given in the book takes us to the other end of the sociotechnical arc—the creation of cyborgs.

The last condition listed in the ICD is: Z99 "Dependence on enabling machines and devices, not elsewhere classified"; with the very last entry, Z99.9, being "Dependence on unspecified enabling machine and device." By some standards we all now qualify for the Z99.9 condition.

The original sequence produced by William Farr (1885, 232) is reproduced in the latest ICD:

The ICD is a variable-axis classification. The structure has developed out of that proposed by William Farr in the early days of international discussions on classification structures. His scheme was that, for practical, epidemiological purposes, statistical data on diseases should be grouped in the following way:

- epidemic diseases
- constitutional or general diseases
- local diseases arranged by site
- developmental diseases
- injuries

This pattern can be identified in the chapters of ICD-10. It has stood the test of time and, though in some ways arbitrary, is still regarded as a more useful structure for general epidemiological purposes than any of the alternatives tested. (ICD-10, 1: 13)

This classification scheme, then, makes no exaggerated claims to timeless truth. To the contrary. Its designers have attempted to paint a fluid picture of the world of disease—one that is sensitive to changes in the world, to sociotechnical conditions, and to the work practices of statisticians and record keepers.

### There Are Many Aids to Storytelling in the ICD

The classification system that is the ICD does more than provide a series of boxes into which diseases can be put; it also encapsulates a

series of stories that are the preferred narratives of the ICD's designers. Certain attributions of intentionality are easy to make; others are
rather difficult. Some ways of life are clearly considered to be well led,
others are called into question. Sometimes context is important; sometimes it can be ignored. Stories also come and go, narratives fade in
importance (the example of AIDS moving, in medical terms, from a
specifically gay-linked disease to a more general one). If one should
have doubts about how to encode a given story, one can turn to volume
2 of the classification, which gives an extensive set of rules for the
interpretation of causes of death. In this section, we will look at the
various aids to storytelling to be found within the ICD.

### The Setting

Frequently, when diseases have first been named, they have taken on
the name of their first scientific describer, of a famous victim, or of the
place where they occur. Each of these kinds of naming strategy tells a
simple story to accompany the classification. Throughout the history
of classification systems over the past 200 years such specifications have
progressively been winnowed away to make way for new kinds of
context and new kinds of description now considered more interesting
and relevant.

What many sufferers of amyotrophic lateral sclerosis know as Lou
Gehrig's disease is coded by the ICD-10 as G12.2: motor neuron
disease. (With the famous physicist Stephen Hawking now suffering
from the disease, it may in future be more well known to the lay public
as Hawking's disease, though baseball player Lou Gehrig first brought
it to public awareness.) In the index to the ICD, the Parisian neurologist
Charcot can lay claim to an arthropathy (tabetic), a cirrhosis, and a
syndrome. In the body of the text, the great doctor tends to slip away;
Charcot's syndrome becomes I73.9 peripheral vascular disease, unspecified, and there is no mention of Charcot. The I73s (other peripheral
vascular diseases) are an interesting category. They show the various
forms of modality: I73.0 is still proudly Raynaud's syndrome, I73.1 is
thromboangiitis obliterans [Buerger], I73.8 is other specified peripheral
vascular diseases, and includes acroparaesthesia—simple [Schultze's
type] or vasomotor [Nothnagel's type]. In general, as the modalities get
deleted the name of the person goes from being the name of the disease
to a bracket after the name to an entry in the index, until finally it slides
gracefully out of the index onto the scrap heap of history. A similar

process occurs with deletion of detail and the uncertainties of discovery in any scientific publication, as Latour and Woolgar noted in their classic *Laboratory Life* (1979; see also Star 1983).

Places follow a similar path to abstraction and formal representation. The ideal ICD disease is not tied to a particular spot, it is rather identified with a particular causal agent. Up to and including ICD-9, however, leishmaniasis was a classification that told a travelers' tale; not only do we know what you got sick of but where you got sick:

085   Leishmaniasis
  085.0   Visceral [kalaazar]
        Dumdum fever Leishmaniasis:
        Infection by Leishmania:    dermal, post-kala-azar
          donovani            Mediterranean
          infantum            visceral (Indian)
  085.1   Cutaneous, urban
        Aleppo boil         Leishmaniasis,
        Baghdad boil       cutaneous:
        Delhi boil          dry form
        Infection by Leishmania  late
        tropica (minor)      recurrent
                          Ulcerating
              Oriental sore
  085.2   Cutaneous, Asian desert
        Infection by Leishmania tropica major
        Leishmaniasis, cutaneous:
          Acute necrotizing
          Rural
              Wet form
          Zoonotic form
  085.3   Cutaneous, Ethiopian
        Infection by Leishmania ethiopica
        Leishmaniasis, cutaneous:
          Diffuse
          Lepromatous
  085.4   Cutaneous, American
          Chiclero ulcer
          Infection by Leishmania mexicana
          Leishmaniasis tegumentaria diffusa
  085.5   Mucocutaneous (American)
          Espundia
          Infection by Leishmania braziliensis
          Uta
  085.9   Leishmaniasis, unspecified
(ICD-9-CM, 16)

Similarly, for ICD-10, we can still find "Delhi boil" in the index, but the main entry itself is a svelte:

B55        Leishmaniasis
B55.0      Visceral leishmaniasis
           Kala-azar
           Post kala-azar dermal leishmaniasis
B55.1      Cutaneous leishmaniasis
B55.2      Mucocutaneous leishmaniasis
B55.9      Leishmaniasis, unspecified

(ICD-10, 1: 166)

So we go from primacy being given to a place (Baghdad boil) to primacy being given to a kind of place (urban cutaneous) to primacy given to a universal agent. Gradually the narrative of travel inscribed in the disease code and thus on the patient's form, present earlier, is deleted.

The loss of eponymy and place markers can of course be read as a story of the advance of science: the replacement of the local and specific with the general; the thing with the kind; the mutable immobile with the immutable mobile, and the concrete instance with the formal abstraction. Another line of argument, however, also deserves attention. As we have already seen, the ICD reflects historical states of the world, and the world has changed. With the huge increase in international travel over the past century and a half, it is more rare for a disease to be tied to any one particular location; rather diseases themselves tend to spread to *kinds* of location. The malaria map of the world hanging on the wall at the WHO headquarters in Geneva shows the expected tropical venues, and it also shows small red circles around major airports as mosquitoes are transported from the tropics. We are as a world becoming more abstract in this way.

Similarly, research now is not attributed to single great figures who can claim sole responsibility for a discovery. Medical work was always done in teams, but they have become larger, involving complex social and institutional relationships of attribution as Gallo and Montaignier would be the first to remind us (Grmek 1990). A typical scientific article has so many authors that the death of the individual scientific author appears certain. In general, the ICD has gone from being the holder of a set of stories about places visited, heroic sufferers, and great doctors to holding another set of stories.

### The Context of Disease

As people and places have moved out of eponymous and loconymous classification, these specific categories are replaced by a general set, what we call the kindness of strangers. By this we mean that the classification system operates a shift away from our being individuals experiencing the world to our being kinds of people experiencing kinds of places. The constructions of social and natural science and of the legal world have moved in. Broken legs and ski resort locations coevolve, as do cancer rates and toxic waste dumps. The classification system, as we shall see in this section, has become a site that holds these constructions together and, through excluding other kinds of story, makes them more real. With the ICD providing the main legitimate means for describing illness, the social, economic, and political stories woven into its fabric become by extension the main legitimate narrative threads for the science of medicine.

Although particular places have moved out, two places have come to play a more significant role in the classification system, the laboratory and the "sociological home." This latter appears in the extra categories developed for ICD-9 as supplemental codes, which in ICD-10 have become fully integrated, what we might call the context codes. Thus housing is one of the conditions that can be broken down and described as part of the classification. In ICD-9 it is described as follows:

| | |
|---|---|
| V60 | Housing, household and economic circumstances |
| V60.0 | Lack of housing |

|  |  |
|---|---|
| Hobos | Transients |
| Social migrants | Vagabonds |
| Tramps | |

| | |
|---|---|
| V60.1 | Inadequate housing |

Lack of heating
Restriction of space
Technical defects in home preventing adequate care

| | |
|---|---|
| V60.2 | Inadequate material resources |

Economic problem    Poverty NOS

| | |
|---|---|
| V60.3 | Person living alone |
| V60.6 | Person living in residential institution |

Boarding school resident

| | |
|---|---|
| V60.8 | Other specified housing or economic circumstances |
| V60.9 | Unspecified housing or economic circumstances. |

(ICD-9, 1: 267)

The related code in ICD-10 is expanded to include discord with neighbors and lack of adequate food (ICD-10, 1: 1,152). In both, the name of the city gives way to the name of the social category and social condition.

These context codes define what is considered to be medically relevant in one's material surroundings. They make it easy to structure studies in these terms (for example, what effect does poor housing have on the incidence of tuberculosis?). At the same time, these codes do make it much more difficult to deal with unrecognized contexts (what effect does conspicuous consumption have on cholesterol levels?). It is not impossible to do these latter studies, but the information is not at hand in the way that it is for medically sanctioned contexts. The reason we stress this point is that it can be taken as a sign of the correctness of allopathic medicine: it has isolated the basic variables that need to be taken into account in the development of public health policy or medical science. Although the ICD is a powerful tool in this sense it also, as infrastructure, enforces a certain understanding of context, place, and time. It makes a certain set of discoveries, which validate its own framework, much more likely than an alternative set outside of the framework, since the economic cost of producing a study outside of the framework of normal data collection is necessarily much higher.

This sort of convergence is an important feature of large-scale networked information systems. Convergence, again, is the double process by which information artifacts and social worlds are fitted to each other and come together (Star, Bowker, and Neumann, in press). On the one hand, a given information artifact (a classification system, a database, an interface, and so forth) is partially constitutive of some social world. The sharing of information resources and tools is a dimension of any coherent community, be it the world of homeless people in Los Angeles sharing survival knowledge via street gossip, or the world of high-energy physicists sharing electronic preprints via the Los Alamos archive. On the other hand, any given social world itself generates many interlinked information artifacts. The social world creates through bricolage, a (loosely coupled but relatively coherent) set of information resources and tools. Thus people without houses also log onto the Internet, and physicists indulge in street gossip at conferences as well as engage in a whole set of other information practices. Put briefly, information artifacts undergird social worlds, and social worlds undergird these same information resources. We will use the concept of convergence to describe this process of mutual constitution.

With these processes of convergence, the site of the medical work itself has gained in importance. The classification of tuberculosis, canonically difficult to diagnose accurately (see chapter 5; and compare Latour, forthcoming) retains the story of what has been done in the laboratory as well as what has occurred in the body. (In chapter 4 we discuss the intersection of these different forms of time.)

A15   Respiratory tuberculosis, bacteriologically and histologically confirmed

A15.0   Tuberculosis of lung, confirmed by sputum microscopy with or without culture
      Tuberculous,

     • bronchiectasis   }

     • fibrosis of lung   }

     • pneumonia     } confirmed by sputum microscopy with or
               } without culture

     • pneumothorax   }

A15.1   Tuberculosis of lung, confirmed by culture only
      Conditions listed in A15.0, confirmed by culture only

A15.2   Tuberculosis of lung, confirmed histologically
      Conditions listed in A15.0, confirmed histologically

A15.3   Tuberculosis of lung, confirmed by unspecified means
      Conditions listed in A15.0, confirmed but unspecified whether bacteriologically or histologically

(ICD-10, 1: 113)

In this case, the disease itself is always classified in terms of the work that has been done in the medical laboratory. Again, as new technologies are invented, historical shifts occur, as seen with the relationship between epilepsy and the EEG machine as diagnostic many decades ago.

 The doctors themselves enter the story at the moment of classification, while the patient rarely does. This comes out clearly if we compare migraine and epilepsy in ICD-9. Epilepsy is a condition that is defined by the doctor in the context of laboratory and so is a real condition:

345   Epilepsy
The following fifth-digit subclassification is for use with categories 345.0, .1, .4–.9:
0     without mention of intractable epilepsy
1     with intractable epilepsy
(ICD-9CM, 80)

So here the question is whether or not the patient objectively has intractable epilepsy in the opinion of the doctor. The determination of intractable migraine, however, relies on the voice of the patient and so is marked as a suspicious designation:

346        Migraine
The following fifth-digit subclassification is for use with category 346:
0          without mention of intractable migraine
1          with intractable migraine, so stated

(ICD-9CM, 80)

The laboratory context then is the 'real' context of the disease; the classification serves to reinforce the separation of the patient from ownership of their condition. We should note at this point that we are not arguing that this makes the ICD a tool for evil and oppression. On the contrary. What we are trying to do is work out what kind of a tool it is, what work it does, and whose voice appears in its unfolding narrative.

The legal context is often enfolded into the classification system. Thus the classification of blindness takes account of the American system of medical benefits:

369        Blindness and low vision
    Note: visual impairment refers to a functional limitation of the eye (e.g., limited visual acuity or visual field). It should be distinguished from visual disability, indicating a limitation of the abilities of the individual (e.g., limited reading skills, vocational skills), and from visual handicap, indicating a limitation of personal and socioeconomic independence (e.g., limited ability, limited employment.)
    The levels of impairment defined in the table on page 92 are based on the recommendations of the WHO Study Group on Prevention of Blindness (Geneva, November 6–10 1972, WHO Technical Report Series 518), and of the International Council of Ophthalmology (1976).
    Note that definitions of blindness vary in different settings.
    For international reporting WHO defines blindness as profound impairment. This definition can be applied to blindness of one eye (369.1, 369.6) and to blindness of the individual (369.0).
    For determination of benefits in the United States, the definition of legal blindness as severe impairment is often used. This definition applies to blindness of the individual only.

369.0      Profound impairment, both eyes
                369.00  Impairment level not further specified
                        Blindness:
                              NOS according to WHO definition
                              both eyes

369.3    Unqualified visual loss, both eyes
            Excludes: blindness NOS:
                legal [U.S. definition] (369.4)
                WHO definition (369.00)
369.4    Legal blindness, as defined in United States
            Blindness NOS according to U.S. definition
Excludes legal blindness with specification of impairment level 9369.01–369.08, 369.11–369.14, 369.21–369.22)
(ICD-9CM, 91)

Note in the above example that "blindness of the individual" might be psychogenic, due to brain damage, or other organic cause outside the eye itself. The problem of localized versus whole organism conditions creates serious coding challenges. For example, depending on one's theory of cancer, it would be an immune disorder affecting the whole person, or a localized phenomenon to be surgically removed, and with many gray areas in between for the different types of cancer.

In the example above, the legal definition can take precedence over the cultural and social. Thus cannabis dependence has its own category, while the culturally profoundly different absinthe and glue addictions are lumped together:

304.3    Cannabis dependence
            Hashish        Marihuana
            Hemp
304.6    Other specified drug dependence
            Absinthe addiction        Glue sniffing
Excludes: tobacco dependence (305.1)
(ICD-9CM, 69–70)

Few would argue that glue sniffing and absinthe addiction are similar phenomena. The former leads to more serious physical conditions than 'cannabis dependence' (a category many would challenge), and yet it does not rate its own category. Absinthe addiction is, one suspects, a hangover from earlier days. Because the origins of the ICD were French, and absinthe abuse an important problem in Paris in the nineteenth century, it persists. These accidents of history, practice and crime contain many clues to re-narrativizing the ICD. E970 to E979 in ICD-9 is an interesting set that covers injuries *caused* by legal interventions:

Legal intervention
    Includes:    injuries inflicted by the police or other law-enforcing agents,
                including military on duty, in the course of arresting or

attempting to arrest lawbreakers, suppressing disturbance, maintain order and other legal action legal execution

Excludes:    injuries caused by civil insurrections (E990.0-E999)

(ICD-9CM, 304)

This set includes state executions. Note that civil insurrections, where the definition of legal intervention is on the table, are classified together with war. The definition of legal, of course, may be subject to its own retrospective reconstruction, as in the case of the Rodney King trial.[14]

Types of abortions, which may be to all intents and purposes medically equuivalent, are marked differently in the ICD according to their legality:

635        Legally induced abortion
                Includes: abortion of termination of pregnancy:
                    elective
                    legal
                    therapeutic
                Excludes: menstrual extraction or regulation (V25.3)
636        Illegally induced abortion
                Includes: abortion:
                    criminal
                    illegal
                    self-induced

(ICD-9CM, 154)

Each type of abortion (spontaneous or 634, legally induced, illegally induced, unspecified, failed attempted, or 638) has the same set of complications attached—nine difficulties, each accorded a digit (making it one of the most closely coded category sets in the ICD). When the issue arises, then, the ICD privileges the voice of the doctor and the laboratory over the voice of the patient; and legal discourse over cultural and social discourse. There are no mentions of what Adele Clarke calls "subtle forms of sterilization abuse" (1983) or of the abortions that never made it into any formal record that Leslie Reagan describes in her book *When Abortion Was A Crime* (1997). The controversial "abortion drug" RU486 is not mentioned in the ICD-10. One can read another order of social history from the nature of the silences in the story.

In general, the ICD carries with it its own context. This is a common feature of classification systems. One way of reading them is that they provide a stabilizing force between the natural and the social worlds. They hold in place sets of arrangements that allow us to read the natural as stable and objective and the social as tightly linked to it. For the ICD this means describing disease in a way that folds the socially and legally contingent into the classification system itself, and so naturalizes it. Inversely, the disease entity out there in the world is brought into the laboratory where the social and organizational work of its stabilization can best be guaranteed.

### Cutting Up the World

To tell stories of the sort we are most familiar with, one needs objects in the world that can be cut up spatially (Berg and Bowker 1997) and temporally into recognizable units. Narrative structures are typically formed with a moving time line, protagonists, and a dramatic structure unfolding over time. The ICD does in fact operate this kind of dissection, which we will discuss below. In the last section we saw the constitution of a context within the ICD, in this section we will see the constitution of actants to populate that context and those stories.

### Time Story One: The Life Cycle

Temporally, the classification system provides a picture of acute (temporally bounded) episodes within an otherwise well-ordered life. It is notoriously bad for describing chronic diseases: the interest is in the episode of treatment (Musen 1992).

Let us go through some temporal units presented by the ICD. Birth is extremely important and is very closely defined:

Live birth is the complete expulsion or extraction from its mother of a product of conception, irrespective of the duration of the pregnancy, which, after such separation, breathes or shows any other evidence of life, such as beating of the heart, pulsation of the umbilical cord, or definite movement of voluntary muscles, whether or not the umbilical cord has been cut or the placenta is attached; each product of such a birth is considered live born. (ICD-10, 2: 129)

We will discuss in chapter 4 the political and religious dimensions of this definition, which have been very closely attended to throughout the period of the ICD's development. For our present purposes, it is sufficient to note that time flows very quickly for the newborn, and so temporal units vary accordingly:

The neonatal period commences at birth and ends 28 completed days after birth. Neonatal deaths (deaths among live births during the first 28 completed days of life) may be subdivided into early neonatal deaths, occurring during the first 7 days of life, and late neonatal deaths, occurring after the seventh day but before 28 completed days of life.

The age at death during the first day of life (day zero) should be recorded in units of completed minutes or hours of life. For the second (day 1), third (day 2), and through 27 completed days of life, age at death should be recorded in days. (ICD-10, 2: 131)

Given the bump in mortality that occurs around birth, this is not surprising.

When we get into adult life, things start to slow down. Adults are defined in ICD9-CM (xiii) as people between 15 and 124 years old. If you make it to 125, you are "hors de catégorie!"

In this middle period, there are some indications of what constitutes a good life. It should be well ordered and rhythmic. Things should happen at the right time. Thus sexual development has its own timing:

259      Other endocrine disorders
259.0    Delay in sexual development and puberty, not elsewhere
         classified
             Delayed puberty
259.1    Precocious sexual development and puberty, not elsewhere
         classified PED
         Sexual precocity:
             NOS
             constitutional
             cryptogenic
             idiopathic
(ICD-9CM, 51)

Similarly, problems with temporal regulation of menstruation are well defined—too early, too late, too frequent, not frequent enough—natural rhythms should not be upset.

A relatively recent temporal problem in addition is jet lag:

307.45 Phase-shift disruption of 24-hour sleep-wake cycle
             Irregular sleep-wake rhythm, nonorganic origin
             Jet-lag syndrome
             Rapid time-zone change
             Shifting sleep-work schedule
(ICD-9CM, 71)

The reference to the "nonorganic origin" highlights that this is a situation-bound condition: the context (jet travel or night-shift work) is directly folded into the disease.

To an outside observer, there is remarkably little reference to the process of aging. An adult is a timeless being who should be healthy:

disease is not in general indexed by age. Further, the body is not present as something that gets used up and worn out: such stories have to be superadded. (Indeed, the category of being "worn out" was in earlier additions of the ICD but has since been removed).

If you rent a house, your agreement with the landlord includes a "fair use" or "normal wear and tear" category: it is expected that houses depreciate over time and this is written into the legal and tax codes. There are only two references to normal wear and tear in the whole ICD. First, one can as an adult step out of the well-ordered life and suffer from premature or delayed senility, puberty, birth, and aging. Among the conditions under "delay" are delayed birth, development (including intellectual, learning, reading, sexual, speech, and spelling), menstruation, and puberty. In this case the cycle structure is the same, but the patient is taking the steps too early or too late. Second—and there is only one example of this—you could use your body badly. The only specific instance of this, however, is that you can grind or otherwise mismanage your teeth:

521        Diseases of hard tissues of teeth
521.1      Excessive attrition
                Approximal wear. Occlusal wear
(ICD-10, 1: 125)

In ICD-10, abrasion of teeth carries with it an illuminating set of contexts: dentifrice, habitual, occupational, ritual, and traditional. Occupational abrasion in earlier times included the hazard "tailor's tooth," for example, where the teeth were abraded due to biting off the thread in hand sewing. In principle, the timeless adult could do many things excessively—there are categories for excessive thirst, secretion, salivation, sex drive, and sweating and binocular convergence among others. Such superfluity, however, is indexed only in this one case against an aging body. Note that there are of course diseases associated more broadly (and often implicitly) with excessive wear and tear, for example, cirrhosis of the liver associated with alcoholic excess. But here we are concerned directly with representation in the classification system.

This curious invisibility of aging as wear and tear is one way in which the ICD stabilizes context and disease entity, the human body as the substrate of both is outside the flow of time. The human adult body becomes the unmarked category, the cipher against which laboratory, social, and natural time must be coordinated. Indeed one could go a step further and see the adult male body as the unmarked category, since there are many more diseases restricted to women than restricted to men; there are sixteen categories or clusters of categories that apply only to males and forty-two that apply only to females. (ICD-10, 2: 26). Feminist critics of medicine have long remarked on the relative pathologizing of the female body (for example, Ehrenreich and English 1973).

### Nobody Dies of Old Age

To finish with the life cycle before moving on to other temporal features, we should note that death itself is remarkably poorly defined by comparison with life. One can scarcely die of old age (Fagot-Largeault 1989).[15] Unlike in the earlier editions of the ICD (see figures 2.4a and 2.4b), the closest that one may get comes under a banner disclaimer:

Ill-defined and unknown cause of morbidity and mortality (797–799)

797        Senility without mention of psychosis
           Old age
           Senile:
           Senescence debility
           Senile asthenia        exhaustion
       Excludes:        senile psychoses (290.0–290.9)
(ICD-10, 1: 215)

The ICD's life cycle for humans is as follows: a spurt of intense activity at birth; timeless adulthood, when one is afflicted with a range of woes that carry their own temporalities; and an inglorious, ill-defined end. The effect of this is, paradoxically, to make the individual an undefined, tabula rasa onto which various diseases are inscribed. From this blank sheet one can read various stories (with the aid of the ICD), restoring first context and then interpretation (which we shall deal with in the next section).

### Time Story Two: The Virus

Diseases themselves change over time. HIV, for example, mutates rapidly in the individual sufferer, so that no two people suffer from the same disease, nor may the disease be identical with itself over time even within a person. This extreme variability of the object world is a problem for any classification system. The case of virus classification illuminates many features of categorizing difficulties and the strategies used to control them. We look here at some of the work of the

# TABULAR LIST

## XII.—OLD AGE.

### 154. Senility.

*This title includes:*

Age (70y+)
Asthenia (70y+)
Atony (70y+)
Atrophy (70y+)
    of old age
Cachexia (70y+)
    of old age
Debility (70y+)
    of old age
Decline (70y+)
Degeneration (70y+)
Dementia of old age
Euthanasia (70y+)
Exhaustion (70y+)
    of old age
General atrophy (70y+)
    breaking down (70y+)
    debility (70y+)
    decline (70y+)
    marasmus (70y+)
    senile failure
    weakness (70y+)
Gradual decline (70y+)
Imbecility of old age
Inanition (from disease, 70y+)
Infirmity (70y+)
Malassimilation (70y+)
Malnutrition (70y+)
Marasmus (70y+)
    of old age
Morbus senilis
Old age

Progressive asthenia (70y+)
    weakness (70y+)
Prostration (70y+)
Senectus
Senile asthenia
    atrophy
    cachexia
    debility
    decay
    degeneration
    dementia
    exhaustion
    fibrosis
    heart
    imbecility
    insanity
    mania
    marasmus
    melancholia
    paresis
    prostration
    psychosis
    softening
    vascular degeneration
    weakness
Senility
Vital degeneration (70y+)
Want of vitality (70y+)
Wasting (70y+)
Weakness (70y+)
Worn out (70y+)

*This title does not include:* Senile gangrene (142).—Senile paralysis (66).

**Figure 2.4a**
In 1913 it was still possible to die of being worn out.
Source: Department of Commerce, Bureau of the Census, *Manual of the International List of Causes of Death,* Department of Commerce, US Bureau of the Census, Washington, DC: Government Printing Office, 1913: 131.

| | | |
|---|---|---|
| Waldenström's disease | 23512–911 | Osteochondritis of capital epiphysis of femur |
| Wardrop's disease | 173–100 | Onychia |
| Wegener's disease | 402–931 | Essential polyangiitis |
| Wegner's disease | 2...2–147.5 | Osteochondritis with separation of epiphysis due to syphilis |
| Welch's disease | 461–147 | Syphilitic aortitis |
| Werner's disease | 014–797 | Progeria |
| White's disease | 111–097 | Keratosis follicularis |
| Widal-Abrami disease | 501–9x0 | Normocytic anemia, cause unknown |
| Wilkie's disease | 4614–021 | Displacement of abdominal aorta, prolapse |
| Wilks's disease | 110–123 | Tuberculosis luposa |
| Wilson's disease | 110–966 | Dermatitis exfoliativa |
| Wishart's syndrome | 910–8453 | Neurofibromatosis of meningeal and of |
| | 980–8453 | Perineural tissue |
| Witt's disease | 501–736.X | Anemia, hypochromic microcytic, due to insufficient intake, absorption, or metabolism of iron |
| Wyatt's disease | 930–997 | Tuberous sclerosis |

### List of Terms to be Avoided

The following list includes nonacceptable terms which have been referred to the authors for assistance in classification and coding. Not any of these should ever be recorded as a diagnosis on a patient's chart. No attempt is made here to list cross references to acceptable terms as these terms should immediately be referred back to the clinician for statement of diagnosis.

| | |
|---|---|
| Abdominal adipose | Acute abdomen |
| Abdominal hernia | Adrenal crisis |
| Aborted lochia | Anterior chest-wall syndrome |
| Abortion emesis | Apoplexy |
| Abortus fever | Appendiceal colic |

*Figure 2.4b*
The problem of controlled vocabulary: this list shows terms in common use to be avoided in favor of more technical medical terms.

Arteriosclerotic peripheral
  vascular disease
Athlete's foot
August fever
Barber's itch
Blue baby
Bed sores
Blighted ovum
Brittle nails
Burst belly
Carcinoid
Cardiac asthma
Cardiac cirrhosis
Cardiovascular renal disease
Catarrhal jaundice
Cerebral accident
Cervical occipital syndrome
Chicleros disease
Combat fatigue
Consumption
Coronary infarction
Coughing disease
Cow-horn stomach
Deer-fly fever
Desert rheumatism
Devil's grippe
Dhobie itch
Diver's paralysis
Dust consumption
Dysinsulinism
Dyskeratosis
Engorged breasts
Epicondylitis
Epidemic summer disease
Fetal distress
Fetal erythroblastosis
Field fever
Frozen shoulder
Gastric crisis
Glass-blower's cataract
Grinder's consumption
Gym itch

Hepatic flexure syndrome
Hobnail liver
Hydrocephalus, external
Hydrocephalus, internal
Hydrocephalus, primary
Hydrocephalus, secondary
Housewife's dermatitis
Hydrops fetalis
Hypersplenism
Hypertensive crisis
Hyperventilation
Hypotensive syndrome
Icterus neonatorum
Indigestion
Infantile colic
Intervertebral disc syndrome
Intracranial tumor
Iron-storage disease
Jeep disease
Jitter legs
Jockey itch
Kissing spine
La grippe
Lice infestation
Lipoid nephrosis
Lipping spine
Liver spots
Lockjaw
Louping ill
Low leg syndrome
Low reserve kidney
Lumbar disc syndrome
Lumpy jaw
Miner's nystagmus
Mazoplasia
Milk leg
Miner's asthma
Morbus caeruleus
Mud fever
Myocardial fatigue
Myocardial ischemia
Neurasthenia

Source: Edward T. Thompson, *Textbook and Guide to the Standard Nomenclature of Diseases and Operations* (Chicago: Physicians' Record Co., 1958), pp. 247–249.

Neurogenic bladder
Otolith syndrome
Pale ovary
Pel's crisis
Pelvic congestion, chronic
   or acute
Peripheral vascular disease
Phossy jaw
Poker spine
Pitcher's elbow
Potter's rot
Proctalgia fugax
Pseudohemophilia, hereditary
Puerperal sepsis
Pulmonary coin lesion
Pulseless arterial disease
Rabbit fever
Rash
Rectal crisis
Recurrent neoplasm
Renal colic
Restless legs
Rheumatic pneumonitis
Rheumatism
Rock tuberculosis
Rum fits

San Joaquin fever
Scalenus anticus syndrome
Shipyard eye
Singer's node
Slime fever
Slipped disc
Spinal meningitis
Splenic flexure syndrome
Steer-horn stomach
Stone-hewer's phthisis
Stroke
Struma
Superior vena cava syndrome
Swamp fever
Thyroid crisis
Trench mouth
Unstable low back
Vagabond's disease
Valley fever
War neurosis
Washerwoman's itch
Whiplash injury
Whipworm
Winter disease
Woolsorter's disease
Wrinkles

## Medical Terminology an Interesting Study

A knowledge of medical terminology will make the tasks of the medical record librarian much easier. A knowledge of Greek and Latin is not required, but she should become familiar with some of the more common roots, prefixes, and suffixes. As these number less than a thousand, she should experience little difficulty in learning them. In fact, she should find this learning entertaining as well as rewarding, particularly if she associates stories with words. Many words have extremely interesting stories, legends, or reasons back of them. Take, for example, the term "coccyx." This is derived from the Greek. Herophilus (335–280 B.C.) first called this bone coccyx because the bone resembled the bill of the cuckoo (G. *kokkos*, a cuckoo). Vesalius (A.D. 1514–1564) gave the same explanation.

International Committee on Taxonomy of Viruses (ICTV) so as to see how diseases that present differently in each individual and often vertiginously mutate can be usefully classified.

Throughout the history of virology, there have been acerbic debates over just what are viruses. The great virologist Lwoff declaimed in 1953 that, "viruses should be considered as viruses because viruses are viruses" (Matthews 1983: 7). Viruses themselves have moved from scientific category to category. In the early twentieth century, the central definition of a virus was entirely negative: as Waterson and Wilkinson (1978, 17–18) note, a virus was any disease organism that could be filtered through one of the 'filter candles' developed for the purpose. This was a useful definition in that it excluded all other known disease agents; however, it did not guarantee the homogeneity of the category itself. As Andrewes noted in 1930, when describing animal viruses: "judgment must be suspended . . . in the case of the invisible viruses or so-called 'filter-passing' organisms. Here our ignorance is almost complete; they are possibly a heterogeneous group but in the case of creatures that we cannot see and whose very existence is, in many cases, a matter of inference only, it is idle to talk of classification in the usual sense" (Matthews 1983, 4). So there was no one definition, or rather, the ultimate encompassing residual category. Here be dragons.

Equally, there was no one discipline studying the matter of virus classification. There was no study of virology per se until the 1980s. There was an a priori assumption, entrenched in disciplinary specialties, that animal and plant viruses were not the same. This was disproved in the 1940s when it was shown that some plant viruses could also affect insects (Matthews 1983, 7). Groups who were not used to working together were forced to cooperate—and they did not necessarily like it. As with the numerous and passionate battles between cladistics and numerical taxonomy in biology (Duncan and Stuessy 1984), there were a series of virulent virological arguments that have left their traces in the literature. The arguments can be read in two ways. They are simultaneously about a struggle for professional authority on the parts of the various disciplines involved and an attempt to find a single language with which to talk about the complex temporal and spatial properties of viruses.

The role of the classification systems in knitting together (or not) the specialties is clear in all accounts of virus taxonomy. Matthews (1983, 13) notes: "in the period 1966 to 1970 there was considerable contro-

versy regarding some of the rules, which developed into a serious rift
between most of the plant virologists, and some animal virologists." He
comments on Fenner's presidency of the ICTV from 1970 to 1976:

> In retrospect perhaps the major contribution made by Fenner during his
> Presidency was to keep the plant virologists working within the ICTV organi-
> zation. This really meant stopping the insistence of Lwoff's supporters on an
> hierarchical classification and Latinized binomials, and also, as noted above,
> deleting the rule regarding new sigla. In addition Fenner exerted pressure to
> ensure that following two vertebrate virologists, a plant virologist should be
> the next president of the ICTV. (Matthews 1983, 20)

Murphy notes that even today: "Virus taxonomy is a polarizing subject
when it comes up in hallway conversations"; he goes on to praise the
ICTV for its work of:

> true international consensus building, and true pragmatism—and it has been
> successful. The work of the Committee has been published in a series of
> reports, the *Reports of the International Committee on Taxonomy of Viruses, The
> Classification and Nomenclature of Viruses*. These Reports have become part of
> the history and infrastructure of modern virology. (Murphy et al. 1995, v)

We see then that the development of the classification system also
constructs the community for which that system will act as information
infrastructure. The system is built as a political compromise between
specialties. The kinds of truth and the kinds of stories that it can
contain by their nature recognize this.

As Murphy says, the resulting classification system is in some senses
arbitrary:

> Today, there is a sense that a significant fraction of all existing viruses of
> humans, domestic animals, and economically important plants have already
> been isolated and entered into the taxonomic system. . . . [The] present
> universal system of virus taxonomy is useful and usable. It is set arbitrarily at
> hierarchical levels of order, family, subfamily, genus, and species. Lower hier-
> archical levels, such as subspecies, strain, variant, and so forth, are established
> by international specialty groups and by culture collections. (Murphy et al.
> 1995, 2)

The apposition of specialty groups (professionalization work) and
culture collections (naturalization work) is unsurprising; Murphy
offers it in a different form later in the same work: "Unambiguous
virus identification is a major virtue of the universal system of taxonomy . . . and of particular value when the editor of a journal requires precise naming of viruses cited in a publication" (Murphy et al.
1995, 7).

Thus a first temporality associated with viruses is that the field itself has formed and changed rapidly, much like the organisms that it studies. This is an unsurprising echo, as the fact that the viruses transgress spatial boundaries and mutate extremely rapidly has contributed to the change.

So what is the problem with correlating virus time with laboratory time? The overwhelming difficulty has been that it is extremely difficult for viruses to produce the kind of genetic classification whose genealogy Patrick Tort (1989) has so brilliantly traced across the social and natural sciences of the nineteenth century. A genetic classification is one that classifies things according to their origins—rocks might be metamorphic or sedimentary, say; languages might be Indo-European or Nilotic. Viruses have multiple possible origins, they look and feel the same since they pass the filter test and make you sick, but they got that way along multiple paths (compare Alder 1998). This is an old problem in medical philosophy and diagnosis—a cure does not necessarily reflect a cause, and there may be many paths to a single symptom or cure (King 1982).

Ward gives four theories for viral origins. First, it is possible that some viruses "evolved from autonomous, self-replicating host cell molecules such as plasmids or transposons by acquiring appropriate genes that code for packaging proteins" (Ward 1993, 433). In this picture they are simple chemical combinations that have acquired the replication habit of their material substrate. Second, "some viruses arose by degeneration from primitive cells in a manner similar to that proposed for the evolution of cellular organelles such as mitochondria and chloroplasts from bacteria" (Ward 1993, 434). Here they are complex organisms that devolved. Third, "some RNA viruses are descendants of prebiotic RNA polymers" (Ward 1993, 433). According to this theory, viruses might have coevolved with life itself. Finally, there is the possibility that "some viruses evolved from viroids or virusoids, although it is equally possible that these small RNA, rather than being progenitors of viruses, are recent degenerative products of the more complex self-replicating systems." (Ward 1993, 434) Where you do not have a single origin story, you cannot have a single biological classification system. Viruses have been classed into families and then into increasingly controversial supervenient categories (only one order—the mononegavirales—has been approved to date by the ICTV). The supervenient categories frequently have the inconvenience of separating viruses that had been considered grouped together.

With the lack of a single origin, the central class of virus 'species' has been defined as follows, "A virus species is a polythetic class of viruses constituting a replicating lineage and occupying a particular ecological niche" (Van Regenmortel 1990). In dealing with obligate parasites, it is necessary to assign them to a particular niche. As we saw above, a polythetic class is a class that is defined by the congruence of multiple characteristics, no one of which is essential. This relatively loose definition opens up a space for the professionalization work that needs to be done in conjunction with the alignment of competing temporalities (of the virus and of the laboratory). There has in recent years developed a line of argument that with genome sequencing it will be possible to produce a coherent history of viruses that will make the species concept more historically accurate. This reflects a wider trend across many social and natural sciences to recover origins—in geology the tide has turned against uniformitarianism (Allegre 1992); in philosophy, Foucault's archeology has grown up in opposition to the postmodern denial of origins. Even today, however, a strictly genetic classification of viruses is possibly leading to category death:

If mammalian viruses are descended from mammals, snake viruses from snakes, and honeybee viruses from honeybees, the group "virus" would cease to have any formal classificatory validity. It could be retained as a nonclassificatory group, analogous to the group of "animals with wings," but if it is not a monophyletic group, there is no doubt how cladism would deal with it; it presents no philosophical difficulty: the taxonomic category "virus" should be exploded. (Ridley 1986, 51)

The demotion to a nonclassificatory group would also have professional consequences.

We see with the history of virus classification, then, that there has been a deliberate effort to create something that looks and feels like other biological classifications, even though the virus itself transgresses basic categories (it jumps across hosts of different kinds, steals from its host, mutates rapidly, and so forth). This has partly been a deliberate political decision on the part of the international virus community: one needs such classification systems to write scientific papers, provide keywords for indexing and abstracting, compare results, and so on. Even in this most phenomenologically difficult of cases, the world must still be cut up into recognizable temporal and spatial units—partly because that is the way the world is and partly because that is the only way that science as we know it can work.

### Stories of Carving Up the Body: The Vermilion Border of the Lip

In *Regions of the Mind* (1989a), Star examined the ways in which researchers seeking to localize cerebral functions cut up the brain into meaningful units. The process is a messy one, since brains themselves come in many shapes and sizes. During the early days of brain research, a diagram of a "typical" monkey brain, with minutely localized and labeled regions, was transposed onto a representation of a human brain in an attempt to produce a standardized diagram. (Human brains are of a much different size than monkey brains.) Nevertheless, the need for standardized representations was so urgent that the physiologists overlooked this source of uncertainty, among others (Star 1985). Much the same problem occurs with the cutting-up of bodies for medical purposes. Stefan Hirschauer (1991) has noted this for the practice of the surgeon's trade; Berg and Bowker (1997) have discussed the same phenomenon in the development of medical records.

The ICD bears traces of this sort of uncertainty most notably at liminal sites (those whose borders are unclear, or used in several different categories) and with respect to roving categories like neoplasms (a cancer may overlap ICD categories). We can use the vermilion border of the lip, also known as the 'lipstick area' as a tracer for this. An early appearance in ICD-9 reads as follows:

4.   Malignant neoplasms overlapping site boundaries
Categories 140–195 are for the classification of primary malignant neoplasms according to their point of origin. A malignant neoplasm that overlaps two or more subcategories within a three-digit rubric and whose point of origin cannot be determined should be classified to the subcategory .8 "Other." For example, 'carcinoma involving tip and ventral surface of tongue' should be assigned to 141.8. On the other hand, "carcinoma of tip of tongue, extending to involve the ventral surface" should be coded to 141.2, as the point of origin, the tip, is known. Three subcategories (149.8, 159.8, 165.8) have been provided for malignant neoplasms that overlap the boundaries of three-digit rubrics within certain systems. Overlapping malignant neoplasms that cannot be classified as indicated above should be assigned to the appropriate subdivision of category 195 (malignant neoplasm of other and ill-defined sites):

140.0     Upper lip, vermilion border
              Upper lip:
                  NOS
                  external
                  lipstick area

(ICD-9CM, 26)

The NOS in this classification stands for "not otherwise specified"; a protean modifier throughout the classification that we shall discuss in chapter 3.

If we consider ICD as a prototype classification system, we can see the manner of treating the vermilion border as part of a general strategy of distinguishing central members of certain categories from outliers. The vermilion border is *strictu sensu* part of the skin of the lip, but it is not a good member of that category: "173.0 Skin of lip. Excludes: vermilion border of lip (140.0–140.1, 140.9)" (ICD-9CM, 32). Equally, it is definitely skin, but is a special subcategory:

238.2    Skin
         Excludes:    anus NOS (235.5)
                      skin of genital organs (236.3, 236.6)
                      vermilion border of lip (235.1)

(ICD-9CM, 45)

Or again, it is definitely soft tissue, but is an outlier:

239.2    Bone, soft tissue, and skin
             Excludes:
         vermilion border of lip (239.0)

(ICD-9CM, 45–46)

In ICD-10, its marginality is explicit,

D00.0    Lip, oral cavity and pharynx
         Aryepiglottic fold:

         • NOS
         • hypopharyngeal aspect
         • Marginal zone

         Vermilion border of lip

(ICD-10, 1: 222)

This multiple reference to the vermilion border of the lip is a typical ICD naming strategy. If a region of the body might fall under several categories, its membership in a special category is explicitly marked.

In principle at least, the world itself—that messy, sprawling, sociotechnical system—should be split up into regions of relevant causal occurrence. The ICD's work is necessarily far from complete. Here, however, is one typically precise definition of a liminal zone in the outside world:

A public highway {trafficway} or street is the entire width between property lines {or other boundary lines} of every way or place, of which any part is open to the use of the public for purposes of vehicular traffic as a matter of right or custom. A roadway is that part of the public highway designed, improved, and ordinarily used for vehicular travel. (ICD-10, 1: 274)

As the ICD records accident statistics, including place and mode, such precision is needed for the compilation of effective safety statistics, for example. This drive for precision is in principle unending. How much of the social and natural worlds would have to be described within the ICD to produce an exhaustive system?

The point here is not that these are bad definitions of lipstick areas and streets. It is that they are ineluctably arbitrary ways of cutting up the world. The goal with a classification system is to produce homogeneous causal regions. Homogeneous causal regions are zones without effective subdivision. For the vermilion border, there is no real distinction between upper and lower lip; for streets, there is no real distinction between tarred and gravel roadways. There is no in principal way that such ontologies can be other than a bootstrapping operation. All research work that explores medical causality has the ICD or a similar system as its base referent and so necessarily assumes the ICD's set of homogeneous regions to design its tests, experiments, or projects. It is analytically always possible to act otherwise, to carve the world up differently into other kinds of causal regions. Latour reminds us of this in *Science in Action* (1987) where he posits the thought experiment: How would someone challenge the basic premises of quantum mechanics?[16] No one would deny that it is *possible* that these premises are wrong; nor that an experiment *might* be designed to prove this. The economic and administrative cost of doing so, however, would be huge. Who would fund the proposal? Who would referee the papers? How, in short, would the inertia of the networks involved be overcome? In the same way it is always possible (and somewhat more common than in the quantum mechanics case) to challenge basic ICD categories. It is in practice, however, much easier to hypostatize them and duplicate them for local usage. Exceptions occur when particular categories are linked with social movements and social problems; an outstanding example of this occurred with the demedicalization of homosexuality in the DSM-3, after challenges from the gay community (Kirk and Kutchins 1992).

We have seen in this section that medical classifications split up the world into useful categories. They do not describe the world as it is in any simple sense. They necessarily model it. This modeling within classification systems of all sorts is where the rubber hits the road in terms of the enfolding of social, political, and organizational agendas into the scientific work of describing nature—in this case in the form of disease entities.

### Interpretation Is Also Enfolded into the ICD

We saw in the last section how the ICD cuts the world up into standard Aristotelian unities of time and place and in so doing how it produces favored readings of the body and of the world at large. The WHO goes one step further. It not only provides, through the ICD, a set of possible stories it also provides, bundled up in the classification system, explicit rules for the interpretation of those stories.

To follow this through, we need to look at the form of the standard international death certificate (see figure 2.3 above). Anne Fagot-Largeault (1989) and Lennart Nordenfelt (1983) have produced wonderful philosophical analyses of this document; our own description will not attempt to be as complete. It is the death certificate that constitutes the archetypal use of the ICD; indeed, until ICD-5, the classification only covered causes of mortality and did not seek to represent morbidity. The death certificate itself has as a major heading, "cause of death." It is split into sections, "cause of death," "approximate interval between onset and death," and other contributing factors or significant conditions.

It is a hard job to boil down a complex series of conditions to a single cause of death; and the work of interpretation begins on the form itself. A single cause is favored for very practical reasons. In the first place it is hard enough to compile statistics at all; the task could get overwhelming if multiple causes were allowed. Further, a single cause of death provides the lowest common denominator over multiple collection systems, from medical examiners in a large hospital to medical paraprofessionals in underdeveloped rural areas. Finally, as the ICD's developers point out, the goal of the classification system is not to describe complex phenomenologies, but to prevent death:

From the standpoint of prevention of death, it is necessary to break the chain of events or to effect a cure at some point. The most effective public health objective is to prevent the precipitating cause from operating. For the purpose, the underlying cause has been defined as "(a) the disease or injury that initiated the train of morbid events leading directly to death, or (b) the circumstances of the accident or violence which produced the fatal injury." (ICD-10, 2: 31)

This statement revealingly indicates a recognition by the system's developers that reality is indeed more complex than their registration system can describe. All the analytic points made to date in this chapter can be read into this one statement: the ICD is a pragmatic classification ("the most effective public health objective"); and it segments the world up spatially and temporally into causal zones that underwrite preferred stories ("it is necessary to break the chain of events . . . at some point").

The cause of death as given on the death certificate by the attending physician is frequently not, as Fagot-Largeault points out, the cause of death that enters into the statistical record. The classifications entered on the certificate are themselves systematically recoded so as to constrain the kinds of story that the statistics tell.

One informal algorithm is that precision always beats no precision. (This is an echo of John King's wonderful observation about technical arguments in the policy domain: "some numbers beat no numbers every time.") On a deeper epistemological level, the substitution of precision for validity is often a needed expedient in getting work done (Star 1989a, Kirk and Kutchins 1992). It may also become a kind of gatekeeping tool in theoretically defining a ground of knowledge. It functions as follows in the ICD:

Where the selected cause describes a condition in general terms and a term that provides more precise information about the site or nature of this condition is reported on the certificate, prefer the more informative term. This rule will often apply when the general term becomes an adjective, qualifying the more precise term.

Example 57:    I (a)  meningitis
           (b)  tuberculosis

Code to tuberculous meningitis (A17.0). The conditions are stated in the correct causal relationship. (ICD-10, 2: 48)

This is doubtless a very reasonable rule. It is significant, however, that it sets in train a process that begins putting in mediating layers between what the doctor says and what gets reported.

In general, these mediating layers refashion the story that the act of classification permits. The records clerk is given a license to change the doctor's classification in such a way that it will reflect the best current medical theories: "*Rule 3*. If the condition selected by the General Principle or by Rule 1 or Rule 2 is obviously a direct consequence of another reported condition, whether in Part I or Part II, select this primary condition."[17] Thus, for example: "Where the selected cause is a trivial condition unlikely to cause death and a more serious condition is reported, reselect the underlying cause as if the trivial condition had not been reported. If the death was the result of an adverse reaction to treatment of the trivial condition, select the adverse reaction" (ICD-10, 2: 45). Derrida (1980) reminds us that it is through what is excluded as trivial that we can frequently understand systems of thought by pointing directly at what is important. Similarly, this opening of the door to an undetermined attribution of triviality is one significant moment, hidden in the third volume of a massive classification system, where the work of reifying current categories is done. Only certain causal chains will be permitted at the moment of classification. This in turn naturally affects the interpretation at the other end of "raw data" in the form of epidemiological statistics: "The expression 'highly improbable' has been used since the sixth revision of the ICD to indicate an unacceptable causal relationship. As a guide to the acceptability of sequences in the application of the general principle and the selection rules, the following relationships should be regarded as 'highly improbable'" (ICD-10, 2: 67). After this passage, there follow a series of unacceptable chains. For example, a malignant neoplasm can not be reported as due to any other disease than HIV; hemophilia cannot be due to anything, and no accident can be reported as due to any other cause, except epilepsy (ICD-10, 2: 68).

An acceptable string of classifications in a death certificate is one that fits into an internally consistent chain that reflects current medical knowledge. In the process of crafting such a chain, all qualifiers should be removed: "Qualifying expressions indicating some doubt as to the accuracy of the diagnosis, such as apparently, presumably, possibly, etc., should be ignored, since entries without such qualification differ only in the degree of certainty of the diagnosis" (ICD-10, 2: 88). In the process of achieving this certainty, multiple causality must often be arbitrarily collapsed into unicausality, here by a principle of first come first served: "If several conditions that cannot be coded together are recorded as the 'main condition,' and other details on the record point to one of them as the 'main condition' for which the patient received care, select that condition. Otherwise select the condition first mentioned" (ICD-10, 2: 106). Any working classification system will have such rules of thumb attached. Such rules are theoretically interesting for several reasons. First, the ICD developers have explicitly recognized that it is not enough to control the classification (the name of the disease) they also must attempt to exercise control over the language game in which the classification is inserted. This indeed is the purpose of the rules contained in volume 2. This attention to both the base level and its metalevel is a bureaucratic necessity, which at the same time conjures the wild world of the patient's body into the ordered world of medical knowledge. Second, the rules themselves serve to systematically reduce ambiguity and uncertainty, even where these are integral to the attendant physician's depiction of the patient's situation. Those who see the patients are aware of this uncertainty; those who apply the rules also know of it; those who read the final statistics are shielded from it. The patients live it.

Finally, there is a potential infinite regress in the control of first the name of the disease then on rules for using these names and so forth. The final level at which regress occurs is in the presentation of results. The WHO recognizes that when dealing with small populations, one may get wild fluctuations of information on mortality or morbidity from year to year. To achieve stability and certainty at this level, one needs to sacrifice precision, to go up to broader ICD rubrics, aggregate data over a longer period, use the broadest of the recommended age groupings and aggregate areas (ICD-10, 2: 137). Recommended age groupings and regional groupings are:

<1, 1–4, 5 year groups from 5 to 84, 85+
<1, 1–4, 5–14, 15–24, 25–34, 35–44, 45–54, 55–64, 65–74, 75+
< 1, 1–14, 15–44, 45–64, 65+

(ICD-10, 3: 128)

Classification by area should, as appropriate, be in accordance with:

(i)    each major civil division;

(ii)   each town or conurbation of 1,000,000 population and over, otherwise the largest town with a population of at least 100,000;

(iii)  a national aggregate of urban areas of 100,000 population and over;

(iv)   a national aggregate of urban areas of less than 100,000 population;

(v)    a national aggregate of rural areas.

(ICD-10, 3: 128)

The regress itself to ever higher levels of control marks the fact that the world is always slightly out of reach. It cannot be contained in the classification system, or the (system + set of rules), or the (system +

set of rules for interpretation + set of rules for change or the system + set of rules for interpretation +set of rules for change + set of rules for presentation).

## *Conclusion*

At the start of this chapter we looked at two basic kinds of classification system: Aristotelian and prototypical. We have seen in the course of our analysis that medical classification systems are "naturally" prototypical, and that they nevertheless have to appear Aristotelian to bear the bureaucratic burden that is put on them. This burden is to act as a gateway between the worlds of the laboratory and the hospital (with precisely defined, closed environments) and the workaday world. As we consider the stories embedded in the system, from the point of view of work and practice, we understand that both the intuitive and the technical are always present in systems such as the ICD.

The way in which this gateway function is provided is twofold. First, the Aristotelian classification embeds within itself a set of implicit narratives that align the artificial categories of the ICD with the real world. Second, the rules for interpretation and presentation sit on top of the ICD and nudge its categories along prepared, legitimate pathways. This combination of embedded and supervenient narratives provides the give through which the prototypical classification can be made to look and feel Aristotelian.

# 3

## The ICD as Information Infrastructure

Science is a systematized and classified knowledge of facts. The proposed change in the definition of stillbirth does not appear to be based upon such an orderly classification of known facts. It seems to be based upon misty theory, contrary to established concepts. It is therefore unscientific.

*(C.H./expert stat./46. doc. 43806, doss. 22685, 22 December 1927, 11)*

### Introduction: Histories of Classifications

Over the past 300 years (beginning perhaps with the ineffable Leibnitz) there have been a number of sweeping encyclopedic visions for storing all knowledge in a single form—be this through perfecting language (Slaughter 1982), classification systems (for example, Melvyl Dewey's library and industry schemes), or modes of knowledge organization (for example, Otlet (Rayward 1975)). These schemes have found their historians, but their shadow side appears to be discovered anew each generation. This side is the barrier to complete knowledge systems, notably in the following forms:

• *Data entry as work.* No matter how good the scheme, its scope is limited by the fact that data entry is never an easy task, and there are never enough resources or trained personnel to make it happen. Not only will there inevitably be mistakes with respect to the internal structure of whatever classification one is representing, there will also inevitably be cultural variations with respect to how it is interpreted as well as culturally biased omissions.

• *Convergence between the medium and the message.* Within any society there are a limited number of technologies for storing information (from ledger books to file cards to computer databases). The information that gets stored is at best what can be stored using the currently available technology: the encyclopedia comes to mirror the affordances of its technological base. In this process, people naturalize the historically contingent structuring of information; they often begin to see it as inevitable.

• *Infrastructural routines as conceptual problems.* No knowledge system exists in a vacuum, it must be rendered compatible with other systems.[18] The tricky, behind the scenes work of ensuring backward and sideways compatibility is not only technical work, it challenges the very integrity of any unifying scheme. Such work, however, is itself often classed as "mere" maintenance and deemed unworthy of public, historical pride of place.

To understand the architecture of such schemes, then, we need to look at the traces they leave of their own history and constitution (as we did through reading the ICD). This chapter examines the historical intertwining of medium, message, routine, and data entry in the ICD. One of the challenges here is to understand how the drive for universal languages and databases is reconciled with the pragmatics of practice and the constraints of the installed base. True universality is necessarily always out of reach. At the same time, the vision of global data gathering and sharing is enormously powerful, and it needs to be understood in its own terms. This is one important context for the development and deployment of such systems as the ICD as tools.

During this century the information sciences have grappled with new ways of configuring, storing, and retrieving information. The rise of networked computing, and the extravagant advances in processing capacity have increased the pace and pressure of this struggle. We are clearly at a point today where we are witnessing the birth of an information technology as fundamentally new as was the printing press in its day (see Eisenstein 1979 for the latter). We do not take this as an unproblematic information revolution. Rather, by looking at the more sober, less glamorous aspects of this infrastructural transformation, we hope to discomfit some of the revolutionary hype (Bowker 1998). This chapter examines the historical background of the development of the ICD as information infrastructure.

As noted in chapter 1, there are too few theoretical tools available to the historian for grasping the development of a new information infrastructure. Infrastructure does more than make work easier, faster or, more efficient; it changes the very nature of what is understood by work. Such changes always span multiple disciplines, industries, and lines of work. Forms of automation, for instance, begin in one sphere and spread across lines of innovation and dependency. Scientists say that the natures of their disciplines are changing, in no small part due to these infrastructural shifts. Stephen Hawking in his inaugural lecture as Lucasian Professor at Cambridge (a post once held by Newton) expressed a belief that by the turn of the century, computers would essentially perform work in theoretical physics. Humans would not be able to understand the mathematics, but they could aspire to interpreting its consequences (Hawking 1980). Pure mathematicians have now adopted a method of existence proofs that would have been unmanageable before the development of the computer, such as the solution to the four-color problem. Not only the scientist and the mathematician are affected. Classical scholars had to learn a new set of techniques (dealing with complex searches on a computer) and indeed pose a new set of questions of their data when classical text became available on-line (Ruhleder 1995). The Thesaurus Linguae Graecae houses the complete canon of classical Greek literature in electronic form. Since its inception in the 1980s, classical scholars have changed their working practices, including the definition of text, the value of word searches, and the role of concordances. J. David Bolter, inter alia, has commented that genres of fiction, too, are undergoing radical change with the development of hypertext (Bolter 1991). And more generally, as Beniger (1986) amongst others reminds us, the structure of industry is changing such that "information work" has become the dominant mode of work in industrialized economies (Kling, Olin, and Poster 1991).

Among other things, these changes imply that the worlds of knowledge and of industry are not the same worlds after the development of this new information infrastructure as they were beforehand. To explain what has happened, the historian has to range freely between the "inside" (looking at knowledge within physics, mathematics, classics, and so forth) and the "outside" (looking at changes in work practice and information management that hold over many fields at once).

The story of information infrastructures is not, in this sense, the history of great people. Much of the work has been done offstage by communities of hackers, technicians, and engineers, and in maintenance, upgrades, and integration. Creating an infrastructure is as much social, political, and economic work as it is theoretical. Although in some sense knowledge is its raison d'être, it bursts the bounds of traditional history of ideas. How then to write its history, avoiding both hype and getting lost in the details? As a one of the participants in the 1920s revision of the ICD declared in a frank letter to Dr. Norman White of the Health Section of the League of Nations:

You know, I am a great believer in taking things of this kind slowly. Statistics is a very unexciting field in which to work up a revolution. . . . I am offering you my personal opinion, of course, but it is this: that the Committee on joint causes of death propose, if it wishes, an ideal certificate and an ideal method of classifying joint causes of death, if it desires, in the next year or so, but that as a *beginning* it should propose some things which can be done and which the statisticians of the different countries feel that they *can* and *want* to do. If a few of these can be accomplished, then the Committee of experts and the Health Section can record some achievements which will pave the way for more fundamental reforms. (12B R842/ Doc. 51040, Doss. 22685, letter from Edgar Sydenstricker to Norman White, June 11 1925)

We will not attempt to give an overview of the whole range of infrastructural work. It is constitutive of, to use the unfortunate phrase, an "information age." (Unfortunate because those who write about this as the information age tend to immediately retrospectively define all of human history as the history of information processing, and thus to effectively deny specificity to whatever age they are defining.) The phrase draws attention away from the material bases and work practices that are analyzed in this chapter. Information cannot analytically be released from these contexts.

The infrastructural work entailed in both design and use of this classification system is considerable. As we have seen, the ICD is used worldwide by states (on death certificates), by insurance companies, and within hospitals. The ICD fits perfectly into Star and Ruhleder's (1996) definition of infrastructure (alluded to in the introduction and discussed in more detail in chapter 7 below). It is embedded in a myriad of databases. It is transparent as it invisibly supports medical work, and has wide spatial reach. (All countries in the world operate with a version of the ICD, though not always the same version!) It is learned as part of membership in the medical and medical actuarial professions, and it is linked with conventions of practice in all these domains.

The discipline and practice of statistics grew up during the nineteenth century (Porter 1986, Hacking 1990). As pioneering medical classifier Farr wrote: "statistics is eminently a science of classification" (1885, 252). As the word's etymology indicates, it was a discipline intimately connected with the rise of statehood. Political and economic life in the industrializing countries of Europe was becoming ever more complex. States experienced the need to gather and keep information about their citizenry. Medical statistics emerged as part of this burgeoning information-gathering activity.

## Medical Classification and the State

During this century large modern states have found themselves forced into developing complex classification systems to promote their political and economic smooth functioning; people are travelling further and more frequently and living longer, more information-dense lives. Producing these classifications is tedious, long, committee work. It is nonheroic work, carried out by bureaucrats. For many, such work does not have a history. The archives of the WHO in Geneva preserve in black leather boxes stamped with gold on bright steel shelves the records of the struggle against smallpox. *Western medicine defeats ancient enemy!* The boxes stand proudly on the shelves, a battery of headlines awaiting a chronicler.

In searching for the archives of the construction and revision of the ICD, however, we were unable to find any such centralized and well-archived cache. Typically for an information infrastructure, the achievement of producing and maintaining a standard international list of causes of death—a massive bureaucratic, scientific, technical, statistical, epidemiological, human achievement—is considered beneath archival priority. Until recently, when every ten years a new edition was produced, records of the negotiations leading up to those editions were destroyed. Some earlier information remains as correspondence of individuals or committee associated with the League of Nations, or later the United Nations or WHO. For it usually appeared to the ICD's designers, its record-keepers, and even those involved in implementing it, that what was interesting scientifically was the agreed-upon outcome, not the error-strewn path leading to it. That has for the most part been seen as too boring to bother with (with some notable exceptions that we shall discuss below).

The ICD's lack of formal, boxed archives does not mean that it lacks history, as we have shown in the last chapter. Inscribed in the form and content of the list, as we saw, are a series of technical, social, political, and economic decisions taken at different moments. These decisions, taken at particular times for a given set of reasons, are paradoxically often more entrenched in the otherwise ahistorical ICD than they would be in some other form of historical object. This is due to the inherently conservative nature of reform of large-scale data collection efforts. To maintain comparability of items in the classification from one revision to the next, and thence to carry out large-scale longitudinal public health and epidemiological surveys, changes must be minimized from one edition to the next. Thus the preface to the ICD's fifth revision (1938) notes:

The Conference endeavored to make no changes in the contents, number, and even the numbering itself of the various items, so that statistics based on the successive Lists should be as comparable as possible, and employees of the registration and statistical services of the different countries should have their habits of work changed as little as possible. Many possible improvements in matters of form and order were abandoned to achieve this practical object. (League of Nations 1938, 947)

As smallpox was eradicated from the face of the earth, its archives swelled. On the other hand, as the ICD grew larger, the archives disappeared. The list folded its history in on itself, however, becoming ever more ramified and complex, involving larger numbers of people in the processes of revision. The complexity of the artifact itself can be summarized as follows:

• Increased detail in data collection

• Increased cross-referencing by cause (occupational, disability-related, safety, morbidity as well as classical mortality)

• Conservatism in abandoning categories due to the need for historical comparability, leading to the preservation of anachronistic categories

• Links between the ICD and other state information systems, such as social security

• Preserving the ever more complex concerns of the governments involved in developing the ICD in category contents

The health of the citizen is central to the modern state, as François Ewald (1986) and others have shown:

In the earliest dawn of the nation the English inquired into the causes of death with a view to discovery and prevention. The protection of life was a fundamental principle of their laws. It was as much an object of their political organization as national defense or war. . . . The plagues of the sixteenth century proved that human life is exposed to invisible enemies more deadly than the mechanical forces of nature, the ferocity of animals, or the malignity of manslayers; and toward the end of Queen Elizabeth's reign the London Bills of Mortality were commenced. (Farr 1885, 218)

The equation can be brutal. One doctor responded in 1984 to a questionnaire on missed diagnoses (believed from autopsy analyses to be about 10 percent of all cases) that these "be quantitated on the basis of functional units, for example, number of productive work-years lost or number of symptom-free months lost" (Anderson 1984, 492). Or the equation can be martial: "By studying the causes which are injurious and fatal to men in our countries and in our cities, statistics will contribute to the removal of evils that shorten human life and to the improvement of the race of men, so that Citizens of a civilized State may be made to excel barbarians as much in strength as they do in the arts of peace and of war" (Farr 1885, 218). It can also be richly paternal, as shown here:

In 1974, New Zealand became the first country in the world to accept responsibility for the safety of its inhabitants for 24 hours every day, 365 days every year, from birth until death. At the same time, the Accident Compensation Corporation became the first organization in New Zealand, and possibly in the world, to become responsible for the prevention of accidents to all inhabitants as well as for compensating, and where necessary rehabilitating, those who suffered personal injury by accident. (Heidenstrom 1985, 69)

In each of these cases, the state pits itself against the passage of time and tries, in its own interest, to legislate immortality for its citizens. The benign side of this process is improved health, social justice, and quality of life. Its darker side, of great concern now in medical policy circles about genetic disease risks and conditions such as AIDS, is surveillance and medical discrimination.

Much has been written about the state's role in classifying and monitoring mental health (Kutchins and Kirk 1997); much less about classifying and monitoring physical health. Yet this latter work has been just as politically fraught and just as imbricated in the rise of the modern state. As mentioned above, the ICD was developed following an international event in 1893 in Paris. This conference in turn followed a series of conferences to deal internationally with cholera.

Why cholera? There were a number of cholera epidemics in nineteenth-century Europe; one series in the latter part of the century being caused by pilgrims returning from a pilgrimage to Mecca infected with the cholera bacillus (see figure 3.1). In early years, returning on foot and by sailboat, infected pilgrims had died before they returned to France. As noted in the introduction, now they were returning by rail and steamboat, and they were able to bring the disease with them back into the metropolis before dying. So as

ABLUTION SACRÉE DE ZEMZEM
(D'après le *British medical Journal*.)

**Figure 3.1**
Sacred ablutions in the zemzem, or fountain, at Mecca.
Source: Proust 1892.

communication between the countries of the world increased at the
apogee of the imperialist age in the 1890s, so did the need to monitor
health and welfare on an international scale.

When the ICD classification was first drawn up, it was based on one
of the few preexisting classifications: Bertillon's list of causes of death
in Paris (see Bertillon 1900). The center of the French empire imposed
its own classification scheme on its colonies and other imperial powers
followed suit. This fact was remarked upon at the time by many. For
example, a South African doctor noted that tropical diseases were
underrepresented. This omission remains a sore point to this day.

Doctors in Africa, for example, have complained about the burden of providing AIDS statistics to first world countries. These statistics are of little use to them internally, in those (all too common) circumstances when they have no means to treat patients. The compilation, demanded by epidemiological agencies such as the WHO, takes up valuable medical and epidemiological resources. For them, AIDS is only one of a series of endemic causes of death. It is one that it will only be worth singling out for treatment once appropriate medical services are in place and general urban sanitary conditions are improved. Until then, the need that western doctors have to trace the detailed genesis and development of the disease is not felt so urgently. Thus the concern was raised in a conversation with the director of health statistics at the WHO in Geneva in 1989 that many of these statistics were being collected to further the careers of public health officials in the United States. (This is discussed further in the next chapter.)

A simple agonistic reading of the ICD is that the system was set up in an age of imperialism and helped impose an imperialist reading of disease from the West onto the rest of the world. There is truth in this, and many medical anthropologists and activists have ably told this story (Anderson 1996). Another, more subtle story can be told alongside this one. Management of the ICD played a part in the creation of the modern state, in many protocols for state-to-state negotiations, and in many international organizations. The degree to which it came to constitute medical knowledge is unknown, and that story is yet to be told.

This knowledge-creating role becomes clear when one looks at the highly complex bureaucratic work involved in developing and maintaining the ICD. Numerous groups use it for many different purposes. Medical insurance companies need a standard list of causes of death and of morbidity to work out standardized scales of payment for different treatments. They develop elaborate risk tables for different groups of subscribers (and thence modulate their premiums). Epidemiologists also use the ICD; to track down the causes of a given new disease, one needs a standard terminology and good records. Only with these in place can one determine that the disease only affects those who eat a certain kind of food, or have a certain genetic heritage or occupational history. Government health officials need good records to determine public health policy and services. For example, if tuberculosis is a major problem in a given area, then one might set up a clinic, or organize free x-rays there.

To maintain a good international system of medical classification, a huge amount of detailed information is needed about both the citizens of a particular state and about citizens of countries with whom they are in contact. No information is irrelevant. The state must have better information than the family itself. As noted in the case of New Zealand above, its need for information is effectively infinite. Below, for example, is a wish list from 1985 for a national medical information system in the United States:

The system must capture more data than just the names of lesions and diseases and the therapeutic procedures used to correct them to meet these needs. In a statistical model proposed by Kerr White, all factors affecting health are incorporated: genetic and biological; environmental, behavioral, psychological, and social conditions which precipitate health problems' complaints, symptoms, and diseases which prompt people to seek medical care; and evaluation of severity and functional capacity, including impairment and handicaps. To accomplish this, a series of interlinked classifications would be required, designed so that all of the information is stored in a common database. The entire spectrum of medical terminology would be included, from the layman's language used to describe ill health and terms used by professionals at the institutional level to molecular terms from each of the basic sciences and terms related to causes of death used at the international level. Feinstein, in a recent paper, proposed to capture even more data consisting of the observations and quantitation of such clinical phenomena as the type and severity of symptoms, the rate of progression of illness, the severity of co-morbidity, the functional capacity of the patient, the reasons for medical decision, problems in maintaining therapy, the impact of the ailment and its treatment on familial and interpersonal relationships, and other aspects of the physical activities and mental functions of daily life. (Rothwell 1985, 169–170)

There is no foretelling what information will be relevant. This is what Lemke (1995) calls the dilemma of choosing between typological and topological. The construction of typologies or classes forecloses labeling options and presets the options about the range of possibilities. For scientific and ethical reasons, he argues for a topological approach in classifying persons, visualizing dimensions that may be added to in an expandable matrix (Lemke 1995). Those who gather information for the ICD and related systems face precisely this dilemma. Heidenstrom says, for example, "to classify a chisel, a hand drill, and a spanner [wrench] together as 'hand tools', or the first two as 'cutting and piercing instruments' may be obscurantist, or even misleading. Whereas to one accident researcher it is significant that a chisel is edged, a drill pointed, and a spanner neither, to another it may be more important that the chisel is pushed, the spanner turned and the drill operated by rotary motion" (1985, 76). A topological approach would ideally preserve this multiplicity of meanings.

To record information properly about a given disease at a given time, everything about the social, economic, personal, and physical conditions of the patient can be relevant (some will recognize this as a version of Spinoza's problem). The encyclopedic vision so common in information sciences envisions a preemptive, or open-ended capture of the information attributes of any object. For epidemiology, this would ideally mean that patient information would be captured at the lowest descriptive level (atomic units). In future years, the data so collected could be remined as advances in medical knowledge reconfigure the attributes. For example, the discovery of a new disease could be read backwards into existing data, and entities unknown at the time of data collection could be read out of the data. In practice, as the above examples show, the infinite possible ontologies of objects is limited by the pragmatics of data collection and by the inescapable inertia of categories already in use.

The expanding wave of information gathering practices is a defining characteristic of the modern state, as Foucault (1991) observes. To produce and maintain standardized medical records, state bureaucrats needed to create a uniform set of data-gathering and encoding practices. Without these practices, standardization could not be achieved. These standards entailed a range of governmental activities, including accustoming citizens to the regular collection of information about ever more detailed aspects of their personal lives. Standards also meant enforcing a standardized set of procedures.

These practices, and the standardized information thus generated, meant that information could be rendered comparable across situations. In turn, the development of a professional class that could use the information garnered was fostered (see Abbott 1988 for the medical profession as archetypal modern profession). This process appeared indefinitely expansive.

As the general level of sanitation improved during the nineteenth century in industrializing countries, doctors needed ever finer classification systems to discriminate these infrastructural effects from other disease agents. At least in the western world, more people were living to an age when they died more complex deaths. Infant mortality, appendicitis, or malnutrition no longer killed them in such great numbers at earlier ages. This is the story from within the history of medicine. But by the same token, as the modern state developed its view of legitimate government as the management of a large information system, states produced a proliferation of ever finer classification systems. Along with this, a bureaucracy developed to manage these systems across a wide set of domains of which the medical domain is a chief example. Building the ICD involved building the state as much as developing medical knowledge.

This double movement—building an information system and building the state—is an intricate one. Bootstrapping is always involved. To create something as basic as an information infrastructure or a scientific standard, much of the infrastructure or standard already needs to be in place. (How else does one organize the data?) Thus, from the early days of the modern state, the need for such a chicken-and-egg operation can be traced with the development of hospitals (e.g., in France after the revolution). Until there was a working classification of diseases—such that people with one disease would not be mixed in with those suffering from others—then patients died wholesale. The hospital served as a place in which to share diseases and on that basis was dreaded by most. But a classification could not be developed unless people suffering from a given disease could be isolated. The establishment of working classifications depended on being able to develop specialized information about particular diseases. This in turn could only be obtained through studying cases in a controlled situation where patients were not subject to a wide range of complicating illnesses and infections (Dagognet 1970).[19] To solve this class of problem in establishing and maintaining the ICD, its designers quite explicitly acted as if ICD statistics were *already* accurate. By so doing, they hoped that the future data gathering would conform to this gamble. Thus when the League of Nations began working on morbidity statistics, it did not try to impose a perfect classification scheme with a functioning bureaucracy. Rather, it admitted:

It is fully realized that much of the information called for in this plan is now utterly lacking in international, or frequently even in interurban comparability. This is evidently the case, for example, in regards to the records of school medical examinations, which are frequently not comparable even between two different examiners in the same town. Experience shows, however, that comparability of statistics has rarely, if ever, been obtained before there was a definite demand for it. Rather than omit from the beginning all data which are not yet satisfactory, the authors have hoped, by including them and utilizing them for what they are worth, to create a demand for their improvement and for international definitions and standards which lead to the development of comparability. Wherever possible, checks have been devised to facilitate evaluation of the data. (Stouman and Falk 1936, 904)

There have been many minor methods for cobuilding the ICD and its bureaucracy in this fashion. Laws have been passed in individual countries demanding that all causes of death be reported by the relevant statistical service. A single standardized death certificate (developed in the 1920s) has been adopted worldwide.

Doctors frequently bemoan the (clinical) resources wasted on searching for the one true cause of death, and as discussed below, consider filling out the certificate virtually a waste of time. Many attempts have been made to educate them in the epidemiological value of a good death certificate. None of these measures alone has rendered the ICD a perfect tool, as Kerr White complained of the ICD up to the ninth edition: "There is no coherent conceptual or organizing theme, to say nothing of theory, and yet this classification and its modifications seek to meet the needs of policymakers, statisticians, third-party payers, managers, clinicians, and investigators of all persuasions and preoccupations in a wide range of socioeconomic and cultural settings around the worlds" (Kerr White 1985, 17–18). Both the acceptance of the role of the state in garnering statistics, however, and its bureaucratic competence so to do, has increased drastically over the past 100 years.

Over the past several hundred years, there have been many critiques of the veracity of medical statistics. John Graunt in 1662 (using original spelling) noted:

17   . . . I found that all mentioned to die of the French-Pox were retured [sic] by the Clerks of Saint Giles's, and Saint Martin's in the Fields onely; in which place I understood that most of the vilest, and most miserable houses of uncleanness were: from whence I concluded, that onely hated persons, and such, whose very Noses were eaten of, were reported by the Searchers to have died of this too frequent Maladie.

18   In the next place, it shall be examined under what name, or Casualtie, such as die of these diseases are brought in: I say, under the Consumption: forsasmuch, as all dying thereof die so emaciated and lean (their Ulcers disappearing upon Death) that the Old-women Searchers after the mist of a Cup of Ale, and the bribe of a two-groat fee, instead of one, given them, cannot tell whether this emaciation, or leanness were from a Phthisis, or from an Hectick Fever, Atrophy, etc; or from an Infection of the Spermatick parts, which in length of time, and in various disguises hath at last vitiated the habit of the Body, and by disabling the part to digest their nourishment brought them to the condition of Leanness above-mentioned. (Graunt 1662, 37)

William Farr, almost two centuries later made a similar remark (constituting an example of the assertion that systems have progressive histories but their work-arounds do not):

The French explicitly reject women as informants and thus must in many cases forego the best possible testimony. Women are almost always, except on the field of battle, in attendance or present at death. The wife does not forsake the husband, the mother the child, in the last moments. In marriage and in birth, the two great acts of registration, the woman is indissolubly associated with the other sex, and from men in death they are not divided. On what ground then is the woman rejected peremptorily as a witness? The French principle is inapplicable to English women. But in England we may well avoid rushing to the other extreme. Why should a majority of the informants of some districts be ignorant women who sign the registers with marks and cannot read and check the entry to which their signature is attached in the national records? (Farr 1885, 226)

Farr went on to write that classification was "another name for generalization," which was basic to the natural sciences, and that good classification depended on the "form, character, and accuracy of the observations" (Farr 1885, 233). He recommended that the mode of observation should be recorded along with the cause of death. In the early 1900s in Russia, one priest would have the task of filling in the death certificates for a scattered population of 100,000 rural inhabitants (Fagot-Largeault 1989, 242). These results would in no way be comparable with the meticulous statistics collected in Paris. Such discrepancies, through a slow series of changes, have become less marked, although they have by no means disappeared (see Sorlie and Gold 1987). Comstock and Markush (1986, 180) remark that "most physicians have had no training in the purpose and process of death certification . . . medical information on death certificates is often incomplete . . . diagnoses on death certificates do not necessarily reflect information obtained after death . . . mortality statistics are not published promptly."

The original list of causes of death covered several million people. The ICD's fith edition (1938) was estimated to cover a reasonable proportion of the world population at 630 million (League of Nations 1938, 946). The current tenth edition is not yet by any means universal; several countries have decided to stay with ICD-9.

With the rise of the state and of statistics playing such a role in the creation and maintenance of the ICD, it is no surprise that the list itself—to the casual glance a flat list of causes of death—has inscribed affairs of state onto its representation of the afflicted human body, as shown in the last chapter. From the beginning, the definition of the moment of life has been a key battleground. Catholic countries fought to recognize the embryo as a living being, statistically equivalent to an infant; Protestant countries were far less likely to accord the status of life to embryonic citizens. There were in addition wide ranges of variation even within nations in how long gestation must last to encode a stillbirth. In Maryland, (historically a Roman Catholic U.S. state), life was defined as "all products of conception." In the state of Washington, it was only those advanced beyond the "seventh month of utero-gestation" (C.H./expert stat./46. doc. 43806, doss. 22685, 22 December 1927, 3). Even in discussing breathing as the sine qua non of life, the committee was forced to ask whether the baby breathed or only attempted to breathe.

A compromise position reached in 1930 was that a baby must have tried to breath three times to be ranked as an infant mortality rather than a stillbirth. Various editions of the ICD have had special sections devoted to this topic. Equally, the ultimate cause of death is also state defined. This was made explicit in 1932 when, if there were two equal underlying causes of death (e.g., cholera and leukemia), then the cause that would be most useful to the public health arm of the state (in this case cholera, which was a matter of public health concern) would be taken statistically as the underlying cause.

Categories of accidental death and death by suicide have similarly always inscribed a diverse series of government regulations and local bureaucratic contingencies. Consider this set of categories from the ICD's fifth revision (1938). In this edition there were many categories for suicide, with categories 163 (suicide by poisoning) and 164 (other forms of suicide) being devoted to it. Subcategories of the latter included:

164.   f. Suicide by crushing.
        fa. Suicide on railways.
        fb. Other suicide by crushing.

(ICD-5, 974)

Some chief forms of accident included:

187.   Cataclysm (all deaths, whatever their cause).
192.   Lightning.
193.   Other accidents due to electric currents.

(ICD-5, 976).

This latter is footnoted: "Except accidents from transport, accidents in mines and quarries, agricultural and forestry accidents, or accidents due to machinery, classed under nos. 169–176, and deaths from operations of war, classed under nos. 196 and 197" (ICD-5, 976).

The pattern is clear. The railway authorities needed to keep track of the number of bodies of suicides they had to recover and manage. The London underground introduced "suicide pits" between the rails (still in use today). They wanted to know the efficacy of these pits in capturing and saving the life of the attempted suicide. Equally, mines, quarries, and war were tracked and managed by different government departments. Thus, it was useful for them to keep these statistics separate, even though the lay observer might see no difference between electrocution on a battlefield and electrocution at home. Again, the typological-topological problem of encyclopedic knowledge reappears. Who will be able to recover which knowledge? Finally, categories about which nothing could be done (medical "fait divers" of all sorts) could not demand detailed treatment: "cataclysm" (as a residual category) would do for them all.

This government pressure on medical authorities to develop useful classifications has been a constant theme. To take but one example, in Norway in 1981 the Government Action Committee for the Prevention of Child Accidents and public servants dealing with the 1976 Act of Product Control on working with consumer products asked the health authorities for a registration scheme. In this fashion all consumer products could be brought under a standard classification scheme (Lund 1985, 84). Thus the health authorities got into the business of classifying not only diseases (natural kinds) but also manufactured articles (social kinds) that might become causal agents in morbidity and mortality. The horizon of detail expands again.

It is clear then that a history of the ICD is only in part a history of medical progress, strictly speaking. Indeed, it must inevitably lag behind the field of medicine. To maintain historical comparability of statistics, the ICD is necessarily conservative with respect to changes. Even at ten-year intervals, a new disease entity may take more than twenty years to be included since the pace of medical discovery and the uncertain process of consensus can be very slow. As shown in chapter 1, some diagnoses may only by achieved with advanced medical technology. In turn, this technology may be slow to spread around the world to become available and familiar to revision centers. In the advent of a new epidemic such as AIDS, diagnostic, nosological and epidemiological tangles have persisted for more than a decade, spanning the implementation of ICD-9 and ICD-10 among affected countries.

The history of the ICD is thus inextricably a history of the formation of the modern state—both at the small-scale level óf the development of particular bureaucratic structures and at the large-scale level of the installing of and justification for methods to keep populations under surveillance. After World War II, this development increasingly involved multinational corporations and the computerized flow of epidemiological and medical information across all manner of organizations.

## Medical Classification and Information Processing

The ICD is a complex information-processing tool. As such it is at any one time associated both with a theory of applied historical knowledge and a particular configuration of technology. It must have an historiographic aspect—a theory of historical knowledge—in the sense that it embodies an understanding of what information about the past can and should be retained.

The historical problem is particularly complex since there has been a secular change in the form that death takes. People no longer die the way that they used to at the turn of the century, as noted above. In 1900, the overriding causes of death were the single great epidemic diseases: tuberculosis, pneumonia, smallpox, and influenza. These tended to attack people indiscriminately, from their prime to old age. Nowadays, with antibiotics and other medicines, people tend to live longer and to break down more slowly. As noted in chapter 2, they tend to be carried off not by a single disease but by a complex of diseases (Israel et al. 1986, 161). Matching this trend, the emphasis in applications of the ICD has changed. It has gone from recording a single underlying cause of death to looking for a complex of causes.

Consider for example the standard International Form of Medical Certificate of Cause of Death adopted by the WHO in July 1948 (reproduced in Fagot-Largeault 1989, 72 and discussed in chapter 2). This was the canonical form that was used to apply the ICD so as to produce epidemiological statistics. It locates a single "disease or condition directly leading to death," with space for two antecedent causes ("morbid conditions, if any, giving rise to the above cause, stating the underlying condition last"). There is then free space for "other significant conditions contributing to the death but not related to the disease or condition causing it."

Designed and standardized in the 1930s, the death certificate echoed the positivist analytic philosophy of the time. Now the trend is toward fractured, postmodern, multiple causation. (Nordenfelt 1983 has an excellent discussion of this philosophical history.) The history of philosophy, the history of ways of dying, and the history of death classification—are these three histories or one? This question makes no sense unless we look at the ways the ICD as an information infrastructure knits together temporal, philosophical and scientific concerns. Earlier in this century, historians in general effected closure on the past by bringing what we now see as problematic single historical actors ("great men") before the tribunal of history. Similarly, the original ICD also tried to effect closure: to provide a single, centralized record of the great epidemic diseases. Modern historians constantly point to the openness of the past. The past, we are told, is recreated afresh at each instant in the present; one role of the historian is to honor this openness while telling the best story one can (Serres 1993, C. Becker 1967). Modern medical classification systems, most particularly the ICD rival SNOMED (Système de Nomenclature Médicale) strive in precisely the same way to keep the past open. Ideally, they would become topological, but with an ease of management, data entry, and controlled vocabulary preserved. Thus far, this goal has proved elusive.

To tell the story as one internal to the history of medicine, consider the problem of tracking AIDS through history. AIDS achieved recognition as a disease in a slow process. Gay and sexual politics, medical profit making, and medical research were embroiled together in both its definition and its control. From the public health side, researchers at the Centers for Disease Control began to notice increased requests for a drug used to treat Kaposi's sarcoma—a rare condition previously afflicting certain localized or well-defined population groups, such as elderly men of eastern European descent. Intensive epidemiological work revealed that sufferers were largely male homosexuals. Transmission to hemophiliacs indicated that it could be passed on in the blood. Then Luc Montaignier and others located a virus that is generally believed to cause the disease (although for some both causal and priority questions remain open).

Statistics compiled before the epidemic used ICD categories without reference to AIDS-related conditions. It is thus virtually impossible to search back through the historical record to find earlier instances of AIDS. The old statistics do not record what were believed to be contributing causes of death. As a thought experiment, people with AIDS from the 1920s might have died of any of a number of opportunistic infections now associated with AIDS, such as pneumocystis carinii. Their deaths, by the then current ICD, would be widely distributed throughout the classification system, and some would be buried as simple pneumonia.

Further complicating the historical retrieval problem, "rare" contributing diseases are often deliberately excluded from the kept record. The ICD is after all primarily a statistical classification. Biostatisticians and epidemiologists are most interested in regular patterns of recurrent diseases. The standard death certificate has no room for clusters of diseases to be recorded; and even if two or three are mentioned, they often get reduced to one at the moment of the compilation of statistics. Grmek (1990) discusses at length the evidence for possible former cases of AIDS that remained in the historical record contingently. In some cases, they involved famous figures (for example, Erasmus) whose life was recorded in great detail. In others, they stood out as medical curiosities, such as the unexplained death of a Norwegian family in the 1950s. In that case, the husband had been a sailor who visited East Africa; his wife and subsequent children all died with a set of symptoms akin to those afflicting AIDS patients. Old blood samples have only been kept in aleatory fashion. Even where they have been preserved, they might give false positives to diagnostic tests. It is only through finding a pattern of immune system breakdown that there could be any hope of tracking such diseases through time. That would (would have) require(d) recording all contributing causes, even when it is, at the current state of knowledge, "obvious" what the patients died of. This is infeasible for both practical and epistemological reasons.

The reasons for wishing to maintain an "open past" as told from within the history of medicine are clear, however impossible to implement. But if one focuses through the history of medicine in such a fashion, one is likely to miss both the contemporary pragmatics of the period in question and to isolate medicine as a special case among disciplines.

In fact, medicine shares many commonalties with other disciplines and professions that have adopted new information infrastructures. To tell this story, we can turn to the history of information technology during this century and to the ways in which technological constraints and information processing developments have shaped the ICD.

## The Technological Configuration: Another Way to Think of Epidemiological History

The ICD is at each historical point associated with a particular configuration of technology. Like much modern information technology it bears traces of its past, inheriting the inertia of the installed base upon which it was built (Hanseth, Monteiro, and Hatling 1996). Computer screens tend to be eighty characters wide, an echo of the eighty columns of the preceding punch-card technology (Norman 1988). Similarly the ICD bears traces of its technological ancestry. The 200 headings restriction inherited from the census forms is the most obvious physical example. Both the form and the implementation of the ICD have been influenced by development of information processing technology. For the former, Blois (1984, 124) remarks that the use of numeric codes in the ICD was directly attached to the development of punch-card technology. As an example of the latter, in the United States coding of more than the single underlying cause of death was a failure before 1968 despite repeated attempts. Such coding became standard when an automated computerized system was implemented for the selection of the underlying cause of death (Israel et al. 1986, 165).

Tracing the imbrication of the technological configuration and the form and use of the classification system, the history of the ICD attaches directly to the development of information processing technology this century. The story begins in the nineteenth century, with the rise of large-scale bureaucracies. This development is still under-explored by historians, but one consensus that appears to be emerging is that insurance companies, banks, railway companies, the post office, and the government were at the heart of this development (Chandler 1977, Yates 1994, Campbell-Kelly 1994, John 1994, Friedlander 1995, Bud-Frierman 1994). As companies began to operate over a very large space (railway companies simultaneously created that space and operated within it), a need arose to share information on a standardized form. A mechanical punched-card technology was developed for storing and sorting large quantities of tabulated information. A hole punched on a certain row of a certain column of a card could mean whatever one wanted it to mean: and cards could be mechanically sorted. Among the first applications of this technology was the use of the Hollerith tabulators for the American census in 1890. Without this aid the information gathered at this census would have taken longer than ten years (the period between censuses) to sort using the old methods.

Information stored on punch cards could be retrieved much more quickly than information stored in, say, ledger books. It was difficult to implement the punch card technology, however, which came into its own only for large-scale statistical and accounting applications. It was expensive and cumbersome to go through huge numbers of cards. The cards themselves had to be printed on the finest quality material, an additional expense (Campbell-Kelly 1989). In the case of the ICD, only certain centralized government bureaucracies could afford the necessary technology and personnel to successfully implement the new information processing possibilities. As a result some countries soon adopted this mode of information processing while others never did. For similar reasons, the problem of divergent information technology resources has dogged the ICD to the present day.

In the 1950s, electronic stored program computers began to appear. There was talk both in the popular press and in academic circles of creating an "electronic brain." The dream in medical circles became the integration of all the various kinds of trace that were kept of medical encounters (Blois 1984, 127). First there was the patient medical record: the hospital's central account of what had happened to the patient. Then there were the local versions of that record stored and maintained by the various hospital departments. Then there were the notes kept by the doctor, the reports to health insurance companies, and the reports to government statistical services. If a single standard language (drawing in part on the ICD) could be imposed on all these reports, then all the various services that needed information could draw it from a single central source. All relevant information would be preserved. The most famous resultant record system, still operating today, is COSTAR: the Computer-Stored Ambulatory Record. This was developed at the Massachusetts General Hospital, starting in 1969, where it was first applied to a population of some 37,000 Harvard Health Care Plan patients. The record was designed to be used by researchers, doctors, and government agencies. Its programming was written in a special interactive programming language called MUMPS (the Massachusetts General Hospital Utility Multi-Programming System) (Barnett 1975, 4).

The central challenge in the subsequent period up to about 1980 became integration of the data so collected. It was clear that the new information technology could provide data integration. At the same time, it was not clear just what sort of integration was needed.

The various ways the new information technology would interact with medical practice was hotly debated. One such project was that of making automatic medical diagnostic tools. In some tests, the expert system MYCIN could outperform doctors in clinical tests, rather than in diagnosis. Paradoxically it was never actually adopted, since it tended to be very cumbersome and slow. Berg relates how this kind of expert system was pitted against the production of clinical-decision support systems, which could advise the doctor not so much on the diagnosis as on the course of treatment to follow (Berg 1997). (In a further complication, often the diagnosis itself is ex post facto. That is, the treatment worked, therefore the patient must have had such and such a disease.) A myriad of similar examples litter the history of the ICD and medical language and recordkeeping.

Whatever the form of integration and automation, more categories were needed to manage the range of uses to which the system would be applied. During the post-World War II period, the ICD has increased hugely in size. Sprawling sets of modifications were produced for specific clinical and administrative purposes. Thus ICD-8 was modified by the U.S. Public Health Service to provide greater detail in certain disease categories and was published for use in the United States as the International Classification of Diseases—Adapted (ICDA) 1967. This in turn underwent further revision by the Commission on Professional and Hospital Activities (CPHA) for use in American hospitals, which was published in 1968 as the Hospital Adaptation of ICDA (H-ICDA). Later versions included that of the Royal College of General Practitioners (1972); the International Classification of Health Problems in Primary Care (1975); and the OXMIS Code of the Oxford Community Health Project 1975. Huffman (1990, 346–364) gives a clear summary of all the modifications and modified modifications that were generated.

Again this is a classic story of information processing from 1950 to 1980. More than 100 standard computer languages were created during this time. Each of these standard languages spawned 100, often mutually incomprehensible, dialects (Metropolis, Howlett, and Rota 1980). The WHO attempted to control this process for the ICD by producing guidelines on how to modify the ICD for particular purposes. These guidelines were themselves modified locally, however, a classic problem in decentralized organizational control.

In the post-1980 period, the resultant steely skyscraper is not so different in kind from the Gothic brick construction of the 1890s. There are a thousand "controlled medical vocabularies" for a thousand purposes, many of them having embedded within them some version or other of the ICD. As one article put it: "We are often reminded that medical knowledge has grown to the point where we require the assistance of computers to manage it. One response has been the construction of controlled vocabularies to facilitate this process. We are now at the point where the vocabularies themselves have reached unmanageable proportions and must again call on computers for help" (Cimono et al. 1989, 517). The call now is for a unified medical language system (UMLS) that will provide for automatic, flexible communication among all authorized controlled medical vocabularies. Embedded within the UMLS will be the ICD. Embedded in the ICD will be flexible classifications that will, in principle, allow a reconfiguration of past records. And surrounding all of it are secondary and tertiary analysis and fiduciary industries that audit, monitor, and collect revenues based on their expertise in analyzing the intertwined category schemes. Figure 3.2, for example, advertises a firm with precisely this mission. Readers will recognize herein a familiar chapter in the history of expert systems, with the emphasis moving from faith in a unitary vision of the world as modeled in symbolic artificial intelligence to the management of multiplicity and pragmatic circumstances. By concentrating on the ICD and information technology, we are able to see a new kind of "open past." Rather than searching for disease precedents described before about the internal history of medicine, we may also find them in the history of information technology.

As such, this open past is shared by a number of different disciplines and professions. Note though, as David Levy (1994) points out more generally, that the move to computerization may lead in some senses to less flexibility and local variability than in the past. Jucovy (1982, 467) states that in the medical field, "computers will probably firm up lexical use in much the same way as printing served to fix the spelling of words a few centuries ago." The reconfigurable past of the ideal database meets the installed inertia of the standardized bases—thus the old dialectic is transferred to a new medium.

### Conclusion

A key outcome of the work of information scientists of all kinds is the design and implementation of information infrastructures. In looking at the case of the development of the ICD, a fundamental figure-ground problem emerges in the analysis of such infrastructures. In particular, the medical classification system that underlies a large part

*Figure 3.2*
Monitoring the intertwined classifications: the recovery room. An advertisement for one of the many auditing service firms that track the integration of information.

Source: *Healthcare Financial Management*, December 1996. Courtesy of Leco and Associates, Pittsburgh. (Note: They no longer provide the DRG review service.)

of all medical bureaucracy is historically contingent both with respect to its political origins and technological underpinnings. However we may tell the story of the open past, the dream of an unconstrained encyclopedia is evanescent.

This is not of itself surprising. Towers of Babel are perhaps the rule, not the exception. To classify is human and all cultures at all times have produced classification systems. Modern Western culture has produced more than most, often without realizing it. It is often asserted that Eskimos have fifty terms to describe snow. On close examination, this is an urban legend—Eskimos have only a handful of such terms (Pullum 1991). On the other hand, however, Arctic explorers have hundreds, scientifically laid out in their expedition manuals (Pyne 1986).

A consistent finding of the history of science is that there is no such thing as a natural or universal classification system (see for example Lakoff 1987, Latour 1987). Classifications that appear natural, eloquent, and homogeneous within a given human context appear forced and heterogeneous outside of that context. Borgès gives a wonderful invented list created by the Chinese emperor: "animals are divided into: (a) belonging to the emperor, (b) embalmed, (c) tame, (d) sucking pigs, (e) sirens, (f) fabulous, (g) stray dogs, (h) included in the present classification, (i) frenzied, (j) innumerable, (k) drawn with a very fine camelhair brush, (l) *et cetera,* (m) having just broken the water pitcher, (n) that from a long way off look like flies" (cited in Foucault 1970, 15).

In a similar trope, Bertillon (1895, 263) pointed out the incongruity of Farr's "natural" (for the 1850s) grouping of gout, anemia, cancer, and senile gangrene as a single kind of disease. In like fashion, our own lists can appear strange to outsiders. Thus supporters of the rival schools in modern biological classifications—cladistics and numerical taxonomy—each make rapprochement between species or splits between them that jar common sense perceptions. The ICD as an information infrastructure is an invisible underpinning to medical practice. On close examination it constitutes a classification as strange in its way as Borgès'. But as with many strange things, it has become well adapted to modern bureaucracy. We can tell the story of this adaptation as the integral, costructuring rise of both the modern state and the new information technologies.

As for the ICD, we saw that one could foreground state interests and see the developing ICD as reflecting and partially determined by these interests. In this picture, the ICD is a passive list, molded by outside forces. The ICD can, on the other hand, be brought into center stage as one of the mechanisms developed this century for producing and defining the modern state. According to this position, it is no happenstance that a series of universal classifications emerged in the late nineteenth century (classifications of work, industrial equipment, criminal physiognomies, see Tort 1989). Rather, the development and maintenance of such classifications by increasingly ramified bureaucracies changed what it was to be a citizen of a given state. They provided fundamental tools for communication and control.

Finally, looking at the ICD and information processing, we saw that one could as well tell the history of medical classification internally from within the history of medicine as the story of the development of better and better classifications in tune with the development of medical knowledge. When we emphasize the infrastructure of classified medical knowledge associated with the ICD, we see a classic story of the development of computing infrastructure. Thus the ICD can be understood as one of many classification systems this century that have changed in tune with the development of computing technology: the storage and retrieval devices involved played a large part in shaping the nature and form of the classification system.

Given Star and Ruhleder's (1996) definition of infrastructures as being hybrid creations of work practice and information medium, such figure-ground switches are helpful historically. Working infrastructures like classification systems are deeply embedded both in practice and in technology. Their history cannot be told independently of the work practices that they constitute or the media in which they are inscribed. The work practices associated with the ICD link its history with of a set of classificatory practices defining the modern state and later the modern corporation. The media associated with the ICD link its history with a set of classificatory principles associated with a particular technological base developed for the management of distributed information.

The analysis of information infrastructures forces us to pay close attention to the unit of historical analysis. One might say that typically an historian seeks to examine the change in an historical entity over time—a person as she or he gets older, a state as it goes to war, an idea as it is born, developed, and superseded. In these standard cases one assumes—rightly or wrongly—that what it is to be a person, a state, or an idea does not change in the course of the historical treatment. It is, in other words, the passive backdrop against which the historical drama is played out. Information infrastructures are constitutive of that backdrop; when they are foregrounded, and the historian's standard categories are rendered contingent, they become objects of historical examination.

To bring together this Janus face of infrastructure, we make a double kind of shift. This is the "infrastructural inversion," discussed in chapter 1 (Bowker 1994). The inversion helps provide a framework within which one can consider the filiation among information processing practices and technologies across a range of arenas. This also generalizes the history told here of the ICD. The problems faced by the ICD and its solutions have as much in common with the history of the Dewey classification system in libraries and industry as they do with the history of medicine before the ICD.

To do historical justice to the development of information infrastructures, one must move among stories that historians traditionally tell of people and places and things and those stories that are generally left untold: of the woof and warp of the canvas on which historical dramas are painted.[20]

# 4

## Classification, Coding, and Coordination

### Introduction: Coordination Work

Marx referred to technology as "frozen labor"—work and its values embedded and inscribed in transportable form. Modern information technologies similarly embed and inscribe work in ways that are important for policymakers, but which, as we have shown, are often difficult to see. Where they are used to make decisions, or to represent decision-making processes, such technologies also act to embed and reify those decisions. The arguments, decisions, uncertainties, and processual nature of decision making are hidden away inside a piece of technology or in a complex representation. Thus, values, opinions, and rhetoric are frozen into codes, electronic thresholds, and computer applications. Extending Marx, then, we can say that in many ways software is frozen organizational and policy discourse.

The purpose of this chapter is to explore this idea theoretically, taking stock of some of the issues in the sociology of technology, and to reflect on how they might contribute to research in organizational policy about classification, standardization, and information systems. We continue to use the example of the ICD. The previous two chapters have looked first at the ICD as a text that can be read for its cultural values and as an historically developing infrastructure. Taken together, these processes display convergence among categories, information technology, and people. In this chapter, we will turn to the ICD as an object that facilitates the coordination of work among multiple agencies—an agent for distributed work and cognition.

The ICD, as we have seen, is an important infrastructural component of medical and epidemiological software. It is increasingly important, as well, for the financial and administrative components of medical care, as it is used (in a number of different forms) to encode reimbursement, in cost assessments, and in the allocation of expensive equipment on the basis of diagnostic need. Issues of social justice and policy are very deeply embedded in all these question of due process (Gerson and Star 1986).

The ICD is one of the earliest modern attempts to collect global information across a number of federated suborganizations and feeder sources. Like state censuses and many forms of business and governmental statistical data collection efforts, it is protean in its effects on knowledge in both the private and public sectors. People use the ICD and its knowledge in a myriad of ways, including developing diagnosis-related groups (DRGs),[21] electronic medical records, assessing equipment and personnel needs, and reporting vital statistics. These in turn have a profound impact on health care costs and business policies (see Geist and Hardesty 1992 for a discussion of this in the American case).

The ICD and similar categorizing schemes are examples of the practical and theoretical difficulties and challenges inherent in modeling large-scale knowledge work and collaboration. They are also good examples of how management and decision-making tools become part and parcel of organizational structure. In recent years, burgeoning interest has risen in joining management information systems (including complex modeling), social history of statistics and classification, sociology of technology, and studies of infrastructural development (Hughes 1987, Boltanski and Thévenot 1991, Desrosières 1988 and 1993, Yates and Orlikowski 1992, Star 1995c, 1991b).

As noted, the ICD is about 100 years old and revised nearly every ten years since the end of the nineteenth century. It is distributed as a book, or as a component of medical record-keeping software, to public health offices, hospitals, insurance companies, health accountancy firms, and bureaus of vital statistics throughout the world. It contains numbers that correspond to causes of death or illness of the sort discussed in chapter 2, and algorithms for arriving at those numbers in complex cases involving more than one disease or cause. In a sense, the ICD is the backbone of a sophisticated, very large computer-supported cooperative work (CSCW) system as well as a form of large-scale organizational memory. It is also a decision-making tool for all sorts of policymakers. On the basis of data collected using the ICD system, decisions are made about allocation of resources, whether and how to control epidemics or endemic illnesses, and whether there are shifts in population based on infant mortality rates, and so forth. This chapter considers the nature and design of the ICD seen as a kind of super-CSCW or coordinating and decision-support tool. Large though it is in both temporal and spatial scope, however, it is also worth remembering that the ICD shares many attributes with more modest, contemporary CSCW and decision-support and modeling tools. This includes, frequently, gaps in frame of reference between designers and users, difficulties with data entry quality and speed, and incompatibility of new tools with legacy systems.

The first point in this analysis is a simple one: the ICD is a list and lists may be considered as interesting tools.

### Truth Comes at the Point of a List

List making has frequently been seen as one of the foundational activities of advanced human society. The first written records are lists (of kings and of equipment) (Goody 1971, 1987). Leroi-Gourhan (1965) pointed out that what gets written down first are things that cannot be retained in the head. This is especially true of lists. The earlier feats of memorization by Welsh poets (up to 100,000 lines for professional bards) were of lists within epic poems. The memorization task was aided by numerous cues within the text, and they were embedded in social practice.

Michel Foucault (1970) and Patrick Tort (1989) have, in different ways, claimed that the production of lists (of languages, races, the minerals, and animals) revolutionized science in the nineteenth century and led directly to modern science. The list in this case is both a hierarchical ordering and a practical tool for organizing work and the division of labor. The prime job of the bureaucrat, according to Latour (1987), is to compile lists that can then be shuffled and compared. Yates (1989) makes a similar point about the humble file folder. And so empires are controlled from a distance, using these simplest of technologies.

These diverse authors have all looked at the work involved in making these productions possible. Instead of analyzing the dazzling end products of data collection and analysis—in the various forms of Hammurabi's code, mythologies, the theory of evolution, the welfare state—they have instead chosen to dust off the archives and discover piles and piles of lowly, dull, mechanical lists. The material culture of bureaucracy and empire is not found in pomp and circumstance, nor even in the first instance at the point of a gun, but rather at the point of a list.

List making is foundational for coordinating activity distributed in time and space. Consider an apparently simple problem of coordination that children in many cultures solve routinely: the treasure hunt. In this game a list of objects is made, usually by an adult, and teams of children are each given duplicate lists. The first team to bring back all the items wins. Even a local, improvised list such as this entails judgment calls: objects should be difficult enough to challenge the children's ingenuity, but not beyond their reach; they should not require impossible resources (e.g., no objects requiring use of a car to fetch). Typically they are things that are odd but not impossibly rare— a copy of the front page of the *New York Times* from June 4 1994; a green high-heeled left shoe. Teams may decide to coordinate their internal work by assigning each person an item, by working in pairs, by moving as a group, and so on.

Lists are in themselves a genre of representation (Yates and Orlikowski 1992). Genres are: "typified communicative action performed by members of an organizational community in response to a recurrent situation . . . identified both by their socially recognized communicative purpose and by common characteristics of form" (Yates, Orlikowski, and Rennecker 1997). When lists are used to coordinate important work that is distributed widely over time and space, a correspondingly complex organizational structure and infrastructure evolves. Lists are stitched together with other genres to form what Yates and Orlikowski call genre systems. Genre systems, a very useful concept, encompass both the abstract top-level notion of the genre, in this case the list, and enfold as well the more concrete local variants (such as the list of diseases, the mortality rolls, and the metadata coordinating lists of epidemiological results). "A genre system is an interlocking and interdependent set of genres that, by definition, requires collaboration" (Yates, Orlikowski, and Rennecker 1997, 2).

The ICD genre system includes sets of codification practices, medical nomenclature lists as well as the actual numbered labels themselves. Over time, the genre system may achieve a kind of closure, as routines and communications depend on each other. For instance, in the genre system of university admissions procedures, genres of standardized tests, letters of recommendation, grade point averages, and geographical distribution all contribute to the assessment of the candidate.

In the case of the ICD, negotiations over the content of the list become reified—frozen—and often take quantitative form, especially if the items are numerous, costly, or critical for other operations. Down the line, this obscures the nature of the genres being linked together in the decision-making process. The judgment calls are still present but now involve multiple actors and their routines, including individuals, organizations, and technologies. The decisions about division of labor remain, but now entail bureaucracies as well as spot judgments. As all the authors cited above have concluded, large-scale coordinated work is impossible without lists. As well, those lists considered as genre systems entrain whole series of substantive political and cognitive changes in the classes they inventory.

The ICD, as a functioning means of coordinating information and work highly distributed over space and time, contributes several valuable lessons to understanding the management and use of information technologies in very large multinational organizations and to the study of genre systems:

• First, there is a permanent tension between attempts at universal standardization of lists and the local circumstances of those using them.

• Second, this tension should not, and cannot, be resolved by imposed standardization because the problem is recursive.

• Third, from the point of view of coordination, ad hoc responses to standardized lists can themselves be mined for their rich information about local circumstances: in turn, information technology might be tailored to support those needs, not subvert them.

• Fourth, this type of list is an example of the sort of object that must satisfy members of communities or organizations with conflicting requirements. In its creation, and later in its use, the complex list is a kind of knowledge representation particularly useful for coordinating distributed work that often contains requirements of this sort. Some, ourselves among them, would argue that they are necessarily conflicting or at least divergent (Hewitt 1985, 1986; Star 1989a).

The problems of such knowledge in very large, distributed organizations are increasing as multinational firms confront local variation and definitions of knowledge in their subsidiaries in different countries and with other processes of globalization. The problem here is generic to all such efforts where diversity is the central issue in representing information.

### The Impact of the ICD
To continue in a foundational vein, the ICD can be seen as one of the tools bound up in the origins of the welfare state (Ewald 1986, see chapter 3): the epidemiologists and government statisticians who

originally drew it up were concerned with large-scale public health measures. It has often silently accompanied all major epidemiological work of this century. The power of a classification of disease can be seen, for example, in the debate about Britain's mortality decline in the nineteenth century (Szreter 1988). Three interest groups have at different times claimed the kudos here (and a share of funding and recognition appropriate to their contribution): medical specialists who claimed new forms of treatment rid the country of its major scourges (particularly tuberculosis); public health officials who asserted the value of sanitation in the cities; and laissez-faire economists who highlighted the general rise in the standard of living in a successful economy unburdened by expensive medical welfare. The stakes in the debate were clearly very high. As Szreter's careful revisionist history details, the debate's outcome hinges on a reading of the tables of mortality that listed causes of death by region. These tables show unequivocally that the new forms of treatment developed after the decline in mortality, not before. This is in accord with an earlier, brilliant demonstration by McKeown, in the context of a debate about national medicine (1976). But contra McKeown, who underscored the rise in the standard of living, Szreter shows the changes are in step with local public health measures. The core of Szreter's argument is an interpretation of disease classification in the nineteenth century (particularly the categories of airborne disease).

The ICD has thus played a key (albeit usually silent) role in determining the outcome of epidemiological, public health, and economic arguments. We will look at the way the ICD has been used by different groups, constituting both a common and a customizable object for these groups and a genre system in use. We will look at the tension between the desire to standardize (so as to be able to perform bureaucratic functions such as comparison over time and space, produce algorithms, compute etc.) and the drive of each interested party to produce and use its own specific list. We will also examine the tension between attempts to make a universally standard list and the idiosyncrasies and local circumstances of users. Both these tensions speak to the nature of all knowledge-based informatic policy and management tools.

To develop this analysis further, we first inventory the different classes of informational conflicts involved with building up and using the list and examine the types of informational needs and structures involved in each case. At the same time, we will examine some associated problems drawn from the list's history.

### International Conflicts

One value of a list like the ICD is that it can be used in transnational comparisons, especially where there are radical local differences in belief, practice, and knowledge representation. This is necessary for epidemiology in that one may trace specific environmental and nutritional factors involved in particular diseases, track epidemics, and impose quarantines.

These advantages can only be fully exploited if the various suborganizations agree on how to collect and code information. A continuing problem has been that different countries have sent their information in to the central collection agency more or less promptly (a problem finding its parallel with the reports sent to central office by subsidiaries). During the 1920s, France and Portugal were notably slack. And once the information comes in, it is often of variable quality—countries with large rural populations find it difficult to give the same sophisticated medical treatment of each case as do heavily urbanized, western countries (Réunion de Conseil 1923). At one stage in the USSR, no attempt was made to compute causes of death in places with less than 10,000 inhabitants (CH/experts stat/78).

Different states have different bureaucratic structures. In the nineteenth century, for example, the statistical system was run by a central service for the whole country in Italy but was broken down by province in France (Bertillon, 1887). The regulations regarding the use of death certificates have made an appreciable difference to the type of results the ICD achieved. Thus in Germany during the 1920s there was no separation between the civil statement of the cause of death and the cause of death issued for statistical purposes. In Switzerland, the statistical cause of death was confidential, making it much easier for doctors to cite causes that might distress relatives (and upset insurance companies). When Holland switched over to the confidential system in 1927: "There was a considerable increase in Amsterdam of cases of death from syphilis, tabes, dementia paralytics, aneurysm, carcinoma, diabetes, diseases of the prostate, and suicide, while deaths from benignant tumors and the secondary diseases such as encephalitis, sepsis, peritonitis, and so forth showed a falling-off" (League of Nations 1938, 10). More recently, a similar artifact was reported as a result of physician terminology preference in ICD-9 on coronary heart disease mortality, varying significantly across states within the United States (Sorlie and Gold 1987).

Further, different cultures place differential emphasis on causes of death. A recent example is a controversy about Japan's low rate of fatal heart attacks. A traditional reading of the list suggested that this statistic is due to nutritional or environmental factors peculiar to that country—level of fat in the diet and so on. Recently, however, some epidemiologists have suggested that the cause may well be that disease is a very low status cause of death within Japanese culture, suggesting as it does a life of physical labor and a physical breakdown. Accordingly, what Americans would call heart attacks often get described as strokes, since an overworked brain is more acceptable there. When this is factored in, they suggest, there is no discrepancy in Japan's figures. These national differences are complicated by the facts that some diseases present differently in different countries. AIDS is one such disease; malaria is another. For the latter, E.J. Pampana noted in an article entitled "Malaria as a Problem for the WHO" that "At a first glance, malaria does not appear to have an international character at all; one could almost say that no other disease is so strictly dependent on local conditions. Malaria might, in fact, almost be called a nationalistic disease, because it takes from the country its very characteristics, as does its folklore. These very local aspects of malaria epidemiology are the bricks with which the science of malariology is built. . . ." (WHO archives 453–1–4).

Different national schools of medicine may disagree about issues such as simultaneous causes of death. One WHO committee noted early on that there were indeed such differences; and that if there were no agreement by "reason," then countries would vary according to "facts of pathology (or) clinical medicine, (or) public health importance (WHO archives 453–1–4, 11)." It recommended that the different countries produce a table of contributory causes so that a comparison could be made; however, the problem proved unwieldy. In the Census Manual of the International List of Causes of Death there were 8,300 terms, which represented 34 million possible combinations. If half of the terms could enter into combination, then an assignation of priority in all possible cases would involve sixty-one volumes of 1,000 pages each (CH/E Stats/34 1927, 10–11).

Finally, handling of the ICD has been politically charged in terms of its internal bureaucracy. Originally, it was run by the French Office Internationale d'Hygiène Publique (OIHP) and was for the French government a sign of its natural diplomatic leadership. When the

League of Nations started to gain control of the production of the list, one British diplomat noted that "an influential clique in the French Foreign Office is moving heaven and earth to retain the Office Internationale unaltered" (Société des Nations, box R822, 1921). The United States became a key participant when it refused to join the League of Nations, leaving the OIHP as broker between the United States and the League of Nations. They tried to squeeze out the International Institute of Statistics as advisors. This led the director of that organization to complain that: "The new masters of the world are laying down their law, without any consideration for the rights of others and for an international organization that has received universal respect to this time" (WHO archives 455-3-3).

Relationships among developed and less-developed countries figure large in the construction of the ICD. For the former, with access to the latest computer equipment, some kind of state-of-the-art expert system could handle more data and detail, more flexibly, than has ever been possible in the past. For other members of the WHO, however, lacking a computer infrastructure capable of implementing the sophisticated software, the list would be useless. Or, it could be seen as an administrative burden imposed by colonialist interests.

Even if it were possible technically, this level of granularity is unnecessary for many countries. As the director of health statistics at the WHO in the 1980s explained to us, death in his country (Indonesia) is overwhelmingly caused by infant diarrhea via contaminated water supplies. Why spend precious resources codifying at a finer level when the problem is so obvious? Until these issues are solved, who cares about the incidence of rarer diseases? The question is not rhetorical—other member nations *do* care, since they want to be able to trace the etiology and development of epidemics that are likely to affect *their* populations (flu, AIDS, etc.) throughout the world.

In this fashion, international cooperation has been hampered within each nation by the diversity of ways of recording and reporting, by cultures with varying stigma and prestige for certain diseases, by local medical cultures and by the different "national character" of some diseases. It has been hampered among nations by the issue of control of the prestigious ICD and by the medical and epidemiological needs of the different nations. The public health policymakers involved before this apparently simple, homogeneous list could be compiled and implemented included government officials, statisticians, anthropologists, medical analysts, epidemiologists, and diplomats. We can easily see parallels with power struggles, control, and containment in the multinational firm and its information management—classic problems of decentralized control in the post-Fordist era.

### Government: The State versus the Individual

Another series of actors emerges when we turn to the relationship between the state and the individual. There are a number of moral and political categories here that directly affect the structures of information. The classification of death by suicide is a good example. Early in this century, many doctors complained about the detailed breakdown of this category, which had "no prophylactic value." Statisticians responded that the details should be recorded "for their sociological interest and for the police," defined by the judiciary not by medicine. This incorporated some moral and political distinctions. Thus: "In the case of collective suicides, you have to count as many suicides as there are people over the age of majority. Minors have to be considered victims of murder. Death by starvation was said to be a "crime" if children suffered it; a "misfortune" if an adult cause of death (Commission Internationale Nomenclature Internationale des Maladies 1910, 116–118).

Similarly, when criminal abortion was defined in a fairly undifferentiated way as homicide (whereas legal abortion had its own category), it was difficult to obtain statistics about it. Similarly, stillbirth was a political and religious category that varied by nation and by brand of Christianity. Should a fetus that had never breathed (or tried to breathe) be recorded as a death? If so, it would contribute both to infant mortality statistics and have a soul; if not, the miscarriage would simply be recorded under the morbidity tables.

### Conflicting Needs of Doctors, Epidemiologists, and Statisticians: Questions of Data Accuracy

How accurate does information need to be? The question is not a trivial one as the opportunity and transaction costs involved in collecting information multiply with precision. In the case of the ICD, clinicians saw the work of collecting data as trading off against patient resources, while statisticians wanted as much accurate information as possible. The task of filling in the death certificates ordinarily falls on the doctor who does not necessarily see the value in filling in a complex form to the degree of accuracy required. After all, this patient is dead; is the time not better spent on the living? As we noted in the last chapter, this creates an impossible situation from the point of view of data quality. John Carter, in a paper discussing "the problematic death certificate" recommended that: "The American Medical Association should be asked to restate annually its resolution of 1980 that 'the American Medical Association encourage physicians to give thoughtful attention to more accurate completion of death certificates'" (Carter 1980, 1,286).

In a study of death certification, Cameron and McGoogan found that: "diagnostic accuracy bore an inverse relationship to the patient's age" (1981, 273). That is, the practicing physician does not see accurate recording of the death of an old person as a high priority. This is a bureaucratic extension of Sudnow's (1967) "social death" and of Glaser and Strauss' "social value of the dying patient" (1965).

When it comes to use of the tables produced with the ICD as a basis, in general: "practicing specialists want more categories and urban statisticians want less" (Société des Nations, Organisation d'Hygiène, Commission d'Experts Statisticiens, CH/experts stat./1–43 1927, 1–2). Here, specialists wish to know the breakdown of each disease strain, whereas the public health urban statistician wants broader, action-oriented categories like nutritional deficiencies, environmental factors that could be changed, and so on. This has at times led to a double bind: "So-called administrative statistics have no value in the eyes of practitioners, who as a result are completely uninterested in it; whereas, unless these practitioners provide exact data, then the scientific value of administrative statistics has to be called into question" (Société des Nations, Organisation d'Hygiène, Commission d'Experts Statisticiens, CH/experts stat./1–43 1927, 2). The ICD does not speak to general practice. Froom ascribes the need for an international classification of health problems in primary care to the fact that attempts: "to use the . . . ICD to classify health problems encountered by general practitioners have often been unsuccessful" (Froom 1975, 1,257). To continue to draw the parallels with decentralized control of distributed work in firms, one hears clear echoes here of the infamous tension between R&D on the one hand and marketing on the other about the need for precision vs. speed.

The different groups have spoken to issues at the core of the ICD. Statisticians, for example, wanted the first ICD to have only 200 categories, since a statistical table as used in censuses could only be approximately that many lines long. For them, lists had to be stable over time and space for comparability. "This is why diseases must be classed according to their seat and not their nature or their cause.

Because the seat is much more easy to determine than the nature"
(Commission Internationale Nomenclature Internationale des Maladies 1910, 11). Statisticians stressed that the role of the list makers was
not to produce "philosophy" but to make a "truthful" and "comparable" list. The Spanish authorities wanted the list of general diseases to
be set out according to how public authorities could react, breaking
them down as follows:

- General and sporadic
- Epidemic
- Imported
- Common to people and animals
- Professional intoxications

(Commission Internationale Nomenclature Internationale des Maladies 1910, 17).

Another set of statisticians wanted to give precedence to social-
biological factors (CH/experts stat/80, 4). "Violent death" should move
up the list, since this would: "settle various doubts . . . as to whether
consequences due to visible external causes are to be classified here,
or, for example, under infectious diseases (a case in point is infectious
diseases of wounds)." Or again, it was argued that there should be a
subdivision for diseases for which statistics were required under international conventions, for example, lead poisoning (CH/experts stat/80, 4).

As we have gone through the different categories, we have been
getting closer and closer to seeing the list as entirely heterogeneous.
The ICD is not so much a list of causes of death as a series of dynamic
compromises among a wide range of players across a number of
different venues—perhaps like an organization chart or a labor contract. Or, as one observer noted: "In short, the nomenclature of diseases and of causes of death established for the needs of statistical
organization constitutes a sort of contract between the two organizations who are charged with statistical works—that is to say the service
who makes the observations and that which produces statistics with the
help of these data" (CH/expert stat/43 1927, 3).

### Industrial Actors

The above discussions indicate that many people from diverse social
worlds had a stake in how the ICD was compiled and used. Three
other significant groups were also involved.

*Insurance companies*   The many insurance companies with a stake in the ICD wanted a breakdown of the ICD statistics in such a way as would be useful for them: "For example, there should be groups corresponding to the age at which direct compulsory sickness insurance begins, and the age at which compulsory old-age insurance starts" (CH/experts stat/80, 3). Since this rule was different for different countries (and nonexistent in many), this would have been impossible to apply.

*Industrial firms*   Some of the first groups to produce lists of causes of death were from the vast German chemical companies of the late nineteenth century. For them, relevant variables included whether the deceased had touched or not touched certain compounds, had worked inside or outside, and so forth. Again, we have here a different set of variables from those of interest to other groups.

*Pharmaceutical companies*   The claims that can be made for different drugs are in part a function of the list of diseases. For example, a classic case occurs in the Spanish pharmacopoeia (Bijker and Law 1992). Due to Catholoic religious restrictions against contraception, this handbook redefines what are commonly described as birth control pills. These pills may have a (typically undesirable) side effect of high blood pressure. In some cases, the pills may be prescribed as a treatment for hypotension. In a figure-ground switch, the technical side effect becomes the inhibition of birth. (Note what is likely to happen to the statistical records of incidence of hypotension in Catholic countries in such cases.)

This process may work inversely in some cases. One of us formerly had a student who was a national representative for a large drug company. A major part of her job was interviewing doctors about whether any of their patients had gotten better from one disease while taking one of the company's medications for another. If yes, that disease might potentially be added to the list of indications for the illness. The representative said that she was constantly pressured by her superiors to "broaden her indications" in this fashion. Here again, there is a trade-off between market pressures, frames of meaning, and regulation that require conflicting levels of restriction (see also Gerson and Star 1986).

No attempt will be made here to continue to list all the various actors who have been involved in compiling and implementing the ICD, but

**Table 4.1**
Some conflicting needs of ICD users

|  | Information needs | Problems |
|---|---|---|
| International Public Health Data Collection | precision; thorough coverage by case; timeliness; consistency | religious and cultural customs; incompatible medical systems ownership and administration of data; different granularity needs of users |
| Government | legality; vital statistics for planning | matching legal and medical categories; crimes unreported for various reasons |
| Doctors, epidemiologists, and statisticians | diagnostic; preventative; predictive | statistical vs. clinical approaches different hierarchies of multiple causes; early detection vs. clear clinical case |
| Industrial | targeting special groups; industrial pathogens; drug impacts and indications | different aggregations of data; shifting market needs |

the matrix shown in table 4.1 summarizes what should be obvious: something has to give. The list cannot be homogeneous, neutral, and appeal simultaneously to all parties. This is always the case for tools and objects that inhabit a number of different social worlds (Star and Griesemer 1989, Star 1989b). King and Star (1990) have examined this problem for the decision-making process in organizations and its implications for designing organizational-level decision support.

### Policy Inscribed into the ICD

The discussion now turns to the solutions to the problems of multiple membership and thus heterogeneous definitions and goals that have been explored through the ICD's history. A number of very bright people have long been working on the difficulties posed by the ICD that we adumbrate above. This section inventories working solutions to the above problems at the level of the negotiations about the design of the ICD. The concluding section then draws some connections with how the list is used in practice and how it comes to inform policy downstream and locally. The solutions proposed here are generic ones commonly appearing wherever diverse information sources must be reconciled into categorical schemata.

### Distributed Residual Categories
A first solution—spreading out garbage categories—might appear to be no solution at all, but rather a studied avoidance of the problem. It does, however, offer some interesting insights. Garbage categories include an array of categories where things get put that you do not know what to do with—the ubiquitous "other."[22] In mid-nineteenth century Paris, more than 10 percent of causes of death were ascribed to "other causes" (Bertillon 1906). In Berlin at the turn of the century, doctors were reluctant to provide valuable morbidity information. Thus one table gave acute bronchitis 1,571, chronic bronchitis 225, bronchitis, without any other qualifier, 12,844 (CH/experts stat/88 1929, 8). There were two general causes for the creation of garbage categories. The major subcategory "undefined diseases" was used "either because there was not enough information or because the disease was badly characterized or finally because the doctor failed to formulate a complete diagnosis" (Commission Internationale Nomenclature Internationale des Maladies 1910, 128).

It would be extremely difficult to envisage a time when there would be no need for these categories. Their management has been a constant thread throughout the history of the ICD. A major feature of this management has been their distribution throughout the list. Thus at the time of the first revision of the ICD, the U.S. representatives suggested getting rid of the categories "eclampsia" (nonpuerperal) and "children's convulsions," since they were ill-defined (pun unavoidable). The committee rejected the suggestion since it would lead to the attribution of too many "unknown causes . . . and this would discredit the statistics" (Commission Internationale Nomenclature Internationale des Maladies 1910, 62). Or again, the vague "hemorrhage" was kept with a view to "not overinflating the figures concerning badly defined diseases" (Commission Internationale Nomenclature Internationale des Maladies 1910, 73). This distribution went to the lengths of distinguishing between two types: "other diseases" and "unknown or badly defined diseases." "Proposed conclusion: Each of these two rubrics is very important. The latter in particular indicates what is missing from the other figures in their approach to truth" (Commission Internationale Nomenclature Internationale des Maladies 1910, 138). The need to distribute was urgent—Jacques Bertillon estimated that over half the causes of death would be "other" in Paris in 1900 if all the residual categories were gathered together (Commission Internationale Nomenclature Internationale des Maladies 1910, 5).

These garbage or residual categories, then, tend to fix the maximum level of granularity that is possible. Their advantage is that they can signal uncertainty at the level of data collection or interpretation under conditions where forcing a more precise designation could give a false impression of positive data. The major disadvantage is that the lazy or rushed death certifier will be tempted to overuse "other." By their nature, forms of this kind are only manageable if there is a zone of ambiguity written into them. In this case, precise definitions would drive a wedge among doctor, statistician and epidemiologist.

### Heterogeneous Lists

Throughout the history of the ICD, there has been continual, endemic debate about whether it constituted a nomenclature or a classification. The difference is that a nomenclature is merely a list of names that does not give any indication of cause. Nomenclatures are not thus necessarily tied to models of disease. A classification, on the other hand, gives causes and arranges them in relation to one another. The advantage of a nomenclature is that it can remain more stable over time. For example, a nomenclature based on the "seat" of the disease can list a series of indications that can then be used at a second degree of analysis to rediagnose in line with current theory. Systemic diseases like AIDS or systemic lupus erythematosis can be tracked this way, even though the category might not have existed at the time the original diagnosis was made. Classification systems are more immediately convenient in that they carry more complex information, but as we have seen, they change every few years with the development of new medical techniques or knowledge.

Intuitively it might appear desirable to have a single, well-defined classificatory governing principle for the ICD. Just as for garbage categories, however, and for the same reasons—the array of actors and opinions involved—the solution that has emerged over time has not been monolithic. Instead, it has incorporated a workable (practically and politically) level of ambiguity (the same issue arises in nursing classifications in chapter 7). The ICD has been as heterogeneous as possible to enable the different groups to find their own concerns reflected. Because different models of medicine hold, they embody different rules for classifying. This has resulted in the fact that, although the list is in appearance homogeneous, there are at least four classificatory principles involved:

1. *Topographical.* This refers to the seat of the disease, which part of the body it manifests in.

2. *Etiological.* This refers to the origin of the disease—genetic, viral, bacterial, and so forth.

3. *Operational.* This refers to the responses to certain tests, without there being a necessary one-to-one correspondence between test results and a given topographical or etiological feature (though in general one or the other is asserted). HTLV versus HIV is a case in point. HTLV was defined as a positive reaction to a test searching for antibodies. When what we call HIV initially produced the same reaction, Gallo classified it as an HTLV even though the virus had not been isolated.

4. *Ethical-political.* We have seen examples of this above. The definitions of stillbirth, abortion, suicide, iatrogenesis, and euthanasia, for example, are the outcome of ethical and political decisions.

### Parallel Different Lists

Frequently over the course of the history of the ICD, different groups have found that the list did not serve their purposes and so they have modified it. This sometimes happened in a country with a different range of medical problems not covered by the European ICD. For example, the first ICD was drawn up partly through a comparison of the tables of mortality of six European countries. Naturally, then, little room was left for the whole range of tropical diseases. African and southeast Asian countries were forced to produce their own modifications. Or again, different users of the list might find that the current one did not meet their exigencies. For this reason, for example, medical insurance companies have often produced their own versions, tailored to populations, reimbursement policies, and the company's software configuration.

As with many other attempts to standardize (computer languages come to mind), each time an international standard is laid down—every ten years in the ICD's case—there is an immediate efflorescence of modifications. Rather than lose control of this whole process, the ICD committee has chosen rather to issue rules for how the list is to be modified. This gives the WHO a degree of control at the second level that it has lost at the primary one. The advantage of this secondary control is that it gives an algorithm for working back from the modified list to the ICD itself. Recoverability, while expensive, is theoretically possible and the open history is maintained in part.

### Full Complementary Localization

In some instances, it has been suggested that the list itself be ignored and detailed local studies be carried out instead. Thus the Registrar General of England and Wales, responding to the call for an International List of Causes of Morbidity to complement the ICD, recommended "large sample investigations into particular groups of morbid conditions . . . instead of international classification, which would impose an order which masked the inherent vagueness of diagnosis" (CH/experts stat/87). Even for notifiable infectious diseases, he noted, intranational (let alone international) comparison is difficult. Furthermore, doctors were too diverse a group to unite internationally around a given list. "Dr. Roesle is tacitly assuming that the flagrant noncomparability of existing morbidity statistics is chiefly due to diversity of classification. The cause of the divergence may lie deeper and may reflect important differences in the points of view of the practitioners themselves" (CH/experts stat/87). The registrar's conclusion was that time spent on classification was wasted. For example, he wrote of breast cancer: "The fact that this disease does not greatly contribute to the *statistical* incidence of morbidity, is an evil not capable of remedy by any international rules of classification—it can only be cured by raising the standard of hygienic education; that of the public at least as much as that of the medical profession" (CH/experts stat/87).

This solution is a further step from the ambiguity discussed above or of the necessary diversity of these lists. It suggests that no list at all is valuable—local practices should be the focal point. From the point of view of the ICD, however, opening up this denegation in fact served to strengthen the ICD as a boundary object. Through open recognition of the tension between the local and the international-universal, the ICD has been continually tested and its limits set. Boundary objects do not claim to represent universal, transcendent truth; they are pragmatic constructions that do the job required (Star 1989a).

### Convergent Bureaucracy

Not all the work that has made the ICD more applicable has been done internally through modifications to the list. Indeed, one background factor that has had a great impact has been the convergence of international bureaucracy. Throughout this century, in general, people have become more and more used to being counted and classified. Public organizations have become increasingly adept at the necessary procedures. Inhabitants of rural areas and of developing countries are less likely to slip through the net now than fifty years ago. It is much less likely anywhere that it is the village priest who determines the cause of death. WHO resolved in the 1970s to cooperate with developing countries "in their endeavor to establish or to expand the system of collection of morality and morbidity statistics through lay and paramedical personnel" (Kupka 1978). WHO is attempting to achieve this through working with trained lay personnel on a modified, simplified version of the ICD.

We introduce this factor as a reminder of the historical and contingent nature of universally applicable lists. In a related domain, Alain Desrosières (1988) has shown how census breakdowns of the populations of Germany, France, and England have remained closely tied to the history of work, trade unions, and government intervention in those countries. As the ICD "naturally" becomes more universally applicable, this is partly the result of the (often-tacit) spread of western values through the application of modernist bureaucratic techniques. These techniques appear rational, natural, and general to citizens of western states, but when looked at in detail prove highly contingent. Just how contingent comes out clearly when we look at an alternative. Thomas McKeown wrote a thoughtful essay for the *British Medical Journal* in which he proposed "A classification of disease that distinguishes diseases determined at fertilization from those—the large majority—not so determined and manifested only in an appropriate environment. Whereas the latter are in principle preventable, contraception, abortion, treatment must deal with the former, or, some would add, modification of genes or chromosomes" (McKeown 1983, 594). This new classification would be threefold, divided into diseases determined before birth, and those arising subsequently from either "deficiencies and hazards" or "maladaption." Hazards would include, from above and below, predation and parasites. Maladaption would include western diseases associated with technology, for example, diabetes, rare in Kenya in the 1930s and now rampant (McKeown 1983, 595). Of course, this can easily obscure and inscribe the deepest of ethical problems such as the importation of smallpox to Native Americans through the process of conquest. Maladaption appears to be an obscenely mild term for the resulting devastation. Monica Casper

(1998) has written of similar moral dilemmas in the conduct of fetal surgery.

This proposed new grouping of diseases is very different from the grouping in the ICD, yet it is both consistent and possibly useful. It serves to remind us that the "rational and general" is only ever specific to time, practice, and place.

### Computerization

From the early 1920s, with the use of Hollerith cards and Powers machines, the history of the ICD is interwoven with that of computing. The chief advantage that computing offers today to the ICD and similar schemes is the ability to maintain uncertainty at the level of closure on analysis. When the list involved a relative handful of categories arrayed along one dimension, then a whole series of decisions were forced, whether the disease was environmental (e.g., of industrial origin), genetic, or viral, and so forth. Even when the maximal degree of ambiguity was kept, it was impossible to compare large bodies of data because the original wealth of material simply could not be maintained. Now that more numbers can be crunched and more axes added to the disease descriptions encoded by computers, the time of diagnostic decision can be held off. This theoretically brings closer the prospect of true comparability, although the range of practical and even ontological problems are unlikely to disappear even with the most advanced multivalent, object-oriented system.

The growing literature in organizational and managerial computing and its impact on knowledge attests to the importance of ambiguity as an organizational resource (see Kraemer, Dickhoven, Tierney, and King 1987 for an example using computerized information modeling; Pinsonneault and Kraemer 1989 review this literature for decision support). Since March and Simon's first conception of satisficing in the absence of universal, complete, knowledge (1958), increasingly sophisticated models and metaphors have been advanced to attempt to address this issue (see Morgan 1986 for a review). As shown in chapters 7 and 8, maintaining ambiguity may also consist in organizational autonomy, professional legitimation, and other forms of discretion.

### Standardized Forms

The goal of standardizing the ICD is by no means equivalent to rendering it unambiguous. Consider, for example, the following definition of a cause of death produced by a committee seeking to standardize death certificates. "A cause of death is a morbid condition or disease process, abnormality, injury or poisoning leading directly, or indirectly, to death. Symptoms or modes of dying, such as heart failure, asthenia, etc. are not considered to be statistical causes of death" (WHO archives, 455–3–4, 31/3/48). The committee proposed a uniform death certificate with several blanks to be filled in for causes and symptoms.

It is clear that standard forms are essential for the ICD to work and that these standard forms cannot be overprecise or people will not be able to use them. That is the fault of this death certificate—it attempted to make the determination of the real cause of death at the time of certification. This entailed asking busy doctors to do work they had no interest in doing nor often any ability to do. It entailed making choices that were more historically contingent than the ICD itself, which allowed a deal of flexibility by not itself making any causal claims. Standardization procedures must be tailored to the degree of granularity that can be realistically achieved (Fujimura 1987, Star 1991). As Harvey (1997, p. 1) states, "Standards are good. Quality is better."

## Using the ICD: Links between Design and Practice in the Organizational Infrastructure

It is clear that there are many unsettled arguments about the design of the ICD; let us look for a moment at the practices associated with its use by people certifying death and illness. None of this will come as a particular surprise to social scientists involved with quality control and practical survey research methods. For many years they have been exploring the gap between representations and codes, and the practices of filling out forms. Cicourel's (1964) ground-breaking critique of methods in survey research in the early 1960s is one example; Bitner and Garfinkel's (1967) exploration of "'Good' Organizational Reasons for 'Bad' Clinical Records'" extends the analysis, as does the work of Suchman and Jordan (1990). Our favorite is a lovely and extraordinarily honest participatory observation article, by sociologist Julius Roth, that explores some of the practices of coders in survey research:

After it became obvious how tedious it was to write down numbers on pieces of paper which didn't even fulfill one's own sense of reality and which did not remind one of the goals of the project we all in little ways started avoiding our work and cheating on the project . . . We had a special category in our coding system, a question mark, which we noted by its symbol on our code sheets whenever we could not hear what was going on between two patients. As the purgatory of writing numbers on pieces of paper lengthened, more and more transcripts were passed in with question marks on them. (Roth 1966, 190)

Lest one think that this would be picked up and corrected at a later point, he continues:

To ensure the reliability of our coding, the research design called for an "inter-rater reliability check" once every two months. We learned to loathe these checks; we knew that the coding system was inadequate in terms of reliability and that our choice of categories was optional, subjective and largely according to our sense of what an interaction is really about, rather than the rigid, stylized, and preconceived design into which we were supposed to make reality fit. (Roth 1966, 191)

He goes on to describe how the coders conspired to come up with an inter-rater reliability coefficient of .70 on checking days to be able to keep the research going (Roth 1966, 191).

Roth argues that this behavior is not unethical, but an inevitable consequence of delegated, large-scale alienated survey research labor—"hired hand research." Recent studies of ICD-using coding practices, which are also highly delegated from the point of view of the WHO and the U.S. Public Health Service and largely unimportant from the point of view of certifying physicians, appears to highlight the same sort of phenomenon.

Mick Bloor (1991) studied the practice of death certification by physicians. He notes that the practice is low status, isolated work and it is not checked or queried very much at all, even though there are legal provisions to do so. Even where autopsies are performed to check diagnoses, the hospital pathologist does not review the death certificate—his or her job is a clinical or research one! Certification is also unevenly distributed among medical practitioners. Out of 482 doctors in one Scottish city Bloor studied, 31 doctors had signed nearly a third of all the death certificates. He found that there were enormous variations. There was only 61 percent agreement on the diagnosis of the underlying cause of death between clinicians and pathologists; other studies have found that inter-rater reliability varies with the deceased's age and condition, their social class, the practitioners' nationalities, and their ages.

Nicolas Dodier has shown how medical judgments, including various ICD-encoded diseases, are transcribed in a fashion that reflects the values and contingencies of the coders' workplace. Again, as one would expect, he found many points of tension and resistance among the clinicians filling out forms, the administrative needs of the information-gathering bureaucracy (in this case tracking occupational illnesses), and the feelings of those doing the coding. He states:

Occupational doctors' vocabularies are tied to the local histories of the workplace. Occupational physician's access to objects is mediated by the instruments and the terms that people—employers, employees, representatives of personnel—themselves use in the workplace . . . there is no guarantee that local vocabularies for identifying reality coincide with administrative nomenclatures, except in the rare cases where the regulatory language is put into use by the employees and the employers themselves. There is, therefore, a conflict between the attention that doctors give to local universes and the standardized administrative definition of pertinent objects for judgement. (Dodier 1994, 6)

He goes on to say that sometimes the doctors treat the coding schemes as black boxes. Sometimes they argue with them, bringing in medical authority and expertise; and at other times the exigencies of time simply mean they code in an ad hoc, even arbitrary fashion.

### Policy Implications

A not unreasonable response to the combined ambiguities of design and certification practice in the ICD would be to throw up one's hands and walk away from any sense of data quality or certainty about the meaning of the ICD statistics. How can one know who is dying of what or where? And yet from the point of view of very large organizations, information, and diversity, this would be to abandon as well a great deal of rich information. This information includes not only content, but also methodological information about the ways in which software and its attendant categories become "frozen policy." The question is extremely complex. Global-scale, highly integrated computing systems are currently being built and augmented, such as the Web, and every day they transmit vast amounts of information around global networks. These tools, and the situation described here, demand new conceptual approaches for understanding the nature of this infrastructure. Two major lessons have emerged from this examination of the ICD as a coordination device:

1. It is unrealistic and counterproductive to try to destroy all uncertainty and ambiguity in these sorts of infrastructural tools. By their very nature, classification systems need appropriate degrees of both to work—only in a totally uniform world (within a given specialty) would

it be even conceivable to try to impose total precision (Serres 1980, Harvey 1997). Rather than root out all instances of ambiguity, analysts of standardized lists should instead seek clearly and consistently to define the degree of ambiguity that is appropriate to the object in question.

2. No such tool can be defined once and for all. They are always the products of continuing negotiation and change. We have noted three spurs for such change. First, there might be a change within one of the communities of practice that has some say in the definition of the tool. Thus, medical specialists might come up with a new test that causes a reclassification of a number of diseases. Second, changes might occur in the bureaucratic background to increase (or decrease) the tool's applicability. We have called this the phenomenon of convergent (or at times divergent) bureaucracy. Third, technical changes might allow for a better match between the actual degree of uncertainty and that permitted by the standard case. Computerization provides such an example.

In general, compromises are all that we have when we seek standard bureaucratic forms for dealing with heterogeneous groups of people and circumstances. Rather than seek to impose the one true way, we should become more aware of the properties inherent in these objects.

Despite a growing body of evidence from sociology and the history of science, distributed artificial intelligence and distributed cognitive science, images of universal policy and encyclopedic knowledge often invoke the ideal imposition of universal standardization schemes. We argue here that while such standards may emerge in physical systems or under certain sorts of market conditions, for the class of phenomena described above, no universal standard is possible. The number of groups and interests, the different ways they structure information, the moving-target nature of collecting scientific information over time when the science itself is changing—all of these factors and more are true of most important classes of problems presenting themselves in wide-scale coordination. It is often difficult to imagine building tools—whose purpose it is to collect precise, uniform, and complete information from a large domain over a long time—and at the same time invoke the necessity of ambiguity, fuzziness, and plastic meanings for their real use. The initial designers of the ICD certainly did not intentionally build such features into their data collection system; on the contrary, they were devout positivists, bent on intellectual and moral recruitment to one medical truth. Yet as the capital 'T' "truth" remained elusive, they did develop pragmatic, workable compromises, many of which used those same features.

Some guidelines emerge here that are instructive for the analysis of any large-scale, infrastructural system:

1. In the face of incompatible information or data structures among users or among those specifying the system, attempts to create unitary knowledge categories are futile. Rather, parallel or multiple-representational forms are required. So, for example, instead of trying to represent a disorder of energy diagnosed with acupuncture as a nervous disease in western medical terms, a parallel representational scheme will avoid imposing inappropriate categories.

2. Pragmatically, the Occam's razor of the coding of information means that too few categories will result in information that is not useful.[23] For instance, alive or dead, while having the virtues of simplicity and [near] exhaustiveness, do not tell us much about disease in the world. On the other hand, too many categories will result in increased bias, or randomness, on the part of those filling out the forms. An ICD with five million category labels may be more ideally scientifically accurate, but most doctors would not even look at the resulting death certificate. Thus, at the level of *encoding,* tools need to be sensitive to the working conditions of those encoding the data.

3. Imposed standards will produce work-arounds. Because imposed standards cannot account for every local contingency, users will tailor standardized forms, information systems, schedules, and so forth to fit their needs. A good summary of this appeared some years ago on a feminist button proclaiming, "One size does NOT fit all!" Gasser (1986) identified three major classes of such informal responses to imposed standards: *fitting, augmenting,* and *working around.* When designing tools for distributed, organizational decision, and policymaking, a detailed catalogue and analysis of such responses could become part of the designers' tool kit; incorporated in the system, it could point out styles of work-arounds at the level of coding.

4. Identifying the granularity of the problem, then encoding it in the system where appropriate, would complement existing organizational information processing. For example, in natural history work, biologists are often classed as lumpers versus splitters. Lumpers tend to

identify fewer species, lumping together specimens with fine-grain distinctions, and conversely with splitters. Such individual-level habits or tendencies have also been documented among those filling out death certificates. At this level of individual encoding, it is possible to track decision making and to signal stylistic bias in one direction or another (such capacities exist in several domains, both computerized and manual). The monitoring of relatively simple habits and creating mnemonic tools to correct for them, however, becomes impossible at the level of occupational specialties or large governmental bodies. Collective memories and practices have a different structure and require much more complex representations. Thus, the rule of thumb for designers here would be to try to tailor the complexity of the representation to this issue of organizational scale.

5. Match the structure of the information system mediating among diverse participants with information needs, specifically taking mismatches and world-views into account. For example, in the case of the ICD, we have a repository maintained by one group of people "fed" by forms coming in from a widely distributed constituency. There is a good match between the types of information being collected (heterogeneous, nonmatching information structures) and the repository; similarly between the use of forms and the far-flung, disparate encoders of information. Another sort of object or system inserted in the middle of this process could be disastrous. An abstract analytical schema with tightly controlled coding requirements, for example, could severely hamper data collection efforts.

### Conclusion

With the advent of very large-scale information systems and technologies, increasing concern with electronic integration, and coding and coordination across geographically dispersed groups, the issues presented here become pressing. Our contribution to this set of questions analyzes the ways human organizations have historically reached solutions to this class of problems with and without computing technology, and reflects back into the technology and the organizational world the angle of vision of history, information science, and sociology. On a more practical level, we would like to define as precisely as possible the creation, maintenance, and perhaps destruction of decisions in information practices, especially inter-organizationally.

The sociology of science and technology has emphasized opening the black box of technology, a kind of social reverse engineering of the interests and rhetoric inscribed therein. Recent organizational and policy analysis have shown how these black boxes may be opened and closed as circumstances and structural conditions change and rhetorical resources mobilized (see also Yates 1989). Yet here we have a hybrid of these conditions, where the box, if you will, is neither clearly closed nor black. Perhaps the oxymoronic "open black box" (Star 1996) would be a fitting name for this phenomenon, deserving further and urgent investigation in its own right.

# II

## Classification and Biography, or System and Suffering

The last three chapters have looked at classification and wide-scale coordination among multiple organizations. The next two chapters examine the relationship between classification and biography. How do classification systems that intimately interpenetrate our lives—shaping and being shaped by them—affect our experience? Chapter 5 looks at the intimate classification systems developed by sufferers of tuberculosis and their doctors. We develop there the themes of trajectory (the movement through time of lives, diseases, and institutions) and torque (the twisting of that biography in the framework of a classification system). Chapter 6 develops these themes further through an analysis of race classification schemes in South Africa under apartheid. Through this extreme example we explore how difficult it is to operate a simple dichotomous classification scheme and how the lives of those caught in its interstices are torqued.

# 5

## Of Tuberculosis and Trajectories

TB is a disease of time; it speeds up life, highlights it, spiritualizes it.
*(Sontag 1977, 14)*

### Introduction

The further away one stands from the disease of tuberculosis, the more
it appears to be a single, uniform phenomenon. It is associated with
one of the great philosophical breakthroughs in medicine—Koch developed his "postulates" for defining disease agency partly with tuberculosis in mind. Indeed, he could hardly avoid it since epidemiologists
assure us that at the time he wrote them in 1881, one seventh of all
reported human deaths and one-third of deaths of "productive mid-
dle-age" groups were attributable to tuberculosis (Brock 1988, 117,
179–180). Yet this single disease, a holocaust of those in their prime,
has historically proved an elusive thing to classify. The work of classification has involved at many levels a complex ecology of localization,
standardization, and time.

As this story proceeds, the interweaving of myth, biography, science,
medicine, and bureaucracy becomes ever thicker, eluding attempts at
standardization and localization from every angle. Just for this reason,
though, the story of tuberculosis holds some profound insights about
how those threads intertwine, tense against each other, and form the
texture of a landscape of time. As the field of science and technology
studies has moved to crisscross nature, culture, and discourse in a
seamless web (Latour 1993), we would add here a fourth strand:
infrastructure, in the form of classification and bureaucracy (Star and
Ruhleder 1996). We do this by borrowing some tools from medical
sociology: notions of body-biography-trajectory (Corbin and Strauss
1988, 1991) and the temporal lessons of chronic illness (Charmaz

***Doc Holliday***

"Lunger," screams the outlaw at Doc Holliday, "come out and fight. Prepare to die." Holliday, one of the most famous of tuberculosis patients, has appeared twice on the big screen in the Hollywood films "Tombstone" and "Wyatt Earp." In both, his pale face, glistening with sweat, is used as a counterpoint to his devil-may-care lifestyle and his gun-happy camaraderie with the Earps. The six years he spent in a sanatorium at the end of his life in Glenwood Springs, Colorado, go unexplored. His tombstone there reads, "He died in bed." As a final irony in the story of romancing disease, localization, and macho myth, another stone reads, "Here lies Doc Holliday whose body is buried somewhere in this cemetery." It seems the body was hidden from potential revenge seekers after he died; its exact whereabouts was then lost in the records (see figures 5.1a and 5.1b).

1991). We seek here to recenter the ways in which time and infrastructure interact with biography. The texture of this web is crisscrossed with great divides between these features, so much so that in literature and popular myth the whole terrain has taken on a phantasmagoric shape. Popular images of tuberculosis are often surreal, distorted; and those images are unrecognizable to those undergoing the experience. We have heard echoes throughout this research of the ways it is for those living with and researching AIDS (Epstein 1996). We hope here to add to that rich analysis of experience, activism, and research currently taking shape.

### Classification and Biography

As researchers attempt to decode the human genome, an obvious question has been posed: whose body is it anyway that is getting analyzed? This question is as old as the development of statistics (Hacking 1990); Quetelet sought for the "ideal type" in a statistical analysis of a regiment of Scottish soldiers. With tuberculosis, the body is constantly in motion and the disease is constantly in motion. An ideal type is difficult to conjure. The disease may be localized or spread throughout the body. The state or general condition of the body and of the person's life both enter into the treatment regime, which may take months and historically has often taken years, sometimes a lifetime.

***Figures 5.1a and 5.1b***
Photographs of Doc Holliday's grave in Glenwood Springs, Colorado, Rocky Mountain home of many tuberculosis sanitoria.
Source: Photograph by Susan Leigh Star, Glenwood Springs, Colorado, 1994.

Any disease classification system should include both spatial and temporal dimensions, but standardized classifications have tended to exist in pure space. As the problems of time emerge in the lives of patients and the work of classifiers, those spatial compartments break down in interesting ways. The formal hierarchy of mutually exclusive categories becomes a set of overlapping contradictory classes. The thesis of this chapter is that when the work of classification abstracts away the flow of historical time, then the goal of standardization can only be achieved at the price of leakage in these classification systems. Under certain conditions, the shifting terrain between standardized classification and the situated temporal biography of the patient is twisted across an axis of negotiation, scientific work and instruments, patience, and time.

### The Disease Is Constantly in Motion

Tuberculosis is a moving target. It is often presented as the great epidemic disease with a cure, heralding the famous optimism (just as AIDS was developing) that epidemic disease could be eradicated from the planet. Disease is in a sense, however, always local and so is its cure, especially when temporal dimensions are taken into account. Consider the following example. In September 1994 the WHO sent out a world-wide press release about the eradication of polio from the planet (*New York Times*, 2 October 1994). A year earlier sociologist Fred Davis, who suffered polio in his youth, and who was one of the most eloquent analysts of uncertainty in illness (Davis 1963) died of a stroke at the age of 65. Was polio eradicated for him? Was this stroke in part the legacy of his earlier illness? Many of those who had polio in the 1940s and 1950s are now beginning to lose their ability to walk as their overburdened spinal cells, designed for backup purposes, are wearing out after years of tough therapy and rehabilitation. Is the disease thus eradicated or delayed? In the lives of these patients, the answer is not so clear.

As Barbara Bates has pointed out (1992, 320–321) many observers "now attribute the decline of tuberculosis chiefly to socioeconomic changes" (a position that has been argued for many diseases, as discussed in Sretzer's argument in chapter 4; see Prins 1981). A historically fully contingent rise in standard of living accompanied by less crowded conditions in the cities possibly worked the real miracle. Other epidemiologists, Bates points out, offer a more brutal but still completely historical cause for the cure. They argue that it is a matter of natural selection, and that what has happened is simply that those humans most susceptible to the disease are now dead. This sort of "local global" is an increasing problem in much science and medical research (Bowker 1994, Latour 1993, Serres 1990). There is a "global truth" here—tuberculosis incidence has declined. There are several explanations: the "fit survived" (the susceptible members died out) *or* human environments changed (better living conditions) *or* there is now a traditional allopathic cure. It is very difficult—if indeed possible—to decide among these three causes. According to the first two, humanity before the "cure" (whichever cure one backs) is not the same thing as humanity after the cure. Either the race will have changed biologically or the infrastructure that makes us what we are will have altered. To complicate matters further, tuberculosis is once again on the rise as its fate intertwines with that of AIDS, global poverty, viral mutation, and international travel.

Not only is humanity in motion, carrying the disease with it along this broken and contested path. The disease has its own history. In a series of works put out by the National Tuberculosis Association from 1950 to the present, tuberculosis often figures wryly as an actor in the text; much as Roy Porter (1994) has noted that gout in the eighteenth century had a character of its own. Thus the 1961 edition of the *Diagnostic Standards and Classification of Tuberculosis* noted that:

For our present purposes, therefore, tuberculosis is defined as that infectious disease caused by one of several closely related mycobacteria, including M. tuberculosis, M. bovis, and M. avium. It usually involves the lungs, but it also involves and sometimes produces gross lesions in other organs and tissues. The clinical and pathologic pictures may range from acute to chronic. Its increasing predilection for middle-aged males makes careful study for differential diagnosis [between the strains of TB] more necessary than ever. (vi)

There is no need to comment on the irony of the passage's naive faith that middle-aged men were naturally the most important. Tuberculosis taken as an agent traverses history and human bodies, taking hold in some and leaving others in a contingent historical progression. The way it is treated and represented affirms the uneven, hierarchical value given to different patient's lives (Glaser and Strauss 1965). Primary infection used to be associated with children, but now that the disease has become less common it has become rather an adult phenomenon. The disease can hit at different points in the life cycle. It becomes here a mirror composed of nature, culture, discourse, and infrastructure.

### The Body Is Constantly in Motion

The development of x-rays was perhaps the most significant breakthrough in the detection and diagnosis of tuberculosis. Unfortunately,
the body itself is constantly in motion and varies by individual, so the
ideal measurement is always a projection from a moving picture onto
a timeless chart:

> The perfect chest roentgenogram, repeatedly obtained, is the aim of those
> who practice roentgenology. The very nature of the problem prevents the
> realization of this aim. The chest is a moving, dynamic part of the body and
> cannot be completely still. It varies from person to person. In some it is thin
> and easy to penetrate. In others it is thick and heavy from fat or muscle and
> hard to penetrate. Some lungs are stiff and hard to inflate. Others are made
> full and voluminous without great effort. To register lungs satisfactorily with
> these variables is at all times difficult. (*Diagnostic Standards* 1955, 71)

Further, each body subjected to tuberculosis is going through its own
biographical and physiological, historical development, and as it develops tuberculosis changes. Thus, "the clinical picture of serious ne-
crotic lesions of primary tuberculosis and widespread dissemination
from them is observed more often in infancy than in later life and
more frequently in nonwhite than in white persons." (*Diagnostic Standards* 1955, 17). Thomas Mann describes one of the tuberculosis sanatorium patients in *The Magic Mountain* responding to another's new
diagnosis of a moist spot. "You can't tell," Joachim said. "That is just
what you never can tell. They said you had already had places, of
which nobody took any notice and they healed of themselves, and left
nothing but a few trifling dullnesses. It might have been the same way
with the moist spot you are supposed to have now, if you hadn't come
up here at all. One can never know" (Mann 1929, 192). There are
several intertwined puzzles involved here.

### Experience in Motion

Not only the disease and the body, but also the patient's experience
has been constantly in motion. Thus Bates points out that institutionalization in a sanatorium may well have worked cures for reasons not
usually recorded in the medical archive. Successful recovery may be
due to good relationships with nurses, doctors, and other patients,
together with removal from bad home conditions. She summarizes:
"Psychological factors have long been thought to alter the course of

tuberculosis, but their actual impact on outcomes is not known" (Bates 1992, 320). And as they were not accounted for, and embedded in other treatments, they may never be known.

## Classification: A Still Life Constantly in Motion

With all these historical trajectories being inscribed into the course of tuberculosis at whatever unit of analysis (humanity, the disease, the body, or the experience of the patient), it will come as no surprise that the work of classifying tuberculosis has generally had very complex temporal ramifications that have often led to problems in classification. They have also led to a sense by those in sanatoria of tuberculosis as inhabiting a phantasmagoric landscape, a borderland filled with monstrous experiences and distortions of time and self (Mann 1929, Roth 1963). The reasons for this are not simply the physical horrors of the disease, though those are terrible enough, but the ways in which our four strands play out against each other in imagination. We are not saying they are separate, but the fact that they are treated as such emerges as very important in the gap between experience and myths so well explicated by Mann and Roth.

One wants to classify tuberculosis first and foremost to say whether or not a particular patient has the disease. This information can be used to suggest a treatment trajectory for the patient, and a trajectory for officials in public health. Said trajectories depend on the current theory of the disease treatment—quarantine, isolation, mountain air, antibiotics—as well as for symbol makers and writers. Tuberculosis has been the poetic illness, the disease of the "sensitive" during the nineteenth century often thought an ideal, ladylike disease for middle-class women.

Tuberculosis classification work is not easy. In the first place, the disease itself is, according to the official handbook, protean and possibly becoming relatively more so over time: "When faced with a difficult diagnosis, the clinician does well to keep tuberculosis in mind, for its mode of onset and course are protean. This needs to be urged all the more now that tuberculosis is becoming relatively less frequent" (*Diagnosis* 1955, 23).

Further, the disease does not have a single cause. Most tuberculosis in humans, according to official accounts, is caused by mycobacterium tuberculosis, but one should not forget mycobacterium bovis and mycobacterium avium. It does not appear in a single place; generally the lungs are affected, but it could produce lesions in other organs and tissues. Star (1989) notes that the disease's tendency to spread through the body implicated it in all investigations of nervous and brain disease in the nineteenth century. Whether a patient had a brain or spinal tumor or tuberculosis was often unclear; the disease, in addition to the cough, may cause seizures, paralysis, lameness, or dementia. Thomas Mann poses of one of his characters "the question whether the disease would be arrested by a chalky petrifaction and heal by means of fibrosis, or whether it would extend the area, create still larger cavities, and destroy the organ." (Mann 1929, 447)

And indeed even pulmonary tuberculosis—its most common form and one of the greatest killers in the history of humanity—cannot be simply classified:

The lesions of tuberculosis are highly diverse in appearance, and their mani-festations are numerous. No single system of classification can give information that completely describes the lesions. Certain classifications and descriptions are needed, however, for records and for statistical purposes. These essential categories may be termed basic, and they should include subdivisions which describe the status of a patient's disease at the time of diagnosis, and at any time in the months and years thereafter. These basic classifications should be used for all cases. (*Diagnosis* 1961, 39)

And this basic classification should, it is recommended, tell a story, detailing the extent of disease, status of clinical activity, bacteriologic status, therapeutic status, exercise status and other lifestyle variables as they are called nowadays.

Medical classification work as based on the ICD does not, however, give a context, it records a fact (one died of the disease or not). There is, as Fagot-Largeault (1989) has pointed out, a complex narrative written into the death certificate that is the primary product of the ICD. Doctors and other health workers must sift through multiple causes of death to determine proximate, contributing, and underlying causes. There can be only one true underlying cause and only a small range of contributing causes. The ICD cannot contain this protean disease.[24] It is oriented toward a cause-and-effect that resembles a set of slots, bins, or blanks on a form, even where it is multivalanced and multislotted; it is not, like disease and diagnosis, messy, leaky, liquid, and textured with time. Indeed, the problem of tuberculosis has been a long-standing problem for the ICD, leading to the convening of several special committees to produce a standard. It remains a problem for the System for Medical Nomenclature (SNOMED), a rival to the

ICD. "The ability to combine terms in SNOMED provides multiple ways to represent the same concept. We need look no further than the favorite example of the College of American Pathologists: pulmonary tuberculosis can be expressed as either D0188 or as T2800 + M44060 + E2001 + F03003 ('Lung' + 'Granuloma' + 'M. Tuberculosis' + 'Fever')" (Cimino et al. 1989, 515). So standard medical classifications, though they may leak at the edges and become configurationally complex, do not reflect the temporal complexity of the disease itself. They do not represent its composite, amodern nature: culture, nature, discourse, and infrastructure. They posit a single answer to the question of whether this person has tuberculosis or not. As Desrosières (1993, 296) has pointed out with respect to all statistical work, this kind of difficulty leads to a contradiction between field workers and bureaucrats. Those in the WHO supported a formal Linnean classification system; those in the field supported Buffon's practical working system. The latter engage in local classification practices (all the traits that one needs in a given situation, with an overall recognition that no classification is inscribed in nature, see Clarke and Casper 1992). It is one of the purest forms of the deduction versus induction debate (compare here the debate between Susan Grobe and the Nursing Interventions Classification designers described in chapter 7).

The classification of the disease of tuberculosis does not stand alone; it is inserted into a shifting terrain of possible classification systems and cultural symbols. For most of the nineteenth century—and part of the twentieth—tuberculosis was believed to be hereditary, and so what was classified was a tuberculoid kind of a person, a temperament: romantic, melancholy, given to emotional extremes, hot cheeked, and so on. Sontag notes, "In 1881, a year before Robert Koch published his paper announcing the discovery of the tubercle bacillus and demonstrating that it was the primary cause of the disease, a standard medical textbook gave the causes of tuberculosis: hereditary disposition, unfavorable climate, sedentary indoor life, defective ventilation, deficiency of light, and 'depressing emotions'" (Sontag 1977, 54).

Sontag also writes of the literary and popular cultural images of tuberculosis, noting that many writers have referred to tuberculosis as ethereal or chaste, somehow pure and mental, not physical. "TB is celebrated as the disease of born victims, of sensitive, passive people who are not quite life-loving enough to survive" (Sontag 1977, 25). In some circles in the nineteenth century, this became a romantic image, especially for middle-class women. "The recurrent figure of the tubercular courtesan indicates that TB was also thought to make the sufferer sexy" (Sontag 1977, 25). Eventually, this romance bled over into a more diffuse concept of style of life and crafting of self. Sontag even states "The romanticizing of TB is the first widespread example of that distinctively modern activity, promoting the self as an image. The tubercular look had to be considered attractive once it came to be considered a mark of distinction, of breeding" (Sontag 1977, 29).

The person with tuberculosis became viewed as a romantic exile: "The myth of tuberculosis provided more than an account of creativity. It supplied an important model of bohemian life, lived without or without the vocation of the artist. The TB sufferer was a dropout, a wanderer in endless search of the healthy place" (Sontag 1977, 33). She also remarks, however, "Agony became romantic in a stylized account of the disease's preliminary symptoms. . . . and the actual agony was simply suppressed" (Sontag 1977, 29).

The work of finding a cure for TB thus involved myriad classificatory activities inserted into a shifting ecology of metrologies and images about temperament and constitution. Bates (1992, 28) notes that members of the Climatological Association in the 1920s compiled measures of altitude, humidity, temperature, sunlight, dampness of the soil, ozone in the air, and emanations from pine and balsam forests to uncover and classify the ideal placement for sanitarium situation. As Bates notes, though, a skeptic "might notice that many of the otherwise disparate conclusions shared one characteristic: physicians tended to discover health-giving attributes in their own locales" (28). Note here a formal similarity between arguments about the viral origin of AIDS and the idea of "cofactors" in lifestyle, including vague notions of stress, sexuality, and community (Shilts 1987).

It was and still is not clear when to *stop* classifying tuberculosis. As the following report on Bergey's manual of determinative bacteriology notes, there is a need to bring order into the classification "unclass-ified" when talking about tuberculosis:

*Unclassified strains* (formerly also know as atypical or anonymous). These species are not adequately differentiated at present. The term unclassified is generally reserved for the following strains isolated from human material which differ from the named species:
*Photochromogens* (M. kansasii, M. luciflavum, the yellow bacillus, Runyon Group I): these strains become yellow- pigmented only after exposure to light.
*Scotochromogens* (Runyon Group II): the yellow-orange pigment of these strains is not completely light-conditioned.

*Nonchromogenic strains* (the 'Battey' type, Runyon Group III): characteristics include variable pigmentation, late in appearance, not light-conditioned.
*Rapid growers* (Runyon Group IV): rapidly growing photochromogens. (*Diagnostic Standards* 1961, 17)

So one can have an "other" or residual category, but at some point even the garbage can will have to be ordered when it becomes large enough.

Further, the committee on the classification of tuberculosis was forced to recognize in general that: "all classifications are ephemeral" (*Diagnosis* 1955, 6, quoting the 1950 edition). The committee fully recognized the temporary, agreed-upon nature of its classification work (it is worth noting in passing how often in infrastructural work like the development of classification systems there is much greater sensitivity to such factors than appears in the published scientific papers). "Complete agreement with respect to the classification of pulmonary tuberculosis, even among the most experienced clinicians in the country, is impossible. . . . The classification presented represents a well-considered compromise of the views of outstanding clinicians." (*Diagnosis* 1955, 5, quoting the 1950 edition).

Indeed, the historiography presented by the texts of diagnostic handbooks was a mixture of pure Whiggish progress tinged with despair ("without roentgenology the fight against TB would be back where it was in the nineteenth century") (*Diagnosis* 1961, 67). A cyclical view of history that Vico would not have been ashamed to espouse: "Readers will note another of those shifts in emphasis that have characterized expositions of the pathogenesis of tuberculosis for thirty-five years. The concepts presented in the current edition are more closely allied to those of former years than to the views expressed in the last edition" (*Diagnostic Standards* 1955, 7). Or, from the 1961 edition: "The one item of change upon which all of our consultants agreed was the need for a classification to include the increasing number of cases which are neither truly 'active' nor 'inactive,' and, chiefly, cases of the 'open negative syndrome.' In defining such a new class and seeking a suitable name for it, we have reached back ten years and reinstated the once-retired term, 'quiescent,' which was previously applied to an intermediate class" (*Diagnostic Standards* 1961, v).

This example moves us into the terrain of tuberculosis and activity, considered in the next section. Here, though, it underscores the situation of the classification act in an historical flow, where time,

*Tuberculosis Test*

Tuberculosis is an infectious disease that primarily involves the lungs. This simple test can show if you may have been exposed to tuberculosis. The test consists of injecting a small amount of fluid under the skin on your forearm and is harmless to your body. This fluid contains a protein derived from the organism of tuberculosis. It is rare that you would have an allergic reaction to this fluid.

The test result must be examined by a Health Center clinician within 48 to 72 hours and the results are determined by a visual inspection of your forearm. It will be necessary to repeat the test if you are not examined within this 72-hour period. Redness, which may occur at the test site, does not indicate a positive test. If swelling (also called an induration) is present, this area is measured. A nurse will determine if this measurement indicates a positive or negative test.

A positive skin test does not mean that you have tuberculosis; rather, that you may have been exposed to the organisms at some time in the past. In this case, a chest x-ray must be obtained in order to be certain there is no active disease. Additionally, if the reaction is positive, we will want to review your history and talk to you about what you should do in the future.

Tuberculosis testing is performed in the Preventive Medicine Clinic on Monday, Tuesday, Wednesday, and Friday during the hours 1:00 to 2:30 p.m. or by appointment (333–2702). No tuberculosis testing is performed on Thursday or on weekends.

—Information given to students by the health center at the University of Illinois.

Source: http://www.uiuc.edu/departments/mckinley/health-info/dis-cond/tb/TB.html

biography, and institutions transmogrify the pure progress of natural science and myth.

## *Freeze Frames: Snapshots of a Disease in Progress*

Throughout the history of tuberculosis classification, one of the key problems has been how to convert a progressive, protean disease to a single mark on a sheet of paper. Many categories have been experimented with. One suggested hallmark was whether or not one tested positive to the tuberculin test. But it was decided that those who tested positive did not *have* the disease, they were: "considered to have tuberculous infection but not disease" (*Diagnosis* 1955, 25). Only those who could bring other evidence of disease to the table would be considered worthy of the classification of pulmonary or nonpulmonary tuberculosis. This other evidence, examined below, is inextricably intertwined with classification and standardization.

Those who did have the disease could be lumped into the categories inactive, active, or activity undetermined. If a "provisional estimate of the probable clinical status is necessary for public health purposes, however, the terms 'probably active' or 'probably inactive' should be used. Every effort should be made to classify cases and to avoid this category" (*Diagnostic Standards* 1955, 28). By 1961, it was agreed that a classification somewhere between active and inactive was needed: this would be the "open negative syndrome" and would, as we have just seen, have the word "quiescent" attached to it. "Inactive" would be redefined to include "constant and definite healing." Ironically, and to underscore the attempt to separate disease from biography, "dead" was also recognized in this classification (41), presumably to stand as a cross between highly active and completely inactive! A supplement to the ICD was developed to serve epidemiological purposes, one that assigned a fourth code number to the given three for any disease. Tuberculosis classifiers took the chance to add categories for "cured or arrested pulmonary tuberculosis" (Y03.0) and tuberculin sensitivity without clinical or roentgenographic symptoms (Y01).

Leaking out of the freeze frame, comes the insertion of biography, negotiation, and struggles with a shifting infrastructure of classification and treatment. Turning now to other presentations and classifications of tuberculosis by a novelist and a sociologist, we will see the complex dialectic of irrevocably local biography and of standard classification.

## Moving through Tuberculosis and Its Classification

The next sections rely on detailed readings of two classic studies of tuberculosis sanatoria and hospitals. The first is Thomas Mann's *The Magic Mountain* (1929) chronicling a Swiss hospital and the seven-year sojourn of a young German engineer there, Hans Castorp. The account was based on Mann's experience as a visitor to a similar institution, when his wife was incarcerated for lung disease. The second is Julius Roth's (1963) *Timetables,* a comparative ethnographic analysis of several American tuberculosis hospitals in the late 1950s. This volume, too, has a strong experiential base in Roth's own hospitalization as a tuberculosis patient while he was collecting data for his doctoral dissertation.

### The Texture of Time: Lost to the World

When Hans Castorp, the hero of Mann's novel, arrives in the Alps as a visitor to his tubercular cousin, one of his first lessons in local culture is the way that values about time change for those "up here." Everything normal appears to change for him, and the whole place seems macabre and oddly humorous. Later in the novel he will explain to another newcomer, "I have no contact with the flatland, it has fallen away. We have a folk-song that says: 'I am lost to the world'—so it is with me" (Mann 1929, 614). This lost-ness first takes on the form of time passing very slowly, but in chunks that appear unimaginable to the newcomer. An old-timer says: "We up here are not acquainted with such a unit of time as the week—if I may be permitted to instruct you, my dear sir. Our smallest unit is the month. We reckon in the grand style—that is a privilege we shadows have" (Mann 1929, 59).

Roth compares the commitment to a tuberculosis sanatorium with having an "indeterminate sentence" for one's years in jail. One does not know how long one will be incarcerated. There are no milestones or turning points that make sense. Time thus also seems endless and distorted with respect to known landscapes, both inner and outer. "Where uniformity rules; and where motion is no more motion, time is no longer time" (Mann 1929, 566).

The patients in both Mann's and Roth's hospitals begin to speculate on the meaning of this lost time, this time out. Is time real, objective, something that can be measured externally—or subjective, illusory? Hans originally opts for a relativist explanation: "After all, time *isn't* 'actual.' When it seems long to you, then it is long; when it seems short, why, then it is short. But how long, or how short, it actually is, that nobody knows" (Mann 1929, 66). His cousin Joachim, a rather hard-nosed soldier who wants only to get off of the mountain and back to his regiment, disagrees. Joachim says "We have watches and calendars for the purpose; and when a month is up, why then up it is, for you, and for me, and for all of us" (Mann 1929, 66). Hans proceeds to demonstrate how slowly seven minutes can go by while taking one's temperature. We indeed feel the seconds creep by in Mann's precise language. What is "the same?" he asks. "The schoolmen of the Middle Ages would have it that time is an illusion; that its flow in sequence and causality is only the result of a sensory device, and their real existence of things in an abiding present" (Mann 1929, 566).

As time goes on, up on the magic mountain, and in each of the hospitals studied by Roth, people inside begin to develop a sense of how to fragment, break up this unbroken monolith. "We are aware that the intercalation of periods of change and novelty is the only means by which we can refresh our sense of time, strengthen, retard, and rejuvenate it, and therewith renew our perception of life itself" (Mann 1929, 107). In one of his many meditations on the nature of time, Mann argues that time and action and space are not separable—nothing fills up time in a platonic-container sense, but these facets are only knowable with respect to each other:

What is time? A mystery, a figment—and all powerful. It conditions the exterior world, it is motion married to and mingled with the existence of bodies in space and with the motion of these. Would there then be no time if there were no motion? No motion if no time? We fondly ask. Is time a function of space? Or space of time? Or are they identical? Echo answers. Time is functional, it can be referred to as action; we say a thing is 'brought about' by time. What sort of thing? Change? (Mann 1929, 356)

At the core of this theory of action is the development of what Roth calls timetables, which are alluded to in more symbolic terms by Mann. Timetables are breaks in space-time that give meaning to action. When will I get out? What will become of me? How will I survive the boredom and the uncertainty of incarceration? Such questions are asked against the specter of unbroken time or eternity, or as Roth's patients and doctors put it for the hopeless cases, "a rather horrifying tubercular Siberia—a seemingly endless waster (of time) without any signposts along the way" (Roth 1963, 21). Or in Mann's words, "Only in time was there progress; in eternity there was none, nor any politics or eloquence either" (Mann 1929, 479).

Gradually a sense that there is in fact no such thing as unbroken time comes about for the patients: "Can one tell—that is to say, narrate—time, time itself, as such, for its own sake? That would surely be an absurd undertaking. A story which read: 'time passed, it ran on, the time flowed onward' and so forth—no one in his senses could consider that a narrative. . . . For narration resembles music in this, that it *fills up* the time. It 'fills it in' and 'breaks it up,' so that 'there's something to it,' 'something going on'" (Mann 1929, 560). The patients begin to fill their days with measurement. On the magic mountain, people walk around with thermometers in their mouths, measuring their temperatures several times a day. In both books, patients are conversant with the details of diagnosis and measurement, the myriad of ways in which the monolithic diagnosis may be broken up and measured. Roth says, "Everyone is frantically trying to find out how long *he* is in for. The new patient questions the doctors, nurses, and other hospital personnel in an effort to discover how may years, months, and days it will take *him* to be cured" (1963, xvi).

## Metrology

One woman has been a patient on the magic mountain for the better part of her life. Eventually she is cured of the disease, but knowing no other life, panics at the thought of leaving.[25] She sabotages her release: runs out in the snow, jumps in the lake, and sticks her thermometer into her tea to make herself appear feverish. When discovered, she is given a thermometer without any marks on it, which can only be read by a doctor with a measuring stick, thus she cannot calibrate her faking. The patients come to call this device the "silent sister."[26] The silent sister becomes the symbol for the ways in which the world of the asylum acquires its own bizarre culture of metrification.

Roth notes that patients are quite systematic in creating measurements for the blocks of time they will spend in the asylum. They begin to construct timetables for themselves (I will get out in six months; I will have surgery in two weeks, and so forth). "After they have been in the hospital for some time, they find that 'mild' and 'bad' are not very meaningful categories." Much more detailed matching categories develop (Roth 1963, 19).

Patients begin observing how other patients are treated. There is a complex edifice of privileges in tuberculosis hospitals based ostensibly on how well the person is perceived to be. If one is making good progress, for example, one is allowed out on brief shopping trips, and so forth. "He divides the patient group into categories, according to his predictions about the course of their treatment. He can then attach himself to one of these categories and thus have a more precise notion of what is likely to happen to him than he could from simply following the more general norms" (Roth 1963, 16–17).

Roth goes on to describe an elaborate system of observations and comparisons made by all the patients about their own bodies, the length of time served, the predilections of the individual doctors, and the technical diagnostic material such as x-rays. Not surprisingly, much of the information available is partial or misleading. Reference points may be more or less clear-cut and stable. If they are prescribed in detail and rigidly adhered to, as they tend to be in the career of pupils in a school system (at least ideally), one's movement through the timetable is fairly predictable. As the reference points become less rigid and less clear-cut, they must be discovered and interpreted through observation and through interaction with others of one's career group. The more unclear the reference points, the harder it is for members of a career group to know where they stand in relation to others and the more likely it is that they will attend to inappropriate clues and thus make grossly inaccurate predictions concerning future progress. The degree of stability is related in part to historical changes in institutional timetables through time (Roth 1963, 99–100).

Managing this instability increases the intensity of comparison and a sense, often, of bewilderment, unfairness, or even madness. Hans Castorp says to his cousin, "I cannot comprehend why, with a harmless fever—assuming for the moment, that there is such a thing—one must keep one's bed, while with one that is not harmless you needn't. And secondly, I tell you the fever has not made me hotter than I was before. My position is that 99.6° is 99.6°. If you can run about with it, so can I" (Mann 1929, 176). "Give me a standard, give me something to hold on to, something clear"; in the face of uncertainty, patients become positivists. Mann describes the rebellion of Hans' cousin again the system of metrification in the hospital, the "Gaffky score" which is a composite score for each patient's progress based on a number of measures:

Yes, the good, the patient, the upright Joachim, so affected to discipline and the service, had been attacked by fits of rebellion, he even questioned the authority of the "Gaffky scale": the method employed in the laboratory—the lab, as one called it—to ascertain the degree of a patient's infection. Whether only a few isolated bacilli, or a whole host of them, were found in the sputum analyzed, determined his "Gaffky number," upon which everything depended. It infallibly reflected the chances of recovery with which the patient had to reckon; the number of months or years he must still remain could with ease be deduced from it. . . . Joachim, then inveighed against the Gaffky scale, openly giving notice that he questioned its authority—or perhaps not *quite* openly. . . . (Mann 1929, 357)

This questioning of authority appears inevitable in a landscape so filled with uncertainty. One character attempts a triage reminiscent of recent attempts on the part of American hospital administrators to quantify health care costs and tradeoff. "Even in the matter of the operation he took a business view, for, so long as he lived, that would be his angle of approach. The expense, he whispered, was fixed at a thousand francs, including the anesthesia of the spinal cord; practically the whole thoracic cavity was involved, six or eight ribs, and the question was whether it would pay. . . . he was not at all clear that he would not do better just to die in peace, with his ribs intact" (Mann 1929, 315).

In the absence of metrics, however, the relationships between doctors and patients come under considerable strain. Patients strive to assign themselves to the proper categories, and then to see whether the doctors agree with them. In *The Magic Mountain,* Settembrini, a slightly satanic character, whispers constantly to Hans about how subjective the reading of the objective measures such as X-rays really is. "You know too that those spots and shadows there are very largely of physiological origin. I have seen a hundred such pictures, looking very like this of yours; the decision as to whether they offered definite proof or not was left more or less to the discretion of the person looking at them" (Mann 1929, 250).

Both physicians and patients struggle to find a standard and to localize it, in the face of a constantly shifting interpretive frame:

The physician finds it difficult to carry out the medical ideal of an individual prescription for each case when at the same time he recognizes the fact that his timing of a given treatment event for a given patient is to a large extent a highly uncertain judgment on his part. If you are going to guess, you might as well make the process more efficient by guessing about the same way each time, especially if you are in a situation where your clients are likely to think that you do not know what you are doing if you change your guess from one time to another. (Roth 1963, 24)

This uncertainty leads to the struggles and negotiations that are at the heart of Roth's analysis. Whose timetable will prevail?

### Classification Struggles

"The TB patient conceives of his treatment largely in terms of putting in time rather than in terms of the changes that occur in his lungs" (Roth 1963, xv).

The length of time one has been inside, combined with patients' observations about where they belong in the general scheme of things, acquires a moral character:

A classification system contains within it a series of restrictions and privileges. When no rigid classification system exists, these privileges themselves become part of the timetable. . . . How long is it before he is allowed two hours a day 'up time' [out of bed]? . . . *these privileges are desired not only in themselves, but for their symbolic value.* They are signs that the treatment is progressing, that the patient is getting closer to discharge. (Roth 1963, 4)

Timetable norms differ from hospital to hospital and from patient to patient. Trust, often in the form of moral condemnation or approval, may play a big part in structuring the timetable negotiations between doctor and patient. For example, alcoholic patients are often refused outside passes, or sometimes a patient with a recalcitrant attitude is refused a pass simply to convince him or her that they *are* very ill. These moralizing attitudes, well documented within medical sociology, add another texture to the landscape we are examining here, twisting it a little away from a simple formal-situated or realist-relativist axis.

Doctors as well as patients may hold the deserving attitude toward those who have "served their time." Roth notes that in treatment conferences, how long the patient has been in is always taken into account in deciding the timetable, "this in itself is given considerable weight entirely aside from the bacteriological and x-ray data" (1963, 27). Even those who appear to be getting better much faster, according to these tests, are kept in longer because "TB just isn't cured that fast" (1963, 27).

Patients know almost to the day when which privileges will arrive. "This relative precision of the timetable results from the emphasis placed upon the classification system by the staff, the consistency in the decisions of the physician in charge, and the physician's explicitness in telling the patients what they can expect in the future" (Roth 1963, 7).

There can be a failure to be promoted in severe cases, and the reaction to this differs among patients. "The subjective reaction to failure varies greatly among TB patients, just as among engineers some of the failures are emotionally disorganized when they do not make the grade while others accept their inferior position with relative equanimity. Some patients regard a few days' delay as a tragedy" (Roth 1963, 15). Bargains are made. "Patients *are* sometimes given regular and frequent passes to induce them to remain in the hospital" (Roth 1963, 53).

Uncertainty plays an important role in negotiations about classification in the hospital. When a patient tries to estimate what classification he or she belongs to, and the physician disagrees, "In effect, the physician tries to get Jones to change his criteria for grouping patients so that his categories will be closer to those of the physician" (Roth 1963, 39). The doctor will provide the patient with examples of others like him or her and relates details about other similar cases. But the physician too is caught in a double bind: ethically he or she is not allowed to give too many details about others' cases. The doctor is thus ultimately reduced to vague generalizations like "no two cases are alike" (Roth 1963, 39). For the patient, this contributes to a house-of-mirrors effect:

Most physicians . . . vary their approach from one patient to another accord-ing to their own judgment of what the patient can take. These judgments, which are usually based on extremely limited information about the patient, are often wrong. . . . the physicians do not know with any precision how long it will take the patient to reach a given level of control over his disease. To allow themselves a freer hand in deciding what the best time is for the patient to leave the hospital, the doctors try to avoid being pinned down to any precise estimates by the patients. (Roth 1963, 45).

The twisting effect of these silences is especially clear where the norms about timetables are also shifting, either due to changes in medical practice, technology, staff, or organizational change. One patient in these circumstances said, "You never seem to get anywhere because people here don't pay too much attention to the classifications. I've been here now since November and I'm still in Group 1. My husband comes to visit me and looks at this tag and thinks I'm never going to get promoted. He wonders what's going on. Then when you *do* get promoted to Group 2, you don't know what it means, anyway. You have no idea what additional privileges you have. . . . *It's like an ungraded school room*" (Roth 1963, 10).

The ungraded schoolroom, combined with uncertainties, shifts in bureaucracy, and changes in the person's biography, begin to form the substance of a kind of monstrous existence.

### Borderlands and Monsters: Time's Torquing of Standards and Experience

"There were those who wanted to make him 'healthy,' to make him 'go back to nature,' when, the truth was, he had never been 'natural'" (Mann 1929, 482).

On the magic mountain, or in any of the hospitals analyzed by Roth, the sense of unreality, of being outside of normal time, and of making up an idiosyncratic timing is very strong. Furthermore, the very insides

---

### The Romance of Tuberculosis

Greta Garbo as Camille drifts across the screen in a cloud of white organza.[26] She is alternately cruel and flirtatious, vulnerable and powerful. She plays with the affections of her lovers, a baron and a struggling young diplomat, from her position as a farm girl who came to Paris. Early in the movie, we understand that she has been ill; from time to time she discreetly covers her mouth with a handkerchief, or seems to swoon (always artistically). At times she recovers, and in a rhythm complexly played out against her wardrobe, she moves from white to black in dress, from sick to well, from powerful to powerless, from country to city. As the movie progresses she becomes more and more ill, and more and more pure—thinner, whiter, more in love with the worthy poor man and less with the nefarious rich Baron. During the whole course of the movie, no one speaks the name of her illness, any prognosis or diagnosis, nor do we see any blood, sputum, feces, or other despoiling of the purified background. Of course, she has tuberculosis—and she is the ideal type, the shadow puppet against which both the medical story and the rich cultural criticisms of tuberculosis have been played out.

---

and outsides of people become mixed up in an almost monstrous way; Hans carries around his love Clavdia's x-ray in his breast pocket so that he may really know her. External time drops away as does one's biography:

(The inhabitants) accorded to the anniversary of arrival no other attention than that of a profound silence. . . . They set store by a proper articulation of the time, they gave heed to the calendar, observed the turning-points of the year, its recurrent limits. But to measure one's own private time, that time which for the individual in these parts was so closely bound up with space— that was held to be an occupation only fit for new arrivals and short-termers. The settled citizens preferred the unmeasured, the eternal, the day that was for ever the same. (Mann 1929, 427)

This sense of time begins to blur important distinctions between life and death, time and space. "But is not this affirmation of the eternal and the infinite the logical-mathematical destruction of every and any limit in time or space, and the reduction of them, more or less, to zero? Is it possible, in eternity, to conceive of a sequences of events, or in the infinite of a succession of space-occupying bodies?" (Mann 1929, 356). As we approach the zero point in the story, Mann notes in an afterword that time-space relations are shifted so that "the story practices a hermetical magic, a temporal distortion of perspective reminding one

of certain abnormal and transcendental experiences in actual life"
(Mann 1929, 561).

We are reminded here of Michel Serres' (1980) invocation of the
passage between the natural and humanistic sciences as indeterminate,
twisted, and full of ice floes; of the images of cyborg and monster
pervading feminist theory about technology (Haraway 1992; Casper
1994a, 1994b, 1995). The following section offers a model for how such
a monstrous borderland terrain is constructed and maintained.

### Trajectories and Twists: The Texture of Action

No one can ever know for certain just when tuberculosis becomes active or
when it becomes inactive. For that matter, one can never be certain that the
disease is inactive, and a patient could logically be kept in the hospital for the
rest of his life on the assumption that some slight undetectable changes might
be occurring in his lungs.

*(Roth 1963, 30)*

The same train brought them as had Hans Castorp, when years ago, years
that had been neither long nor short, but timeless, very eventful yet "the sum
of nothing," he had first come to this place.

*(Mann 1929, 520)*

### Body-Biography Trajectory: Strauss and Corbin

We present our model of the TB landscape in three parts of an
ongoing conversation among all the authors analyzed. First is a model
developed by Anselm Strauss and Juliet Corbin to describe what happens in the course of a chronic illness (1988, 1991). They posit that
bodies and biographies unfold along two intertwined trajectories (the
body-biography chain), nestled in a matrix of other structural and
interactional conditions. For example, a heart attack may temporarily
interrupt work, home life, creativity, dragging down the trajectory of
biography. This in turn is contingent on a number of other circumstances such as access to health care, living in a war zone, or having
another illness that makes recovery longer.

The chain may be viewed geometrically as a topography emerging
from the interplay of these factors. Many illnesses do not have such an
acute nature; during a long chronic illness there is a back-and-forth
tugging across the trajectory of the disease-body and of the person's
biography within the conditional matrix. The title of Kathy Charmaz'
*Good Days, Bad Days: Time and Self in Chronic Illness* throws this relation

into relief (Charmaz 1991). A long, slow downswing may only very gradually affect biography. Personal and family resources may compensate for a brief acute phase, experienced over and over, so that the overall trajectory of the biography remains fairly smooth. Many possible shapes are envisioned: a looping shape in the case of a comeback after a serious, debilitating illness; a very gradual progress of the disease that slowly erodes the biographical trajectory.

This model does not seek a Cartesian mind-body dualism, but rather seeks to find a language for the ways in which two (or more) different processes become inextricably intertwined into one thick chain or braid. It makes more complex the sick-well, able-disabled dichotomies, and it brings in people's active conversations with and work for their ill bodies as a central concern (see figure 5.2).

The body's trajectory and the self's are bound together, but not completely tightly coupled. Careers, plans, work, and relationships may continue in spite of, around, and through illness; or, a sudden illness may interrupt plans and biography and reshape the topography. The background landscape is a nested set of contingent possibilities and structural features which in turn act upon the shape of the trajectory.

### Multiple Identities along a Body-Biography Trajectory in Sudden Illness or Death

Timmermans has suggested emphasizing multiple identities in a dialogue with the Strauss-Corbin BBC model. He studied more than 100 cases of attempted resuscitations (CPR) with victims of cardiac arrest, in the emergency room (in press a, in press b, in press c). He uses the trajectory model to explain the sequence and flow of events as people were brought in by the ambulance crews, worked on by staff, and either declared dead or saved (1998). (The vast majority of people die.) Each patient who undergoes CPR has multiple intertwining identities outside that of heart attack victim, each with its own trajectory. At the moment of the resuscitation attempt, these collapse into a single identity: that of the body-machine (Timmermans 1996). The nurses and doctors and technicians focus down to a single attribute of the person. After resuscitation, if this is successful, the multiple identities restart from the same baseline, but each identity will have been differentially altered by the experience (see figure 5.3).

Here the biographical trajectories—selves—move from complex, multiple activity to a single focus: life-saving. At death, the identities

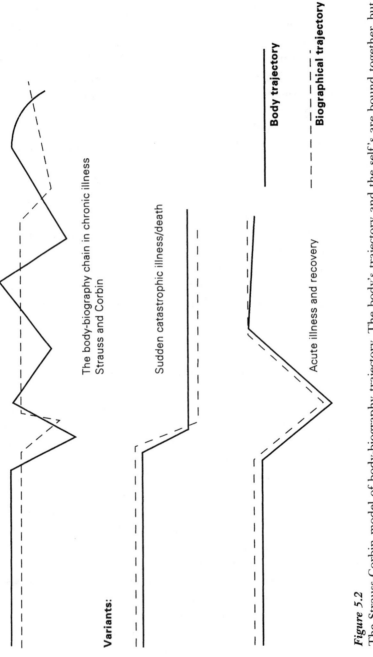

*Figure 5.2*

The Strauss-Corbin model of body-biography trajectory. The body's trajectory and the self's are bound together, but not completely tightly coupled. Careers, plans, work, and relationships may continue in spite of, around, and through illness; or, a sudden illness may interrupt plans and biography and reshape the topography. The background landscape is a nested set of contingent possibilities and structural features that in turn act upon the shape of the trajectory.

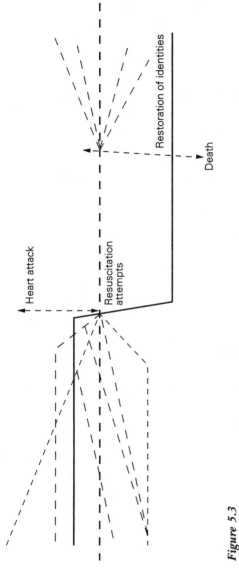

*Figure 5.3*
Multiple identity trajectories along the body-biography trajectory. Timmerman's modifications to trajectory in acute, severe illness (such as cardiac failure). Here the biographical trajectories, selves, move from complex, multiple activity to a single focus: life-saving. At death, the identities are restored. If the patient survives, they are the same but different: an isomorphic transformation has occurred.
Source: Timmermans 1996, in press a.

are restored. If the patient survives, they are the same but different: an *isomorphic transformation* has occurred (Timmermans 1996, in press a).

### The Twisted Landscape: Adding Texture to Multiplicity and Standardization

Next, we have add a third trajectory dimension to play off against the interacting trajectories of bodies and multiple identities: the trajectory of classification systems themselves (as part of infrastructure). In looking at an extreme case temporally, where the "time" of the body and of the multiple identities cannot be aligned with the "time" of the classification system, we have suggested that the latter gets twisted by the former. A variety of monstrous classification schemes bubble through the rift in space-time. In the case of tuberculosis, there is a chronic illness that necessitates withdrawal for a prolonged period from normal life, sequestering with others with the disease, in an uncertain time frame that partly depends on the ways classification schemes are perceived, negotiated, and used by health personnel. It also draws on a matrix of possibilities for the basis for these negotiations, including how medical knowledge is represented by public health agencies, classifications modified in the hospital, and images from literature, film, and popular science about "what people with tuberculosis are like." The rich topography of body and biography intercalates with a bureaucratic infrastructural typology (classification scheme; see figure 5.4).

When standard classifications are added to the scheme, patients try to fit their experiences along both body and biographical trajectories to a standard picture of metric. Changing definitions, local arrangements, and complex relations of all three trajectories contribute to "torquing" the typology-topography via the dotted lines, which represent negotiations.

### Twists and Textures: Classification and Lived Experience

Time morality is not cut and dried.
*(Condon and Schweingruber 1994, 63)*

The information infrastructure that deals with tuberculosis, as with other diseases, operationalizes a classification system that is purely spatial. The disease is localized in this body or not; in this region of the body and no other; it is present among this population but not

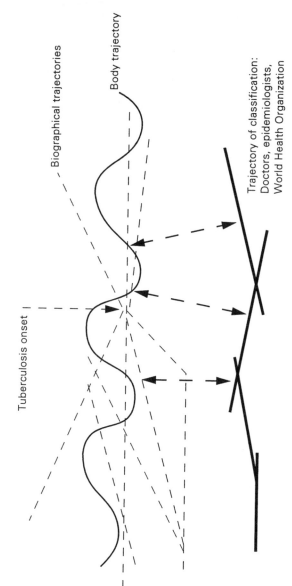

**Figure 5.4**
The topology-typology twist. Tensions among bodies, biographies and classification—the twisted landscape of multiple and classifications. The topology created by the body-biography trajectory is pulled against the idealized, standardized typology of the global classification of tuberculosis, itself a broken and moving target. When standard classifications are added to the scheme, patients try to fit their experiences along both body and biographical trajectories to a standard picture of metric. Changing definitions, local arrangements, and complex relations of all three trajectories contribute to torquing the typology/topography via the dotted lines, which represent negotiations.
Source: Star and Bowker.

that one; the cure can be found in this place but not that, and so forth. The closest that one gets to the flow of time is the description active or passive, latent or virulent; and at these key points the classification system itself breaks down in numerous ways. The information infrastructure dealing with medical knowledge abstracts away from process to produce ways of knowing defined as being true "for all time" about its subject and so able to abstract contingent historical and biographical flow to uncover the underlying reality.

But there is not just one kind of classification in the world, as we have amply demonstrated in the case of the ICD. Classification work is always multiple. As we move further from medical knowledge and closer to the suffering patient, time seeps into the classification systems that get used: how long does group 1 stay here ? (Roth) ; how can I get reclassified so that I can pass more time on the magic mountain? (Mann). Camille's morality tale unfolds in time in binary oppositions of good-bad, fit-ill, black-white; promotion or demotion from class to class occurs in a continually downward career trajectory. Tuberculosis is the archetypal disease of time: chronic, recurrent, progressive.

So what happens when the disease of time meets the classification of space? As we have shown, the formal, spatial classification twists. "Other" categories run rampant, each seeking a way of expressing the elusive, forbidden flow of time (words like "quiescent" and "nonactive" abound). A macabre landscape is born. And the historiography of the classification system twists too: in stunning contrast to most medical scientific texts, tuberculosis classifiers speak of a cyclical flow to their own historical time (not linear progress). From the other point of view, that of the patient, orthogonal classifications are developed that do not interact with medical categories. "I have put in my time here, and I am a good person, so I deserve to be better and to leave." The disease is given a temporal texture at the price of becoming purely local; abstracted away from the standardized language it becomes once again temporally textured and immediate.

This way of framing the problem introduces the idea of *texture* as an important one in conceptualizing the relationships between representations, work, body, and biography. Michael Lynch's (1995) work on topical contextures implies a similar direction: the look and feel of being in a place and using a genre of representations. Kari Thoresen (personal communication), a former geologist, is developing a model and vocabulary for different aspects of texture in organizations and technological networks, examining layers and strata, crystallization processes (a term also used by Strauss and Timmermans), and other metaphors to examine how wires, people, and bits are put together by a large organization.

What is interesting about such twists and textures? Through them we can move on from exploring the seamless web of science and society, of nature and knowledge to an analysis of the information infrastructure that acts as matrix for the web. The web itself is textured in interesting ways by the available modes of information storage and transfer. Medical classification work, typified by the ICD, deals in spatial compartments; and these compartments cannot hold when biography and duration are a necessary part of the story. In general, the information infrastructure holds certain kinds of knowledge and supports certain varieties of network; we believe that it is a task of some urgency to analyze which kinds of knowledge and network. This textural metaphor is explored in detail in chapter 9.

Much of this book concerns itself with the relationship (first conceptualized as a kind of gap) between formal systems of knowledge representation and informal, experiential, empirical, and situated experience; however, it is never the case of "the map OR the territory." One may try to hold a representation constant and change practice to match it, or vice versa. Using the example of medical classifications, however, both coconstruct each other in practice. Thus we have "the map IN the territory" (making the map and the territory converge). It is not a case of the map *and* the territory (Berg 1997).

This chapter attempts to examine one kind of map in a territory marked by severe biographical interruptions, solitude and aspects of total institutions, and in dialogue with a compelling infrastructure (both informatic and managerial). We see the map in this case as a warping factor, not in the sense of deviating from any putative norm, but in the sense of reshaping and constraining other kinds of experience.

Finally, before turning to a second example of classification and biography (the example of race classification) we draw attention to the appearance both here and in chapters 2 and 4 of time as a problem for many classification systems. In chapter 2 we saw two examples: the difficulty of producing a stable classification for rapidly evolving species (for example, viruses) or the difficulty of expressing duration (wear and tear on the human body) in the ICD. In chapter 4, we saw the difficulty of maintaining long-term comparability of epidemiological results if the classification system changed too fast. In this chapter we have seen related difficulties in capturing a disease that itself changes over time (leading to large-scale reclassification) using tools that need to capture a body in motion (x-rays, for example), and one that has a profound temporal effect on the biography of the sufferer. The representation of time is a site of tension within most classification systems used in bureaucracies and in science precisely because when things are put into boxes, then a set of atemporal, spatial relationships are produced—and duration tends to be folded into the interstices. The inverse problem occurs when things are ordered too much in terms of temporal boxes. Fossil classification (Galtier 1986), for example, is often difficult because it is unclear what is a function of space (a species varying between two sites) as against time (a species evolving).

The next chapter takes up another example of the intersection of biography and classification, this time on a mass scale applied to groups of people. In race classification under apartheid in South Africa, a racist classification system was used to divide people into crude racial groups. This practice torqued the biographies of thousands of people, including those caught in between.

# 6

## The Case of Race Classification and Reclassification under Apartheid

Sir de Villiers Graaff asked where the sudden danger to the white group was that had caused the minister to decide to close off the human stud-book he had tried to create. He was endeavoring to classify the unclassifiable.
*(Horrell 1968, 27)*

The stubborn survival of racial categories attests to the enduring power of the old race paradigm, as well as the fact that new insights and methodologies take time to be fully incorporated and internalized.
*(Dubow 1995, 106)*

As information scientists, the theoretical and practical issues of racial and ethnic group categorization, naming, and meaning can be viewed as empirical data about problems associated with the organization of knowledge, representation, classification, and standards setting.
*(Robbin 1998, 3)*

### Introduction: The Texture of Classification

The last chapter examined the detailed interactions among people, institutions, and categories about tuberculosis. Each has a trajectory, and the trajectories may pull or torque each other over time if they move in different directions or at different rates. The threads that tie category to disease, to science, to bureaucracy, and thus to person, often become twisted and tangled in the long process of the disease. The texture of classification here is composed of thick filiations, encompassing much of a person's life, imposed from outside, and filled with uncertainty and contradiction.

This chapter examines another similarly torqued group of filiations between people and classifications, that which tied racial categories to persons under apartheid in South Africa. Here, race classification and reclassification provided the bureaucratic underpinnings for a vicious racism. Here too the attempt to create a normalized, systemic book-keeping system was embedded in a larger program of human destruction. There are enduring lessons to be drawn about moral accountability in the face of modern bureaucracy. The ethical concerns are clearly basic questions of social justice and equity; at the same time, their very extremity can teach us about the quieter, less visible aspects of the politics of classification. We walk here a line similar to that of Hannah Arendt in her *Eichmann in Jerusalem: A Report on the Banality of Evil* (1963). The quiet bureaucrat "just following orders" is in a way more chilling than the expected monster dripping grue. Eichmann explained what he was doing in routine, almost clerical terms; this was fully embedded in the systematic genocide of the Holocaust.

One of this book's central arguments is that classification systems are often sites of political and social struggles, but that these sites are difficult to approach. Politically and socially charged agendas are often first presented as purely technical and they are difficult even to see. As layers of classification system become enfolded into a working infrastructure, the original political intervention becomes more and more firmly entrenched. In many cases, this leads to a naturalization of the political category, through a process of convergence. It becomes taken for granted. (We are using the word naturalization advisedly here, since it is only through our infrastructures that we can describe and manipulate nature.) We emphasize here the stubborn refusal of "race" to fit the desired classification system suborned by its pro-apartheid designers. Thus, we further develop the concept of torque to describe the interaction of classification systems and biography.

### Background

From the early days of Dutch settlement of South Africa, the de jure separation and inequality of people coexisted with interracial relationships. In the mid-nineteenth century charter of the Union, it was simply stated that "equality between White and coloured persons would not be tolerated" (Suzman 1960). Various laws were enacted that reinforced this stance. When the Nationalists came to power in 1948, however, a much more detailed and restrictive policy, apartheid, was put into place. In 1950 two key pieces of legislation, the Population Registration Act and the Group Areas Act were passed. These required that people be strictly classified by racial group, and that those classifications determine where they could live and work. Other areas controlled de jure by apartheid laws included political rights, voting, freedom of movement and settlement, property rights, right to choose the nature of one's work, education, criminal law, social rights including the right to drink alcohol, use of public services including transport, social security, taxation, and immigration (Cornell 1960, United Nations 1968). The brutal cruelty, of which these laws were the scaffolding, continued for more than four decades. Millions of people were dislocated, jailed, murdered, and exiled.

The racial classification that was so structured in the 1950s sought to divide people into four basic groups: Europeans, Asiatics, persons of mixed race or coloureds, and "natives" or "pure-blooded individuals of the Bantu race" (Cornell 1960). The Bantu classification was subdivided into eight main groups, with Xhosa and Zulu the most numerous. The coloured classification was also complexly subdivided, partially by ethnic criteria. The terribly fraught (and anthropologically inaccurate) word Bantu was chosen in preference to African (or black African), partly to underscore Nationalist desires to be recognized as "really African."[28]

State authorities, touching every aspect of work, leisure, and education obsessively enforced apartheid. In a bitter volume detailing his visit to South Africa, Kahn notes:

Apartheid can be inconvenient, and even dangerous. Ambulances are segregated. A so-called European injured in an automobile accident may not be picked up by a non-European ambulance (nor may a non-European by a European one), and if a white man has the misfortune to bleed to death before an appropriate mercy vehicle materializes, he can comfort himself *in extremis* by reflecting that he will most assuredly be buried in an all-white century. (Nonwhite South African doctors may not perform autopsies on white South African corpses.) (Kahn, 1966, 32)

"Separate development" was the euphemism used by the Nationalist party to justify the apartheid system. It argued from a loose eugenic basis that each race must develop separately along its natural pathway, and that race mingling was unnatural. This ideology was presented in state-sanctioned media as a common-sense policy (Cell 1982).

Despite that fact that it was required by law, it often took months or years for blacks to acquire passbooks, during which time they were in danger of jail or being deported to one of the black homelands (Mathabane 1986). Horrell recounts a story about the early years of apartheid, and a group of black people waiting for hours outside the registration office. "The Native Affairs Department official tried to

**Table 6.1**
Charges and conviction under the immorality act during the year ending June 1967

|          | Men     |           | Women   |           |
|----------|---------|-----------|---------|-----------|
|          | Charged | Convicted | Charged | Convicted |
| Whites   | 671     | 349       | 18      | 11        |
| Coloured | 20      | 5         | 264     | 126       |
| Asians   | 11      | 4         | 20      | 13        |
| Africans | 8       | 5         | 338     | 180       |

Source: Adapted from Horrell 1969: 36.

pacify them. He told them to come next Monday, next Wednesday. But no, they said. They had waited yesterday, and now losing their pay for today, and the rent would soon be due, and if they were not classified they would be arrested—'Hulle sal ons optel. My vriend was gister opgetal,' they said. The word came from every quarter of the yard,—'optel, optel'"[29] (Horrell 1958, 59).

Any form of interracial sexuality was strictly forbidden by a series of immorality laws, some of which predated apartheid. Ormond states that "Between 1950 and the end of 1980 more than 11,500 people were convicted of interracial sex; anything from a kiss on up" (1986, 33). These sexual borders were vigorously patrolled by police. Ormond continues, "Special Force Order 025A/69 detailed use of binoculars, tape recorders, cameras, and two-way radios to trap offenders. It also spelled out that bedsheets should be felt for warmth and examined for stains. Police were also reported to have examined the private parts of couples and taken people to district surgeons for examination" (Ormond 1986, 33). The South African Institute of Race Relations (Horrell 1968) shows a typical year for charges and convictions under the Immorality Act in 1966–67. The racialized gender biases speak for themselves (see table 6.1).

For black South Africans, the system of segregation included a legal requirement to carry a pass book, a compilation of documents attesting to birth, education, employment history, marriage, and other life events (see figure 6.1). The books were over fifty pages long. No black was allowed to be in a white area for more than seventy-two hours without special permission, including government authorization for a work contract (such as that for a live-in servant). The consequences of

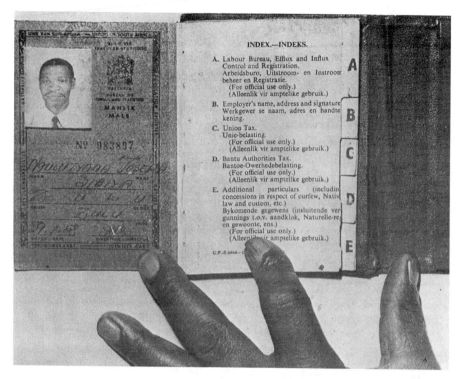

**Figure 6.1**
A passbook required of black South Africans over the age of sixteen, under
the apartheid regime.
Source: The Hoover Institution pamphlet collection, Stanford University,
"The Fight for Freedom in South Africa and What It Means for Workers in
the United States," produced by Red Sun Press publications, 11.

transgression were severe, as Frankel notes. "The inestimable number
of 'illegals' in the urban areas live a life of harassment that is
Kafkaesque in its proportions, yet even those fortunate enough to
qualify for urban status are faced with a harsh and insecure daily
existence where the loss of a document, some technical violation of the
mass of administrative decrees, or some arbitrary (and often vindictive)
stroke of the bureaucratic pen, can mean condemnation to perpetual
displacement" (1979, 205). A Foucaldian system of control of all people
except whites ensued (although by law the restrictions applied to
whites entering proscribed areas, this was rarely enforced and whites
did not carry pass books) (Black Sash 1971, Mathabane 1986). Blacks

were the major targets of scrutiny, and the pass book system allowed for comprehensive surveillance of their actions. "The whole system has been extended and rationalised over the years by widening the categories of officials who can formally demand the production of passes, and by linking this up with sophisticated computer technology centred on the reference book bureau of the Department of Plural Relations in Pretoria" (Frankel 1979, 207).

These data were entered into a centralized database that was cross-referenced across the different domains. Kahn notes:

Every African over sixteen must have on his person what is called a reference book, a bulky document measuring five-by-three and a half inches and containing ninety-five pages. As a rule, it is only Africans who are stopped by the police and asked to produce their passes. "The African must be a collector of documents from the day of his birth to the day of his death," says a publication issued by the Black Sash.[30] His passbook must contain particulars about every job he has had, every tax he has paid, and every x-ray he has taken. He would be well advised, the Black Sash has suggested, not to let himself get too far away from his birth certificate, baptismal certificate, school certificates, employment references, housing permits, hospital and clinic cards, prison discharge papers, rent receipts, and, the organization has added sarcastically, death and burial certificates. (1966, 91)

Horrell (1960) relates a story of an illiterate man "D.L." of Natal, who was arrested for having removed pages from his passbook. He was fined £10 or two months in jail (a huge sum for a black man at that time); unable to pay the fine, he went to jail. After being released, he could no longer find work, as he now had a prison record. A sympathetic literate friend investigated the case and found that the printers of the passbook had by accident eliminated pages 33 to 48 and instead had produced two sets of pages 49 to 64. D.L. had to appeal the conviction up to the Supreme Court level, again a costly and time-consuming business, where it was finally set aside.

In addition to the pass book system regulating the lives of black South Africans, the state attempted to enforce many other forms of segregation. Christopher Hope, in his novel *A Separate Development*, writes of petty apartheid such as the segregation of buses and benches:

One lived, of course, surrounded by such signs and notices. Most of them, however, served some clear purpose, the point of which everyone recognized as being essential for their survival: WHITES ONLY on park benches; BANTU MEN HERE on nonwhite lavatories; or INDIAN BENCH; or DEFENSE FORCE PROPERTY: PHOTOGRAPHS FORBIDDEN; or SECOND-CLASS TAXI; or THIS PLAYGROUND IS RESERVED FOR CHILDREN

OF THE WHITE GROUP. And, of course, people were forever being prosecuted for disobeying one or other of these instructions. (Hope 1980, 20)

For apartheid to function at this level of detail, people had to be unambiguously categorizable by race. Despite the legal requirement for certainty in race identification, however, this task was not to prove so easy. Many people did not conform to the typologies constructed under the law: especially people whose appearance differed from their assigned category, or who lived with those of another race, spoke a different language from the assigned group, or had some other historical deviation from the pure type. New laws and amendments were constantly being debated and passed (see, for example, *Rand Daily Mail* 1966). By 1985, the corpus of racial law in South Africa exceeded 3,000 pages (Lelyveld 1985, 82).

Both the scientific theories about race and the street sense of terms were confused. Prototypical and Aristotelian senses of categorization were used simultaneously, as with the example of the ICD shown in chapter 3. The original official sorting by race after the 1950 Population Registration Act derived from the categories checked on the 1951 census returns. An identity number was given to each individual at that time (Horrell 1958, 19). The census director was in charge of deciding everyone's racial classification, on the basis of the census data, and, where necessary, other records of vital statistics. Horrell notes, "But this classification is by no means formal. Section *Five(3)* of the Population Registration Act provides that if *at any time* it appears to the Director that the classification of a person is incorrect, after giving notice to the person concerned, specifying in which respect the classification is incorrect, and affording him or her an opportunity of being heard, he may alter the classification in the register" (1958, 4). So in the case of apartheid, we see the scientistic belief in race difference on the everyday level and an elaborate formal legal apparatus enforcing separation. At the same time, a much less formal, more prototypical approach uses an amalgam of appearance and acceptance—and the on-the-spot visual judgments of everyone from police and tram drivers to judges—to perform the sorting process on the street.

The conflation of Aristotelian and prototypical categories for race classification has deep historical roots in South Africa and elsewhere. The concept of racial types took firm hold in the nineteenth century across a range of natural and social sciences, and it was embraced by the architects of apartheid. At the same time, the pure types existed nowhere, and racism existed everywhere. Dubow writes about the scientific history of South African racial theories:

The typological method is at the heart of physical anthropology. It was based on empiricist principles of classification taxonomy originally developed in the natural sciences. The conception of race as "type "encouraged a belief in the existence of ideal categories and stressed diversity and difference over similarity and convergence. This was overlaid by binary-based notions of superiority and inferiority, progress and degeneration. One of the many problems associated with the typological method was its fissiparous character. The search for pure racial types could not easily be reconciled with the evident fact that, in practice, *only* hybrids existed. New fossil discoveries led to a proliferation of variant racial types and ever more theories were developed to explain their affinities. (Dubow 1995, 114–115)

Such difficulties are always present when trying to place people in racial categories (see López 1996, Robbin 1998, Harding 1993). As Donna Haraway says of racial taxonomy in the United States:

In these taxonomies, which are, after all, little machines for classifying and separating categories, the entity that always eluded the classifier was simple: race itself. The pure Type, which animated dreams, sciences, and terrors, kept slipping through, and endlessly multiplying, all the typological taxonomies. The rational classifying activity masked a wrenching and denied history. As racial anxieties ran riot through the sober prose of categorical bioscience, the taxonomies could neither pinpoint nor contain their terrible discursive product. (1997, 234)

Although a vague conception of eugenics and other forms of scientific racism are woven throughout the debates about apartheid, this lack of a scientific definition of race appears repeatedly. Dr. M. Shapiro, at a meeting of the Medico-legal Society in Johannesburg in 1952, wryly noted that:

Where for purposes of legal classification, the question arises whether a person is White, Coloured, Negroid or Asiatic, the policeman and the tram conductor, unencumbered by biological lore, can make an assessment with greater conviction, and certainly with fewer reservations, than can the geneticist or anthropologist. Indeed, the law being traditionally intolerant of uncertainty in matters of definition, the evidence of the scientist on the subject of race can only prove an embarrassment to the Courts if not to himself. (quoted in Suzman 1960, 353)

In a legal article reviewing race classification in 1960, Suzman concludes, "As the present study has revealed, the absence of uniformity of definition flows primarily from the absence of any uniform or scientific basis of race classification. Any attempt at race classification and therefore of race definition can at best be only an approximation, for no scientific system of race classification has as yet been devised by man. In the final analysis the legislature is attempting to define the indefinable" (Suzman 1960, 367). Landis (1961) similarly notes that the definition of the law was inherently ambiguous; she argues this was intentional, and that the ambiguity shifted the burden of proof to the individual. In a case where a person could not be proved to be either European or non-European, the burden of proof would fall on disproving the non-European side.

The lack of scientific definition had no bearing on the brutal consequences of the classification, despite the fissiparous—branching and dividing—nature of the scientific problem.

### Conflicting Categories in South Africa

Different aspects of apartheid law could classify a person differently. Where a woman lived, for example, often depended on her husband's classification, although movement from Bantu to white was not possible this way. So she might be of Indian national origin classified as Asian, married to a man classified as coloured, and live in a coloured zone but only be able to work or go to school in an Asian zone. This could be impossible or very arduous due to distance and the segregated transportation infrastructure:

Under the Population Registration Act, the children of mixed unions are, it appears, generally being classified according to the "lower "of the two layers involved—that is, the group carrying fewer privileges. . . . But under the Group Areas Act, the children of Coloured and African parents, or Asian and African parents, would while they were minors presumably be classified according to the racial group of the father in order that they might live with him and his wife in his group area. The child of an Indian father and an African mother might, thus, be brought up in an Indian environment, but, on reaching the age of sixteen and receiving his identity card might be forced to leave his parents and change his mode of living and his associates to those of the African group. (Horrell 1958, 12–13)

Again, the racialized gender structure is prominent here, where patrilinearity and patriarchal definitions of the couple's race are followed. Oddly, at times the multiple, contradictory methods of classifying could be used subversively to work in favor of the individual who lived between the categories. Horrell recounts, "A third case is described to show the absurdities that may arise. Mr. T. is in appearance obviously Coloured, and his sons and daughter are near-White: his sons, in fact, served as Europeans in the army. Both of them now live as Coloured men and were so classified. But Mr.T. trades in an African location and wants to continue doing so. It is said that he asked the official to classify him as an African: certainly this was done" (1958, 53).

In one infamous example a jazz musician, Vic Wilkinson of Cape Town, was born to a coloured man and a white woman, and originally he was classified as white. After apartheid he was reclassified as coloured and then twice more reclassified as he married women of different races and moved to different neighborhoods. (Note that the remarriages took place outside of South Africa for legal reasons.) Finally, both he and his Asian wife Farina were reclassified as coloured, allowing them and their children to live together. At the age of fifty, Vic received a new birth certificate and crossed the race lines for the fifth time (see figure 6.2, *Sunday Times* 1984).

The barriers to movement to a less privileged class were of course more permeable that those to passing "up." The language originally used to encode the classifications was itself inconsistent as well. Officials entering vital statistics in the preapartheid era frequently used the term "mixed." In many cases, this caused later confusion. "It was mentioned earlier that some White people, on sending for the first time for their birth certificate find that their racial group has been entered as "mixed," but that, on further investigation, this may be found to imply nothing more than that one parent was, for example, an immigrant from Sweden, while the other was an Afrikaans-speaking White woman" (Horrell 1958, 73). Again, we see here the conflation of prototypical ("mixed") categories with the attempted Aristotelian definition (the precise, exclusive categories aspired to by Nationalists). In this example, the formal-informal mixture itself produces organizational conditions that favored both structural and face-to-face ad hoc discrimination, the one reinforcing the other. Star (1989a) describes a similar case with scientific anomalies, where anomalies arising in one sector of research may be answered by nonanomalous research drawn from another, thus obscuring the original difficulty. In this case, as with the conflation of prototypical and Aristotelian categories, biases become deeply embedded in both practice and infrastructure. The conflation gives a terrible power of ownership of both the formal and the informal to those in power. The use of both simultaneously is

**Figure 6.2**
Vic Wilkinson and his family of Cape Town, showing the certificate of his fifth
racial classification.
Source: Johannesburg *Sunday Times*, 11/4/84, 21. Terry Shean/*Sunday Times*,
Johannesburg, courtesy of Times Media, Inc.

precisely at the heart of the banality of the evil of which Arendt wrote. It differs from the mindless adhesion to formal rules found in Heller's *Catch –22* (1961). The demoralization that it may produce is more like the Japanese-American *Men of Company K* of which Shibutani (1978) writes so eloquently about a thorough cultural demoralization produced by unbearable role strain.

### The South African Coloured Population and Reclassification

Approximately 1 million South Africans fell into the coloured category at the time of the Population Registration and Group Areas Acts. It was among this group that the majority of borderline cases appeared most often in the form of a person labeled coloured and desiring to be labeled white (or European). Within the internal logic of apartheid, an apparatus had to be constructed to adjudicate these cases. The Registration Act had a proviso that if the person objected to a classification, he or she had thirty days in which to appeal. Several local administration boards were set up to hear borderline cases and reconsider classifications. Their decisions could be appealed up to the level of the Supreme Court, a costly and time-consuming business. The average waiting time for an appeal was fourteen months (many were longer), during which time the person existed in limbo. For example, if someone wanted to be classified white, but was classified coloured, she or he could not go to a white school. If they enrolled at a coloured school, this could later become legal evidence that they *were* coloured. Several took correspondence courses as a solution, as apartheid apparently did not work long-distance (interestingly, without a face-to-face component it was not enforced).

The havoc wreaked in the lives of those in between was considerable. A broadside issued by the South African Institute of Race Relations stated:

While the Population Registration Act of 1950 did not affect the circumstances of the vast majority of the South African population, it created the utmost confusion as to the destiny of the small minority of people whose appearance, associations, and descent do not happen to coincide. The South African Institute of Race Relations pleads with all the power at its command that this small number of persons should be allowed to remain in the racial category in which they feel most at ease. (1969)

But this was not to be allowed by the Nationalist government. "As from 1 August 1966 it became compulsory for all citizens of the Republic

**Table 6.2**
Numbers of objections to racial classifications under the population registration act, 1968

|  | Total number of reclassifications | Of the total, number made as the result of representations by the person concerned |
|---|---|---|
| White to Coloured | 9 | 1 |
| Coloured to White | 91 | 91 |
| Coloured to Bantu | 29 | 3 |
| Bantu to Coloured | 136 | 136 |

Source: Horrell 1969: 42.

over the age of sixteen years to be able to produce identity cards, on which the racial group of the holder is. Until then, large numbers of people on the racial borderline had apparently not submitted themselves for classification" (Horrell 1968, 23).

There were several factors involved in weaving the texture of categories in the lives of those in the borderlands of apartheid. Often there were long bureaucratic delays in assigning a racial classification to those who appeared ambiguous. The Associated Press cites a case in which two preschool children were held in detention for three years while they awaited a government decision about their race (6 April 1984).

Approximately 100,000 people applied for reclassification (Brookes 1968, Horrell 1958). Few were approved. Bamford notes that "The board would seem to have been overstrict against the subject—the court has upheld its decision in only one of the ten reported cases involving the merits of reclassification" (1967, 39). A typical year is shown in table 6.2.

By May 1956 officials had dealt with 18,469 cases "in which objection had been raised to the classification claimed by the person concerned. Of these 1182 had been classified as White, 9,642 as Coloured, and 7,645 as Bantu" (Brookes 1968, 23).

Some years later, the figures had risen slightly but the basic direction of the changes remained the same. In 1981–82, 997 people changed races; in 1983, 690. In 1984, 795 people were reclassified. Of these, 518 went from coloured to white; two whites became Chinese and one became Indian; 89 black Africans became coloured, and 5 coloured people became African (Ormond 1986). A man from Durban won his reclassification appeal against his designation as coloured when a judge declared that he was a "white of the Mediterranean type" (Horrell 1968, 22).

The reclassification process was fraught in myriad ways and was completely internally inconsistent. At first the Race Reclassification Board ruled out descent (or "blood" as it was commonly called) as the determining factor. Instead, it used a mixed criteria of "appearance and general acceptance and repute." This was in explicit contrast with the American one-drop rule (Davis 1991), presumably for the reason that nearly all white South Africans had some traceable black African ancestry.[31] Bamford, in an article in the *South African Law Journal*, attempts to clarify the juridical meanings of "appearance" and "general acceptance" (1967). He notes, "Appearance is a matter of visual observation and assessment, to be undertaken by the tribunal. This observation and assessment should be made at the start of reclassification proceedings. If the subject is obviously white in appearance the presumption in section 19(1) will operate; if he is obviously not white, no further enquiry is necessary since he cannot be reclassified as white; and if he is neither obviously white nor nonwhite, the tribunal must proceed to decide on general acceptance" (41). There was no clear onus of proof about the meaning of general acceptance as white; in ambiguous cases the Race Reclassification Board would decide after conducting hearings and administering a range of tests of race. Like child custody hearings in American courts, such painful (and often shameful) tests were not stable or guaranteed of permanence:

The concept of general acceptance does not preclude a person's movement from one classification to another by virtue of changing association. The acceptance need not be absolute or without exception, so that the fact that a subject maintains contact with relatives or remains friendly with a Coloured family is not in itself fatal. In such cases: '[The tribunal must] decide whether the nonwhite history and associations were so overshadowed by the acceptance as white as to constitute general acceptance of the [subject] as white.' (Bamford 1967, 41)

Acceptance is an ambiguous, highly subjective prototypical concept sitting uneasily in the middle of the attempt for Aristotelian certainty. The *New York Times* reports the story of one Johannes Botha, a mail carrier living in Durban in 1960. Botha returned home to find his wife in tears following a visit from two investigators from the group areas board whose mandate was to seek out those living illegally outside of their category. The sleuths had "pounded on the door and then pushed past her and seated themselves in the living room without invitation. They had demanded identity cards, birth certificates, and a marriage license and had asked whether her husband was white. Mrs. Botha told them to wait for her husband's return, but they persisted in the interrogation" (Bigart 1960, 14).

The men questioned neighbors and the couple's four-year old son, focusing on a visitor to the house on the two previous Sundays. A coloured man had been seen on their doorstep. Botha, a scrap-metal dealer, was interested in purchasing a used car from him. Neighbors had concluded that he was a relative, and thus that the Bothas were passing for white and hiding their "coloured blood." The network of suspicion, spying, and the search for purity implied here affected every aspect of South African life for those in all racial categories.

The actual reclassification hearings were usually done in camera. The procedure was kept highly secret by those at the Population Registration Office. "No observer is in any circumstances allowed to attend. Legal representation is permitted; but as an inquiry by the Board is not analogous to a law suit, the ordinary rules of court do not apply. The officials may ask any question they wish" (Horrell 1958, 31).

Like the tuberculosis patients discussed in the previous chapter, people's biographical trajectories were severely disrupted by the re-classification process. Many lived for years in limbo; it could take months or years for the appeal to be heard and the person to be reclassified. It was, as one of Roth's respondents in the previous chapter declared the case to be with waiting and negotiating for a tuberculosis classification: "an ungraded classroom." A time out of time:

Even if appeals succeed, the people concerned have often suffered much anxiety and hardship before their cases are settled. In 1961 a family in Cape Town was classified Coloured. The eldest daughter had to postpone her marriage to a white man, while the second daughter had to leave a white school. For eight years she studied by correspondence: her parents did not want to prejudice their case by sending her to a Coloured school. The son's job was threatened. They all agreed to commit suicide if they could not get the classification altered. On reading their story in a newspaper, someone in Johannesburg made them an anonymous loan of R500 to cover the expense of an appeal, and this proved successful. (Horrell 1969, 27)

Once heard, the process was an open degradation ceremony (Goffman 1959). The process stripped people of identity, of uniqueness, and publicly humiliated them. "These officials question the people at the head of the queues and fill in forms —there is no privacy as those behind can hear the questions and answers. On average, the investigation of one case appears to take about twelve minutes" (Horrell 1958, 66).

### Technologies of Classifying

How impracticable it is to try to classify human beings, for all time, into definite categories, and how much suffering has resulted from the efforts made to do this.

*(Horrell 1958, 77)*

Apart from the categories themselves, the technology associated with the reclassification process was crude. Combs were sometimes used to test how curly a person's hair was. Horrell (1968) notes that barbers were sometimes called as witnesses to testify about the texture of the person's hair. One source mentioned expert testimony from the South African Trichological Institute (presumably an organization for the scientific study of hair). Affidavits were taken from employers, clergy, neighbors, and others to establish general acceptance or repute. "The official may summon any living relative, including grandparents, and question them in a similar way" (Horrell 1958, 32). Complexion, eyes, hair, features, and bone structure were examined by board officials, and they could summon any relative and examine them in this way as well (see figure 6.3). Horrell (1958) notes, "It is reported that some were even asked 'Do you eat porridge? Do you sleep on the floor or in a bed?' Some Coloured people said that they had been told to turn sideways so that the officials could study their profiles" (62). Folk theories about race abounded; differences in cheekbones, even the notion that blacks have softer earlobes than whites, were taken seriously. A newspaper account notes that some coloured people had reported that "the officials fingered the lobes of their ears—the theory is that Natives have soft lobes" (*Sunday Times* 1955). The same article reported that a coloured man was stopped by the police in the street and asked to which soccer club he belonged. He named a coloured team, and then was told, "only natives play soccer, not coloureds."

The "pencil test" was recounted by many who had undergone the reclassification ordeal. Sowden gives us the following passage, quoting at first from an old black woman describing apartheid to him:

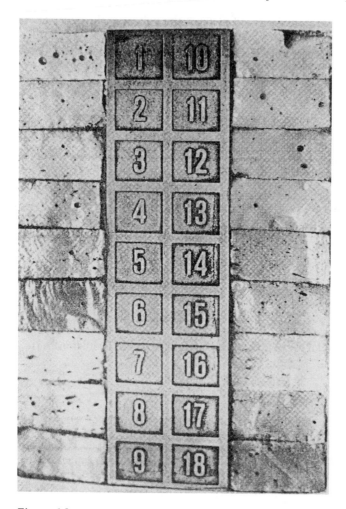

**Figure 6.3**
A scale for comparing the color of skin, one of the technologies used for race discrimination.
Source: Bruwer, J., J. Grobbelaar and H. van Zyl. 1958. *Race Studies (differentiated syllabus) for Std VI*, Voortrekkerpers, Johannesburg.

"If you're black and pretend you're Coloured, the police has the pencil test."
"The pencil test?"
"Oh, yes, sir. They sticks a pencil in your hair and you has to bend down, and if your hair holds the pencil, that shows it's too woolly, too thick. You can't be Coloured with woolly hair like that. You got to stay black, you see." (Sowden 1968, 184)

Because of the ambiguous nature of both the notion of general appearance and of general acceptance, the burden of evidence fell on the person desiring to be reclassified. At the same time, the Population Board fostered the system of informers where someone trying to pass (typically as white or coloured) could be turned in to the classification board for reexamination. Horrell notes the case of Mr. A, who was turned in to the Population Registration Office by an informer and called before them to prove his whiteness. Questioners asked if he knew of any coloured blood in his family and noted their hair, eyes, and skin color.

Mr. A said, "It was a terrible shock to me, and more so to my sons. The whole future of my family now rests on a decision from Pretoria." The worst part, he added, was that the very act of trying to prove himself European suggested that there might be some suspicion in the matter" (Horrell 1958, 34).

### Passing

They are both White and not White at the same time. They are in a White school and there they "must" be White: the law is witness to that. Yet "everybody" knows that they are not White, not really. They are something in between. But the law, which is an ass, knows no in-betweeness. It dichotomizes inflexibly, imposing a clumsy disjunction upon the subtly variegated flux of reality.
*(Watson 1970, 114)*

In the early preapartheid days, it was easier to change race category than it became later. Kahn notes that "between 1911 and 1921 . . . some fifty thousand individuals disappeared from the colored population rolls" (1966, 51). Many families living in the categorical borderlands went to great lengths to establish themselves as white, keeping photos (sometimes fabricated) of white ancestors (Boronstein 1988, 55).

Under apartheid, merely associating with someone of the wrong group could become evidence of membership and thus of race. Horrell writes of the collusion among members of coloured families being torn apart at the same time they collaborated to help some members pass for white:

There are in South Africa many thousands of people who cannot be classified according to a rigid system of racial identification . . . The lightest coloured members of these ["borderline"] families often "passed" as whites and went to live in separate homes. Their darker relatives have been referred to as "Venster-Kykers" ["window lookers"] because, in order not to embarrass those who had "passed," they made a practice of looking studiously into shop windows in order to avoid greetings should they happen to meet on the streets. (1958, 4)

Someone's racial classification could be challenged at any time. This was particularly important to the apartheid government in the case of people trying to pass for white, and a crucial location for the operation of the system of informers. Doman notes:

Soon after the introduction of the legislation many people asked for reclassification, with the result that there are today many families split down the middle. The offspring of the 'across-the-line' marriages are not always as white as their parents, and many families have emigrated rather than risk exposure. Today, there is no concerted effort to unearth the skeleton in the family cupboard. Coloured mothers avoid embarrassing their "White" daughters and do not see them even though they live in the same town. Yet the legislation also lends itself to spite—aggrieved people can get their own back on enemies or people they dislike by exposing a "mixed" marriage, or informing the police about a couple having an immoral (in terms of law) relationship. (1975, 151)

In an extraordinary study of a school in the suburbs of Cape Town, Graham Watson (1970) wrote of the complex negotiations, subterfuge, and balancing acts performed by parents, students, school principals, and the local Race Classification Board in managing "pass whites." Cape Town is the area in South Africa with the largest population of coloured people. Over the years, thousands chose to pass for white (or tried to). To do so, they changed their primary language from Afrikaans (used at the time by most coloured people in Cape Town) to English. They changed their social affiliations, as noted by Horrell above. Some passed for white during working hours, and returned to live with their coloured families in the evening.

In *Passing for White: A Study of Racial Assimilation in a South African School*, Watson drew a vivid picture of Colander High School based on his ethnographic participation. The high school was one of many buffer schools, which meant that "they admitted as pupils children described in the local School Board minutes as 'slightly coloured' but colloquially known as 'borderline cases'" (Watson 1970, 57). These schools predated apartheid, and the practice of accepting light-skinned coloured children into them was well-established before 1948.

Many borderline students were admitted to Colander High through a process full of ad hoc decisions and negotiations spanning years. This was facilitated in the first instance by the contradictory and arms-length relationships among those parts of the government charged with classifying people. To attend a school, a pupil received a classification from the School Board. Before apartheid, the Education Ordinance of 1921 stated that parents must prove European heritage for the child to be defined as white. Prior to 1963, the Superintendent-General of Schools was not bound by the decision of the Director of Census and Statistics (who managed the Population Registration Act). Thus, it was possible for a child whose parents both carried white identification cards, but who was dark-skinned, to be rejected by a white school. Similarly, the Superintendent-General, via the school principals, could act to let a child into the nominally white school. Principals who were in doubt about the race of the pupil could request an interview with both of the parents and with the school committee. Sometimes the parents and even grandparents were asked to produce birth and marriage certificates. The hearings, like those of the Race Classification Board, were held in camera. Watson notes that there was often disagreement among members of the School Committee about the classification decisions. Of the many cases heard by the committee of Colander High School, twenty were rejected. Of these, seven were rejected because one or both parents failed to appear; the rest were refused primarily on grounds of appearance (Watson 1970, 43).

The number of appeals from these decisions were small, as the principals and committee members were loath to confront parents directly. Watson notes that parents were not told of the real reason for the child's rejection from the school, and often the blow was softened by simply saying, "we're full." He continues:

Moreover, the parents of rejectees, whether they have been summoned before the Committee or not, are not informed of the real reason for their rejection—they are normally told simply that the school is full. Not even the School Board—to whom the Principal has been instructed to disclose the reason for each refusal—is told unequivocally that a child has been refused on the grounds of colour: in correspondence addressed to the Board the Principal covers himself by claiming that 'In the first instance, inability to accommodate is the reason for refusal.' In answer to verbal queries from the Board the

Principal is reticent. 'He led me to the brink,' he said, recounting his response to such a query, 'but I wouldn't say it. I told him we were full up with thirty or forty in each classroom and we weren't prepared to take anyone until we got an increase of staff. Mind you, he must have taken one look at the boy and seen there was less milk than coffee and known perfectly well that wasn't my reason, but he couldn't say so.' (1970, 44–45)

Appeals to the board about the committee's decisions, however, were often successful. Thus, there was a delicate invisible negotiation between parents and school principals-school committees. If no real reason was given for rejection, there would have been no grounds for appeal to the Board. The principals were charged with keeping up appearances as a white school, or they would risk far more serious sanctions from the Population Board, as well as complaints from white parents. As one says, "I can accept that child . . . but what do I do when I have a school function and the rest of the family comes along?" (1970, 47). Presumably, he speaks here to the racism of the local white families. In addition, "For the Principal, however, the child represents a sinister threat to the White status of his school, to his ability to attract teachers and pupils of sufficient number and satisfactory quality, and, ultimately, to his own personal prestige" (1970, 48–49). Multiple convergent systems are operating here. One school psychologist held the belief that the school ratings were lowered by coloured children as they performed more poorly on IQ tests. Such was the worry about this sort of status difficulty that principals often rejected those darker skinned children who carried formal white identification cards. The schools also needed to keep their numbers up, however, and some more liberal principals sometimes wanted to help the applicants. Overall, this juggling represents another example, as seen with the ICD, of distributing the residual categories, the "others." For those who do not quite fit the given categories, distribute them around the buffer schools, rather than having them all attend school at one place and thus threaten the white status of the school.

As mentioned, some officials willingly collaborated in the passing process. "It takes two (or more) to complete the process of passing for White" (Watson 1970, 55). Watson writes of a zone of ambiguity in face-to-face decisions. "Is it incumbent upon me, in the circumstances, to decide whether or not this person is White? If I decide that he is White, will others go along with my estimation? And what's in it for me?" (Watson 1970, 55–56). If the zone of ambiguity remains intact, often the amount of trouble incurred by refusing someone the claimed white status is too much. At other times, officials find work-arounds with which they can sidestep any confrontation. At one hospital, for example, a patient came in who was unconscious and looked neither European nor coloured, but was somewhere in between. The hospital authorities wavered on whether to put her in a European or a coloured ward, and finally put her alone in a side ward—in a living architecture of a residual category.

Watson (1970, 59) hypothesizes that those who successfully pass as white interact "segmentally with members of the superordinate group, thus allowing the superordinate-group members leeway in which innumerable ad hoc decisions cumulatively favorable to the aspirant can be made." The person wishing to pass for white manages a kind of shell-game sequence: first obtain employment in a whites-only occupation that is not too fussy about identity cards (such as being a tram director). The next step is to move to a mixed neighborhood, and quietly join local white associations. Working with the fact that even racist whites may find it difficult to confront a person face-to-face as passing, pass-whites are able to manage many face-to-face interactions such as attending white churches. Over time, this establishes a track record that can be used as leverage for reclassification based on general acceptance and repute. With possession of a white identity card, the person has the nominal protection of the law.

The more rigid the system of racial segregation and inequity, the more important passing became to those living in the categorical borderlands. At the same time, with the rise of the black consciousness movement in South Africa in the 1970s, a new ideology of black unity across black African, Asian and coloured lines became powerful there. In an 1975 article, Unterhalter (1975, 61) notes that most coloured people that she surveyed had disapproved of passing. This disapproval was based primarily on the need to remain loyal to the Coloured group. This contrasted with studies done twenty years earlier, where the harmful effect of passing as white on families was the primary reason (Watson 1970, 61). Negative attitudes toward black Africans who try to pass for coloured remained unchanged.

Given the disparities in power and privilege, it is not surprising that so many coloured people wanted to pass as white. Because passing is a partially secret, interactive process, and because it does require ad hoc mixtures of prototypical classifying and confrontations with Aristotelian categories of law, it is a crucible for the issues discussed throughout this book. Another kind of implosion (Haraway 1997) occurs where people try to be *re*classified, or who fall in other ways between the categorical imperatives of apartheid.

## Reclassification and Borderlands

In one family, one twin was classified as coloured and the other as African.
(Horrell 1958, 70)

The ground zero of race categories appeared in the case of an infant of indeterminate appearance who was abandoned on the steps of a hospital in Johannesburg. Because of the requirement of general acceptance and repute as determining one's race classification, it was ludicrous to try to use those criteria to decide the race of Lize Venter, named after the nurse who found her. An article in the *Rand Daily Mail* (7/10/83, 10) in 1983 announced, "Lize? A flat nose and wavy hair could decide her fate." It noted that "the law makes no provision for abandoned, newborn babies." Lize, more than any other person, represented the moment when the gaps created by the enforced mingling of prototype and Aristotelian category are laid bare, and the absurdities of apartheid law made clear.

The case of David Wong displays a more strategic exposure of the laws' internal contradictions. Wong was born in China of Chinese parents, and lived in a white neighborhood of Durban (a city with a large ethnically Asian population). His neighbors, taking advantage of the general acceptance clause of the Population Registration Act, swore a series of affidavits stating that he was White. This was no doubt as well an antiapartheid gesture, read over all. Wong received a white registration card on the basis of the affidavits. Brought to the attention of the press and government, it prompted an outraged reaction by M.P. deKlerk, who thundered:

It now appears, however, that there are certain White persons in this country who, again for reasons of their own, are prepared by means of affidavit to assist a person who admits that he is a full-blooded Chinese by descent, that he looks like a Chinese and who in appearance is obviously a Chinese, to be accepted as a White person by declaring an oath that he is accepted as a White person. This happened in spite of the indisputable fact that that was not the opinion of the community, and I challenge any honorable member on the other side to take this Chinese, David Wong, out of the environment in Durban where he is living and to placing in any other environment in Cape Town or Pretoria or Johannesburg and to get the verdict of public opinion as to whether he is a white person. (DeKlerk 1962, 10)

An amendment to the Population Registration Act was thereby passed, where anyone who claimed to be of a certain racial descent would be so categorized by the Board. The nested absurdities of the search for purity here are apparent, yet the search for purity remained strong in the popular white racist opinion through the 1980s.

### Language and Race as Conflicting Categories

There are thousands of ironic and tragic cases where classification and reclassification separated families, disrupted biographies, and damaged individuals beyond repair. The rigid boxes of race disregarded, among other things, important linguistic differences, especially among African tribal languages. Presented here are a few of the more extreme borderline cases. Collectively, they provide a powerful ethical argument against simple-minded, pure-type categories and for the positive value of ambiguity and complexity when applying racial categories to human beings.

The filiations of appearance and linguistic group become tangled in the case of "Dottie," a girl born to black African parents in the Randfontein area. She "happened to be lighter-skinned than are most Africans and to have long, wavy, copper-colored hair. Because of this she was rejected by principals of African schools and cannot attend a Coloured school because she can speak only Sotho" (Horrell 1968, 21). A similarly cruel situation appeared in the case of the Griqua group, which has a distinctive physical appearance, with "yellowish skin, high cheekbones, hair growing in little curly clusters" (Horrell 1958, 53–54). Many of this group married other native African tribal groups. They were classed by the Population Registration Act as African. This meant that they would be ruled on education by the Bantu Education Act and thus educated in one of the indigenous African languages. This was "completely foreign to most of the Griquas who speak Afrikaans" (Horrell 1958, 53–54).

Layers of invisibility were being enacted here by proapartheid forces. The idea of a separate development required that black people fit into mythic categories of pure tribal groups. The basis on which these groups were established and reported only partly respected actual tribal affiliations and not at all the conditions of people's lives. The hypothetical types were adorned with many natural features such as language and customs. In turn, each hyper-prototypical tribal group must have its own language, its own land, and its own unique customs. There was no room for people or circumstances that did not fit this image.

Again there is resonance with the ways in which Americans have enacted race in different regions. There are thousands of Native

American tribes, and most of them have been displaced from their indigenous lands. Unknown numbers have intermarried outside of and across tribal lines. Yet the registration system of the U.S. Bureau of Indian Affairs for counting who is really a member of a recognized tribe (and thus deserving of government benefits) is as contorted as tribal counting under apartheid. Munson (1997), for example, writes of the laws in the state of New Mexico that seek to protect Native American artisans from non-Indian imitators claiming to be "genuine Indian art." The laws have the ironic effect that a legitimately registered member of a tribe from another state could come to the area, have no prior knowledge of local tradition, but legally sell "genuine" Indian artifacts. At the same time, a local from an unrecognized tribe, having lived in New Mexico all his or her life, would not be able to do so. In an imposed, purified system of categories, both under apartheid and elsewhere, there are many ironies and much individual suffering.

### Sudden Changes

Another ironic twist of the categorical landscape leading to acute torques occurred when race classification was suddenly, unexpectedly shifted. For example, Ronnie van der Walt was a famous boxer in South Africa. At the age of twenty-nine, he was suddenly reclassified from white to coloured on the eve of a big match. One presumes from the *Newsweek* article reporting the case that someone had informed the race classification board, and it timed its inspection to be maximally embarrassing—an object lesson for others. The local race classification board's decision "was based on an inspection of Ronnie, Rachel and their two children." "One man there," Ronnie recalls, "walked around us peering at us from every angle like you do when you buy an animal. He said nothing, just looked . . . Interior Minister P.M.K. Leroux insisted that the ruling on Ronnie would stand. 'He has never been a White person,' sniffed Le Roux. Then, with logic reminiscent of the Mad Hatter the minister added 'And I do not believe he will ever become one'" (*Newsweek* 2/27/67, 42).

Van der Walt's biography and career were suddenly bisected by the revision of his race classification. Other cases were reported that illustrated the precarious nature of race purity. Two white children, Jane-Anne Pepler and Johanna de Bruin, had severe malfunctions of the adrenal glands, which caused their skin to turn brown. Jane-Anne had an operation to remove the glands at the age of fifteen; in a short period of time her skin and hair went from fair to dark brown. Her mother reported that: "only close friends and family who knew her before the operation know she is white" (*Newsweek* 7/3/70, 31). In a statement rich with an unconscious, ironic pragmatism, her mother said, "Some of her school friends have ostracized her completely—just as though she were a real nonwhite" (*Newsweek* 7/3/70, 30). She noted that "All this is particularly embarrassing for us because we are a purely Afrikaner family and strong Nationalists. We believe in white supremacy" (*Newsweek* 7/3/70, 31).

Johanna was an infant when she contracted Addison's disease. "No white school would accept her when she reached school-going age. Her father told a reporter that he intended applying to the Education Department for a tutor to teach her at home until she had passed standard V, after which she would be able to take correspondence courses." (Horrell 1969, 26) Her mother "lives in constant fear that, because of the past difficulties, 'someone' will come and take her daughter away from her" (Wannenburgh 1969). In both cases, the children are stuck with the rigidity of the Aristotelian definition of race—both were born with parents with white identity cards and were thus white—tempered with the prototypical face-to-face judgments of skin color, which would render them coloured.

Perhaps the most famous case of sudden identity change was that of Sandra Laing, who was brought up by white parents and was evicted at the age of ten from her white school for being coloured. The United Nations reported that when she was expelled by school officials under the Population Registration Act, it became illegal for Sandra to attend her Piet Retief boarding school, which was all-white (United Nations Office of Public Information 1969, 4). This reclassification denied her access to all other white institutions. The only way Sandra could continue living with her family was by being registered as a servant.

Scientists explained Sandra's appearance as resulting from a "dormant 'throwback' gene." It was posited that among the six color-determining genes, this throwback was responsible for Sandra's coloring (*Ebony* 1968, 85). Sandra's parents rejected the idea that this made her coloured, however. Sandra's father stated that she had been brought up "naturally" as a white child (*Ebony* 1968, 90). They attempted to tutor her at home, while appealing her case up to the level of the Supreme Court. After two years, she was reclassified again as white, and was legally permitted entry back into white schools. *Ebony* continues, "Out of the fire and back into the frying pan. That's what it all amounted to, because it had been Sandra's *appearance*, not her

---

### The Case of Sandra Laing

"Ten-year-old Sandra Laing slipped unnoticed into the school cloak-room. She made sure she was alone, then picked up a can of white scouring powder and hastily sprinkled her face, arms and hands. Remembering the teasing she had just endured in the schoolyard during recess, she began scrubbing vigorously, trying to wash off the natural brown color of her skin." (*Ebony* 1968, 85)

---

*legal classification,* that had aroused the bigotry of white parents two years before. And they still didn't want their children in school with a dark child. No matter what the statute books ruled, Sandra was still a *kaffir*[32] to them" (1968, 88).[33] The informal categories of racism and the formal classification system meet once again, this time tearing Sandra Laing's biography apart. Sandra's parents, clinging to the formal definition of her race, refused to tell her exactly what was going on or why she was so treated. For two years she "acted out fantasy rather than face the bitter truth. Until recently, she dutifully got up in time for school every morning since her dismissal in March of 1966, dressed herself in her school uniform, then sat around the house and waited, trying not to believe what was happening to her" (*Ebony* June 1968, 86).

In 1983 the *Rand Daily Mail* reported that Sandra had become completely alienated from her family and community. She "eventually lived with a black man and, ironically, applied to be reclassified so she could live legally with her lover" (see figure 6.4. *Rand Daily Mail* 7/23/83, 10).

Christopher Hope's novel, *A Separate Development,* centers on a dark-skinned boy who grew up white and who suddenly, upon reaching adolescence, becomes defined as coloured. A bus conductor throws him off a white bus, calling him a "white *kaffir.*" The boy says bitterly:

The thing is that this entire country has always based itself on two propositions, to wit: that the people in South Africa are divided into separate groups according to their racial characteristics and that all groups are at war with each other. Before you're clear about your groups, you must be sure you're clear about your individuals. As they teach the kids to chant: 'An impure group is a powerless group!' . . . Preserve the bloodlines. That was the rallying cry for generations. May your skin-tones match the great colour chart in the sky. Anyone who broke the bloodlines, who wasn't on the chart, was a danger to the regular order of things. You fought such renegades, mutants, throw-backs

**Figure 6.4**
A photo of Sandra Laing in 1991 at the age of thirty-six. She concludes an interview with the *New York Times* by saying, "my life is over."

and freaks with the power of definition. When in doubt, define. Once defined, the enemy could be classified, registered and consigned to one of the official, separate racial groups which give this country its uniquely rich texture. 'White kaffir': the words have a ring to them. I came to be grateful for them. Up until then I hadn't any proper idea what I was. What the conductor gave me was an identity. Ever since, I've been an identity in search of a group. (1980, 28)

Those who live in the borderlands, as Sandy Stone argues about gender order in her "The Empire Strikes Back: A Transsexual Manifesto," illuminate a larger architecture of social order (1991). Transsexuals, those who cross over from male to female or vice-versa, become in her metaphor a blank piece of paper upon which may be written anyone's fantasy of what a perfect women (or man) should be. Stone herself, a male-to-female transsexual, was initially refused surgery for refusing to wear makeup and high heels and behave "like a woman." Emily Ignacio, writing about Filipino/a identity in diaspora, notes a similar struggle both with and against stereotypes of a "real Filipino" (1998). Wherever ethnic identity exists, such struggles for and against purity exist. South Africa's Nationalists tried to classify in a completely Aristotelian fashion to make racial borderlands and ambiguity impossible. In so doing, the texture of the filiations they created were knotted, twisted, and often torn: another nightmarish texture (Star 1991b).

Earlier, the term torque was used to describe the twisting that occurs when a formal classification system is mismatched with an individual's biographical trajectory, memberships, or location. This chapter has probed more deeply into how this unfolds—the prototypical and Aristotelian are conflated, leaving room for either to be invoked in any given scenario (especially by those in power). The South African case represents an extreme example. For those caught in its racial reclassification system, it constituted an object lesson in the problematics of classifying individuals into life-determining boxes, outside of their control, tightly coupled with their every movement and in an ecology of increasingly densely classified activities. Each borderland case became a projection screen for the stereotypical fantasies of those enforcing the borders themselves. The stories of Sandra Laing, Dottie, and Ronnie van der Walt are ones that help illuminate what can happen when such a classification system is enforced and policed.

In some ways the South African stance is a mirror image of the current American dilemma. In the mid-1990s, a group of Americans held a march on Washington, with the goal of having the the option of choosing multiple racial categories added to the U.S. census.[34] This would replace the vague and to some insulting "other" category. They argued on both scientific and moral grounds that multiracial was the appropriate designation, one which would not force individuals to choose between parts of themselves. Yet many civil rights groups vigorously opposed them. Robbin (1998) recounts the struggle over the decision taken by the Office of Management and Budget (OMB) in 1997 to allow people to choose more than one racial category on the U.S. Census and in other federal government forms. This decision was known by the innocuous name of Statistical Directive 15. Arguments over the nature of racial classification in the U.S. census go back over many decades but with the advent of affirmative action and other similar measures in the late 1960s and early 1970s, race classification became even more consequential and contested. Robbin identifies three major issues involved in arguing for or against changing the categories: the controversy about whether or not to name (and how to name) racial and ethnic groups in government data; the exclusion of minority populations from the decision-making process within the U.S. Census Bureau; and the difficult questions of data quality and measurement. For example, where geographical location and race have been confounded, as with the Hispanic-American groups, people have in the past often simply opted to identify themselves as "other." The size of this residual category, and the potential sampling errors it represents, remains an unsolved difficulty in collecting Hispanic-American census data (1998, 43–46).

In the controversy surrounding the OMB decision, many African-American leaders (among others) argued that if everyone went by the scientific classification and coded themselves as belonging to multiple races, valuable demographic information and resources could be lost for many African Americans (Frisby 1995–96). Regardless of the scientific or genetic basis for the category, they said, racism against people with any black African ancestry was real, and the category was necessary to obtain resources and justice. This stance, sometimes called strategic essentialism by critical race theorists, lives precisely at the pragmatic junction between that which is perceived as real, and the consequences of that perception. Other leaders approved of the category choice change, seeing it as potentially liberating. Keen debate about the nature of these categories continued for months, including issues such as whether the indigenous people of Hawaii should be grouped with Native Americans and how to categorize people of more than one national and ethnic heritage. In October 1997, OMB decided to allow people to identify themselves as of more than one race, that is, to check more than one box. They could not, however, identify themselves as multiracial. The enormous expense and inertia of the decision is striking. It is estimated that it will cost millions of dollars. It is not often that individual categories are championed as social movements; even more rarely does an entire schema come under scrutiny.

### Conclusion

The South African case relates directly to all questions of information systems design where categories are attached to people. It is an extreme case, but at the same time, a valuable one for thinking about the ethics and politics of information systems. Not all systems attempt to classify people as globally, or as consequentially, as did apartheid; yet many systems classify users by age, location, or expertise. Many are used to build up subtle (and not-so-subtle) profiles of individuals based on their filiations to a myriad of categories. In the process of making people and categories converge, there can be tremendous torque of individual biographies. The advantaged are those whose place in a set of classification systems is a powerful one and for whom powerful sets of classifications of knowledge appear natural. For these people the infrastructures that together support and construct their identities operate particularly smoothly (though never fully so). For others, the fitting process of being able to use the infrastructures takes a terrible toll. To "act naturally," they have to reclassify and be reclassified socially.

# III

## *Classification and Work Practice*

These next two empirical chapters look at classification and work practice. Taking the example of the design of a system for classifying nursing work (NIC), we examine how classification systems that represent work embody multiple tensions—notably in this case between control and autonomy (chapter 7) and the representation of current work practice and learning from previous generations of practice (chapter 8). Such tensions are integral to the operation of work classifications. Due attention should be paid to their occurrence in order to evaluate the political and ethical implications of the introduction of new classificatory infrastructures.

# 7

## What a Difference a Name Makes—The Classification of Nursing Work

It's not always suitable to view work as production of information.
*(Bjerknes and Bratteteig 1987a, 323)*

### Introduction: How Work Classifies and Classification Works

To this point, the book has looked largely at the classification of diseases, patients, and of race—entities that are often (highly problematically, as we have seen) claimed to be natural kinds. This chapter and the next are concerned with entities generally seen as social kinds—units of nursing work.[35] Work classification systems are central to the management of a wide range of enterprises: and, we argue, their development is a contested site of great political significance.

Large information systems such as the Internet or global databases carry with them a politics of voice and value that is often invisible, embedded in layers of infrastructure. The "politics of artifacts" of a nuclear bomb or a genetically reengineered organism are more available for public debate then those of information interchange protocols or how insurance data are encoded. Yet these latter decisions and standards may affect markets, differential benefits from particular technologies, and the visibility of constituencies, among other important public goods (Kindleberger 1983). They are important in organizing work, and they are often used explicitly as vehicles for professional and organizational transformation, via accounting and legitimization processes. They appear, as parts of accounting schemes, in technologies of organizational change such as business process reengineering and total quality management; in addition to record-keeping and accounts, they also classify people and their importance in organizations.

For several years we have been investigating this quiet politics of voice, work, and values in information infrastructure, seeking to clarify how it is that values, policies, and modes of practice become embedded in large information systems. This chapter focuses on a classification system directed at nursing work and develops some theoretical notions about the relationship among classification, information systems, work, and organizations. Here we primarily take the point of view of design: what are the problems designers of the system face from their constituencies? The next chapter touches on the implementation of the system in various field sites and its direct impact on nursing work. Here, we examine the upstream dilemmas. These are similar to dilemmas faced by many designers of information systems in a range of application domains.

How does one make a successful, practically workable classification scheme of work practice? The problem of how to produce any classification scheme is an old one in the philosophy of knowledge, from Occam's razor to Quine's objects. Blurring categories means that existing differences are covered up, merged, or removed altogether; while distinctions construct new partitions or reinforcement of existing differences. This mutual process of constructing and shaping differences through classification systems is crucial in anyone's conceptualization of reality; it is the core of much taxonomic anthropology.

The case studies in Douglas and Hull (1992) point to the ways in which a category can be nonexistent (distributed out of existence) until and unless it is socially created. Thus Hacking (1992) talks about the creation of "child abuse" during this century. He argues that it is not that there was nothing in the nineteenth century that we would now call child abuse. Rather, that category per se did not exist then and so tended to go by a disaggregated host of other names.

Once the category was declared a legal and moral one at a particular historical juncture, it could be entered into the historical record (with much the same problematics as with AIDS). Another consequence follows from the canonization of a category: people then socialize themselves to the attributes of the category. Thus, people who abused children could now learn socially how to be a child abuser and what attributes in themselves they might identify as such. Reports in the media would teach them what was expected of the abuser personality. This is similar to Becker's analysis of how to become a marijuana user and what it takes to learn to read the signs of being stoned (1953–54). Naïve users must be taught to read their bodily signs to become intoxicated. Another similarity may be found in how UFO abductees shape their experience to fit the general cultural consensus about flying saucers and medical experiments.

The result of the change in category, and its place in social order, is a shifting of balances of distinctions, a change in the architectural relationships. Every newly constructed difference, or every new merger, changes the workability of the classification in the ecology of the workplace. As with all tools and all knowledge, such classification schemes are entities with consequences, to be managed, negotiated, and experienced all at once.

"Difference"—distinctions among things—is the prime negotiated entity in the construction of a classification system. Differences enter the work stream in a subtle and complex fashion. The practices to be classified do not disappear with new classification schemes. The *work* of categorizing itself, however, may cause shifts that in turn present challenges to the designers of the scheme (faced with decisions about how fine-grained it should be), to users (filling out forms and encoding diagnoses), and consumers (assessing the viability of the scheme). In this process, work itself is neither created nor destroyed, yet may be radically reshaped to fit into the emerging matrix. The larger contexts within which these classification shifts occur commonly include professionalization, automation, and informatization, and the creation of international research and recordkeeping procedures.

There are three main areas of challenge in crafting a classification scheme that will fit the work stream and agendas created by these larger contexts:

1. *Comparability.*   A major purpose of a classification system is to provide good comparability across sites to ensure that there is a regularity in semantics and objects from one to the other, thus enhancing communication. If "injection" means giving medication by needle in one country and by suppositories in another, for example, there is no use trying to count the number of injections given worldwide until some equivalence is reached by negotiation. The more intimate the communication setting, the less necessary are such negotiations for a variety of reasons, including that they may already exist historically or by convention; or they are more private and less subject to regulatory scrutiny.

2. *Visibility.*   How does one differentiate areas of work that are invisible? While they are invisible, they are by definition unclassifiable except as the residual category: "other." If work "just gets done"

according to some, it has found no voice in the classification scheme. Invisibility is not only erasure, though, on this view; it can come from intimacy, as with a team that has worked together for so long they no longer need to voice instructions or classify activities.

3. *Control.* No classification system, any more than any representation, may specify completely the wildness and complexity of what is represented. Therefore any prescription contains some amount of control to be exercised by the user, be it as small as in the most Taylorist factory or prison or as large as the most privileged artists' retreat. Control, like visibility, has good and bad elements, depending on one's perspective. Freedom trades off against structurelessness. The ability to exercise a wide range of judgment is worthwhile only if one has the power and resources to do so safely and effectively. Too much freedom for a novice or a child may be confusing or may lead to breakdowns in comparability across settings, thus impairing communication. Judgment about how differentiated to make the classification must take due consideration of this factor. This balance can never be fully resolved (as novices and strangers are always entering the field of work); the managerial trick is to measure the degree of control required to get the job done well, for most people, most of the time.

From the point of view of design, the creation of a perfect classification scheme ideally preserves common-sense control, enhances comparability in the right places, and makes visible what is wrongly invisible, leaving justly invisible discretionary judgment. It has, simultaneously, intimacy (in its detailed knowledge of the nuances of practice), immutability-standardization, and is manageable. A manageable work classification system works in practice, is not too fine-grained or arcane in its distinctions, and it fits with the way work is organized. It is standard enough to appear the same in every setting and is stable over time as well.[36] Intimacy means that the system acknowledges common understandings that have evolved among members of the community.

Such a perfect scheme, however, does not exist. In the real world, these areas trade off against each other. Maximizing visibility and high levels of control threaten intimacy; comparability and visibility pull against the manageability of the system; comparability and control work against standardization. For a classification system to be standardized, it needs to be comparable across sites and leave a margin of control for its users; however, both requirements are difficult to fulfill simultaneously. A manageable classification system (for whomever) does not only require that the system classifies the same things across sites and times but also that it uncovers invisible work; this affects the recording of data. The combination of these two thus requires compromise. Finally, to keep a level of intimacy in the classification system, control is a trade-off against the requirement to make everything visible. These trade-offs become areas of negotiation and sometimes of conflict.

Because one cannot optimize all three parameters at once to produce simultaneously perfect degrees of intimacy, manageability, and standardization, a real-life classification scheme encompasses a thorough, pragmatic understanding of these trade-offs in their historical context. It places them in the work stream. Here we now situate this process in our observations of the building of a classification system in progress, the Nursing Interventions Classification (NIC).

### The Nursing Interventions Classification

NIC itself is a fascinating system. Those who study it see it as an elegant ethnographic tool. Some categories, like "bleeding reduction—nasal," are on the surface relatively obvious and codable into discrete units of work practice to be carried out on specific occasions. But what about the equally important categories of "hope installation" and "humor" (see figures 7.1 through 7.3)?

Here we look further into the category of humor in NIC. The very definition of the category suggests the operation of a paradigm shift from work as punctual activity to work as process: "Facilitating the patient to perceive, appreciate, and express what is funny, amusing, or ludicrous in order to establish relationships. . . ." It is unclear how this could ever be attached to a time line: it is something the nurse should always do while doing other things. Further, contained within the nursing classification are an anatomy of what it is to be humorous and a theory of what humor does. The recommended procedures break humor down into subelements. One should determine the types of humor appreciated by the patient; determine the patient's typical response to humor (laughter or smiles); select humorous materials that create moderate arousal for the individual (for example "picture a forbidding authority figure dressed only in underwear"); encourage silliness and playfulness, and so on. There are fifteen subactivities, any one of which might be scientifically relevant. A feature traditionally attached to the personality of the nurse (being a cheerful and

## Airway Management                                    3140

DEFINITION: Facilitation of patency of air passages

ACTIVITIES:

Open the airway, using the chin lift or jaw thrust technique, as appropriate

Position patient to maximize ventilation potential

Identify patient requiring actual/potential airway insertion

Insert oral or nasopharyngeal airway, as appropriate

Perform chest physical therapy, as appropriate

Remove secretions by encouraging coughing or suctioning

Encourage slow, deep breathing; turning; and coughing

Instruct how to cough effectively

Assist with incentive spirometer, as appropriate

Auscultate breath sounds, noting areas of decreased or absent ventilation and presence of adventitious sounds

Perform endotracheal or nasotracheal suctioning, as appropriate

Administer bronchodilators, as appropriate

Teach patient how to use prescribed inhalers, as appropriate

Administer aerosol treatments, as appropriate

Administer ultrasonic nebulizer treatments, as appropriate

Administer humidified air or oxygen, as appropriate

Regulate fluid intake to optimize fluid balance

Position to alleviate dyspnea

Monitor respiratory and oxygenation status, as appropriate

BACKGROUND READINGS:

Ahrens, T.S. (1993). Respiratory disorders. In M.R. Kinney, D.R. Packa, & S.B. Dunbar (Eds.), AACN's Clinical Reference for Critical-Care Nursing (pp. 701-740). St. Louis: Mosby.

Suddarth, D. (1991). The Lippincott manual of nursing practice (5th ed.) (pp. 230-246). Philadelphia: J.B. Lippincott.

Thelan, L.A., & Urden, L.D. (1993). Critical care nursing: Diagnosis and management (2nd ed.). St. Louis: Mosby.

Titler, M.G., & Jones, G. (1992). Airway management. In G.M. Bulechek & J.C. McCloskey (Eds.), Nursing Interventions: Essential Nursing Treatments, (2nd ed.) (pp. 512-530). Philadelphia: W.B. Saunders.

*Figure 7.1*

Airway management, NIC. Helping the patient to breathe, including using breathing technologies and medications.

Source: *NIC*, second edition.

---

## Spiritual Support                                              5420

---

DEFINITION: Assisting the patient to feel balance and connection with a greater power

---

ACTIVITIES:

Be open to patient's expressions of loneliness and powerlessness

Encourage chapel service attendance, if desired

Encourage the use of spiritual resources, if desired

Provide desired spiritual articles, according to patient preferences

Refer to spiritual advisor of patient's choice

Use values clarification techniques to help patient clarify beliefs and values, as appropriate

Be available to listen to patient's feelings

Express empathy with patient's feelings

Facilitate patient's use of meditation, prayer, and other religious traditions and rituals

Listen carefully to patient's communication, and develop a sense of timing for prayer or spiritual rituals

Assure patient that nurse will be available to support patient in times of suffering

Be open to patient's feelings about illness and death

Assist patient to properly express and relieve anger in appropriate ways

---

BACKGROUND READINGS:

Fehring, R.J., & Rantz, M. (1991). Spiritual distress. In M. Maas, K. Buckwalter, & M. Hardy (Eds.), Nursing Diagnoses and Interventions for the Elderly (pp. 598-609). Redwood City, CA: Addison-Wesley.

Guzetta, C.E., & Dossey, B.M. (1984). Cardiovascular nursing: Bodymind tapestry. St. Louis: Mosby.

Thompson, J.M., McFarland, G.K., Hirsch, J.E., & Tucker, S.M. (1993). Clinical nursing (3rd ed.) (pp. 1637-1640). St. Louis: Mosby.

*Figure 7.2*
Spiritual support, NIC. Intervention for patients in spiritual distress.
Source: *NIC*, second edition.

supportive person) is now attached through the classification to the job description as an intervention that can be accounted for.

The Iowa group, who are mainly teachers of nursing administration and research, made essentially three arguments for the creation of a nursing classification. First, it was argued that without a standard language to describe nursing interventions, there would be no way of producing a scientific body of knowledge about nursing. NIC in theory would be articulated with two other classification systems: the nursing sensitive patient outcomes classification scheme (NOC) and the nursing diagnosis scheme (NANDA). NOC is a complex classification system in its own right. Since the medical profession has assumed

# Humor

**DEFINITION:** Facilitating the patient to perceive, appreciate, and express what is funny, amusing, or ludicrous in order to establish relationships, relieve tension, release anger, facilitate learning, or cope with painful feelings

**ACTIVITIES:**

Determine the types of humor appreciated by the patient

Determine the patient's typical response to humor (e.g., laughter or smiles)

Determine the time of day that patient is most receptive.

Avoid content areas about which patient is sensitive

Discuss advantages of laughter with patient

Select humorous materials that create moderate arousal for the individual

Make available a selection of humorous games, cartoons, jokes, videos, tapes, books, and so on

Point out humorous incongruity in a situation

Encourage visualization with humor (e.g., picture a forbidding authority figure dressed only in underwear)

Encourage silliness and playfulness

Remove environmental barriers that prevent or diminish the spontaneous occurrence of humor

Monitor patient response and discontinue humor strategy, if ineffective

Avoid use with patient who is cognitively impaired

Demonstrate an appreciative attitude about humor

Respond positively to humor attempts made by patient

**BACKGROUND READINGS:**

Buxman, K. (1991). Make room for laughter. American Journal of Nursing, 91(12), 46-51.

Kolkmeier, L.G. (1988). Play and laughter: Moving toward harmony. In B.M. Dosseyk, L. Keegan, C.E. Guzetta, & L.G. Kolkmeier (Eds.), Holistic Nursing: A Handbook for Practice (pp. 289-304). Rockville, MD: Aspen.

Snyder, M. (1992). Humor. In M. Snyder (Ed.), Independent Nursing Interventions (2nd ed.) (pp. 294-302). Albany: Delmar Publishers.

Sullivan, J.L., & Deane, D.M. (1988). Humor and health. Journal of Gerontological Nursing, 14(1), 20-24.

*Figure 7.3*

Humor, NIC. Cheering up patients—an important part of caring work.
Source: *NIC*, second edition.

responsibility for the cure of diseases, the nurses have to measure more the speed of the cure and the quality of life during the hospital visit and after release (for example, whether patients understood how to deal with the consequences of their bypass surgery).

The three systems could work together in the following fashion. One could perform studies over a set of hospitals employing the three schemes to check if a given category of patient responded well to a given category of nursing intervention. Rather than this comparative work being done anecdotally, as in the past through the accumulation of experience, it could be done scientifically through the conduct of experiments. The Iowa Intervention Project made up a jingle: NANDA, NIC, and NOC to the tune of "Hickory, Dickory, Dock" to stress this interrelationship of the three schemes.

The second argument for classifying nursing interventions was that it was a key strategy for defending the professional autonomy of nursing. The Iowa intervention team is aware of the literature on professionalization—notably Schön (1983)—and of the force of having an accepted body of scientific knowledge as their domain. Andrew Abbott (1988), taking as his central case the professionalization of medicine, makes this one of the key attributes of a profession.

The third argument was that nursing, alongside other medical professions, was moving into the new world of computers and networked information technology. As the representational medium changed, it was important to be able to talk about nursing in a language that computers could understand, else nursing work would not be represented at all in the future. It would risk being even further marginalized than it is at present.

The empirical material for this analysis consists of all the minutes of NIC team meetings and publications of the NIC group since 1987; eighteen open-ended, in-depth interviews with principal investigators, coinvestigators, and research associates; and observations of team meetings.

### Infrastructure and Organizations

There is no simple way to tell the story of the complex theoretical and practical work that goes into the development of an information infrastructure. Star and Ruhleder (1996) argue that an infrastructure has several key properties. Their relationship to NIC is detailed here:

• *Embeddedness* ("it is 'sunk' into other structures").   NIC is embedded into various information practices and tools that are used by hospitals and insurance companies for costing and coding reimbursements and by medical librarians for accessing medical literature. NIC is used in clinical decision-making software, hospital accounting systems, and nursing information systems.

• *Transparency* ("it is ready to hand and does not have to reinvented each time").   NIC is oriented to standard scientific and working practical knowledge and to being ready to hand for the practitioner. Instead of applying idiosyncratic or new labels to diseases, practitioners are asked to turn to classification systems to fill out forms, assign values, and compare results.[37]

• *Having reach or scope* (it is not a "one-off event or one-site practice"). NIC aims to cover U.S. nursing with a slow growth currently into the European and Asian nursing communities. Interest in adapting it has been shown by groups of social workers, occupational therapists, and pharmacists.

• *It is learned as part of membership* ("associated with communities of practice").   NIC is increasingly present in nursing education programs. Because of the ways in which it is propagated, it is closely tied with what it means to *be* a nurse.

• *It is linked with conventions of practice* ("both shapes and is shaped by the conventions of its communities of practice").   For NIC, the informatic conventions are young, but a key design issue is its fit with the conventions of nursing practice. One aspect of NIC user meetings is the developer's insistence that NIC integrate with work practice, and that NIC users share common conventions concerning the system's use. They are currently encouraging the development of clusters of interventions (invisible to the classification system) to represent local practice at specific institutions, for example, at nursing homes.

• *Multifunctionality* ("As with electricity, supports several functions"). NIC supports a wise range of functions, from data collection and basic epidemiological research to accounting by insurance firms and legitimization of work practices. These definitions are touchstones to order discussion here and to help guide the construction of a useful model for organizational analysis. Although there is a rich body of research on computerization, impact, values, and work-place politics, as yet theories of information infrastructure and its evolution, meaning, and values implications are not well developed.

Over the years many innovative applications in information technology have failed due to insufficient consideration of the projected users. When developing new information infrastructures, however, the scope of usage is murky. Users may not know, prior to experience, what they want from the new system and how they will use it. The success of France's Minitel Rose is a prime example where, much to the designers' surprise, personal and pornographic messages and not official information sold the system (Taylor and Van Every 1993). As noted in chapter 3, the work of being a classical Greek scholar has changed with the advent of the Thesaurus Linguae Graecae. The ability to produce concordances on the fly makes previous laborious library work, which had to be carried out over two or three continents, the work of a few minutes. This means that new kinds of questions may, indeed must, be asked of the material the researcher is looking at (Ruhleder 1995). Secretaries now do much less copy typing and professionals do more— the result of a pincer movement between the development of new management philosophies and new information technology. New infrastructures do more than support work that is already being done. They change the very nature of what it is to do work, and what work will count as legitimate.

In this sense, NIC is an actively developing infrastructure. It is fed into a clinical decision support system, directs nurses on which activities to perform, and becomes part of hospital accounting systems. It lays claim to a professional territory for nursing; used as part of ongoing research and teaching programs, current nursing interventions (fluctuating at present) become stabilized. Since research is built around these categories, a feedback loop is set up that stabilizes the current set. (A similar set of events were seen in the development of the international classification of diseases (ICD) in chapter 4.) Political, cultural, ethical, social, religious, economic, and institutional factors each play a role in NIC's development. Thus, for example, the definition of stillbirth has been a site of conflict among states with different religious constituencies; and epidemiologists argue that it is still highly variably diagnosed depending on the beliefs of the attending physicians. Similar coding problems have been documented in the case of AIDS and its associated illnesses, especially during the early 1980s (see Verghese 1994 for a wonderfully compassionate physician's view of the situation).

With NIC, as with the ICD, apparently precise, measurable qualities often prove much fuzzier when looked at closely. And yet as classification systems they present knowledge in a form that is transportable and usable in a wide range of different infrastructural technologies—databases, decision support systems, and so forth. They are complementary, in that NIC concentrates on work practice-information technology and the ICD on information technology-domain specific knowledge (although clearly all three factors are significant for each). NIC is associated with the traditionally "invisible work" which is often gender and status-linked (Star 1991a), while the ICD is linked with highly visible medical knowledge—yet each are being merged into seamlessly integrated infrastructures.

There is a close relationship among inscriptions, work practice, and standards, as Bruno Latour's work demonstrates. As seen in chapter 2, Latour in *Science in Action* (1987) developed the concept of *immutable mobiles* to explore the ways in which scientific knowledge travels from a local, messy field site into the laboratory and out into textbooks. The development of NIC displays the force of both of his analytic points: indeed the work of holding classifications stable and enrolling allies in their use has been central.

Equally important are notions of accounting and quantifying as forms of social order. Foucault's work on "governmentality" (Porter 1994) discusses the rise of statistics as a new mode of government; and following this, Ewald (1986) examined the rise of the welfare state as a form of government of the body and soul. A similar theme arises in the work of Rose (1990), whose argument that accounting systems reflect a moral order and help define the self has been widely adopted in critical social studies of accounting (see, for example, the journal *Accounting, Management and Information Technologies;* Boland and Day 1989; Boland and Hirscheim 1987). Central here is the recognition that statistics and other numbers, invariably based on classification systems—and recognized by WHO and the NIC designers as a key product of their own systems, are socially and politically charged.

The case of NIC is used here to discuss the three dimensions of work classification systems that form pragmatic challenges for designers and users: comparability, visibility, and control.

### Comparability: The Need for Standard Descriptions in Research

The construction of a nursing interventions classification implies a drive to abstract away from the local, the particular—to make nursing the same entity wherever it may appear. Ideally, local terminology and the idiosyncrasies of each ward and each staff nurse should change

### Intimacy and Language

A man and a woman sit in a kitchen. It is early in the morning. He is reading the newspaper intently; she is putting away last night's dishes and preparing breakfast. She pours a cup of coffee and puts it in front of him, carefully avoiding the angle of turning of the newspaper pages. After a moment, he takes a sip of the beverage. "Cold." From this single word, she infers the following: he is still angry over the squabble they had last night; he is feeling apprehensive about his upcoming work review; the dinner they ate together that precipitated the squabble sat heavily on his stomach, and he slept less well than usual. Correctly, she predicts that he will be a little snippy with his secretary in the office and forget to bring his second cup of coffee in the car with him on the way to work, a practice he has recently adopted. This omission will result in a late-morning headache. Psychologist Gail Hornstein analyzes this snippet of conversation as a means to understanding the relationship between intimacy and language. The more intimate the relationship, the more seemingly telegraphic may language become with no loss of meaning.[38]

immediately through an adoption of NIC in hospital administration. Those making the classification examine variability to either eliminate or translate it across settings. This is the strategy of moving toward universality: rendering things comparable, so that each actor may fit their allotted position in a standardized system and comparisons may be communicated across sites.

For the nursing interventions classification, the drive to erase the particular and communicate equivalents is apparent in several strategies the group adopts to further their cause. The developers consider NIC a basis for curriculum development: they reason that only with a complete classification system can one guarantee thorough, standardized, and cross-site comparability in professional training. NIC is being integrated into model course development efforts at Iowa and elsewhere. The basic interventions are part of undergraduate nursing curricula, while the more advanced interventions will be taught to master's students. But NIC is ultimately as well a standardized language for comparability. As one respondent said "The classification is an aspect that makes it a tool, more useable, but it is the standardized language that is really critical." According to the NIC researchers, "a standardized language for nursing treatments is a classification about nursing practice that names what nurses do relative to certain human needs to produce certain outcomes" (McCloskey and Bulechek 1994a, 57). In the eyes of the NIC creators, the classification system provides such a standardized language for nursing treatments that can be used across units, health care settings, and health care disciplines. A classification alone would be useful for costing, recordkeeping, and teaching, but the linguistic aspect is necessary for research and comparability. This intention was clearly expressed in several interviews:

Certainly we are aiming at standardizing nursing languages. So that when we talk among other nurses and other health professionals we all know what we talk about. Because what one nurse might be talking about is very different (from another nurse). What is the difference between therapeutic play and play therapy? And then we need to communicate with parents, consumers, patients, physicians and other health professionals and knowing that they are talking the same language. It is really important that we talk in a language that is not foreign to other groups. Maybe we like to be unique, but sometimes we need to bend so that we talk the same language as families, consumers, and medical professionals.

A hospital administrator asked me a couple of years ago whether nurses could just tell him what they do. You can't say "the nursing process" because everyone does nursing assessment, intervention. That is a model that everyone can apply. Physical therapists can say what they do: muscles and bones. Respiratory therapists can define their tasks. But nurses do all that. Nursing is so broad. The only thing that they know is that they can't work without us. NIC is extremely helpful because it provides a language with a firm scientific base to communicate what we do. (Interview with JoAnne McCloskey 4/6/94).

Thus NIC is seen as providing the means for rendering all nursing work comparable. To study the effectiveness of nursing care, the nursing profession proposed the uniform and routine collection of essential nursing information or a nursing minimum data set (NMDS) (Werley and Lang 1988; Werley, Lang, and Westlake 1986). "The purpose of the NMDS is to foster comparability of nursing care across patient populations, with the ultimate goal the improvement of health care" (McCloskey and Bulechek 1994a, 56). This data set consists of sixteen data elements, including four nursing care elements: nursing diagnoses, nursing interventions, nursing outcomes, and nursing intensity. NIC is promoted by its creators as providing the nursing intervention variable for the NMDS. A standardized language is also necessary to communicate with extant information systems. As a universal, scientific language, NIC is targeting inclusion in the unified medical language system (UMLS)—a spearhead of the drive for a standardized for all health care information systems.

Ironically, NIC's biggest critics come from the same information systems world. Criticism has been directed against NIC's standardized language ambitions. Susan Grobe, a nurse and information scientist at the University of Texas, Houston, criticizes the attempts at creating a universal standardized system as scientifically outmoded and inflexible. Instead Grobe proposes her own nursing intervention system, the Nursing Intervention Lexicon and Taxonomy (NILT) which consists of eight broad categories of nursing interventions. According to Grobe, in NILT "the burden of standardized language is resident in the automated systems and not dictated to practicing professionals for their memorization and adoption" (Grobe 1992, 94). Where NIC expects nurses to learn and use a standardized terminology, Grobe believes that nurses should keep their natural language and computers should be used to standardize language. She argues that having computers decide how terms will be standardized is inevitable and cites researchers who are working on this approach in health care documentation.

NIC researchers defend themselves against Grobe's criticism by specifying how a standardized language increases comparability. They note that although the advent of computers was an impetus for standardized languages, different organizations and agencies developed their own system, "with the result that we cannot collect comparable data from multiple agencies, or even within agencies from one unit to the next." They further quote Sherrer, Côté, and Mandil. "Intelligent documentation systems cannot totally discard classifications. Moreover, the availability of at least one classification is a necessary condition for a good documentation system. Classifications are not a necessary evil but a very effective way of representing knowledge about the domain of discourse" (McCloskey and Bulechek 1994a, 59; see also Bulechek and McCloskey 1993). Thus since a natural language system is at this moment lacking in nursing, the NIC researchers claim that their classification system fills the void and at the same time achieves the goal of comparability.

In their newsletter, the NIC investigators summarize their vision about a standardized language to achieve comparability across sites and professions. "Norma Lang has often been quoted as saying, 'If we cannot name it, we cannot control it, finance it, teach it, research it, or put it into public policy. . . .' We would like to be quoted as saying, 'Now that we have named it, we can control it, finance it, teach it, research it, and put it into public policy'" (NIC newsletter 1994, 2).

Striving for comparability in a standardized language across settings conflicts with the need for visibility within local settings. The nursing intervention architects want their system to be adopted by health care institutions. As a language, its entire vocabulary needs to be available to nursing professionals. Certain institutions, however, will most likely only need part of the NIC taxonomy; for example, nurses in a geriatric hospital would not require "Newborn Care" as an intervention. The results of validation studies with different nursing specialties suggest that between 20 and 80 percent of the terminology would be routinely used by several nursing specialties. This raises the issue of how to limit each institution's modifications. Too much flexibility would obviously undermine the birth of a standard language, but too much control makes a system user unfriendly especially in such a safety critical and busy line of work. As a rule of thumb, the NIC group decided that an institution should adopt the whole classification system at the level of the copywritten interventions, definitions, and labels, but that activity-level descriptions could be modestly changed. Control and enforcement of this rule, however, ultimately rests with the publisher.

This central tension between standards on the one hand and local, tailorable systems on the other is a familiar one in information systems (Trigg and Bødker 1994). It remains a tradeoff—a tension not resolved by resorting to a lowest common denominator, a universal algorithm, or an appeal to universal positivist knowledge (Star 1992).

*Visibility: Legitimacy versus Surveillance*

Comparability rests on the management and mobility of differences and equivalencies across sites. The issue becomes what is local and particular or what do all nurses have in common that can be rendered equivalent across settings and nursing specialties? Then, what does this commonality render invisible? The nursing classification designers employ a definition of nursing interventions as a guideline. "A nursing intervention is any direct-care treatment that a nurse performs on behalf of a client. These treatments include nurse-initiated treatments resulting from nursing diagnoses, physician-initiated treatments resulting from medical diagnoses, and performance of the daily essential functions for the client who cannot do these" (Bulechek and McCloskey 1989, 23). Here, the emphasis is on direct care: that which nurses do to increase the well-being of a patient at the bedside. Direct

> ### *Invisible Categories*
>
> —an anecdote related by literary critic Alice Deck
>
> In the 1930s, an African-American woman travels to South Africa. In the Capetown airport, she looks around for a toilet. She finds four, labeled: "White Women," "Colored Women," "White Men," and "Coloured Men." (Colored in this context means Asian.) She is uncertain what to do; there are no toilets for "Black Women" or "Black Men," since black Africans under the apartheid regime are not expected to travel, and she is among the first African Americans to visit South Africa. She is forced to make a decision that will cause her embarrassment or even police harassment.[39]

care is separated from care that only indirectly benefits the patient. Indirect care includes, for instance, coordinating treatment schedules, discharge planning, and patient supervision. One step further removed from the bedside is administrative care, activities for creating a work environment supporting either direct or indirect care. This includes tasks such as coordinating administrative units and supervising nurses. Initially, the NIC group concentrated on direct-care interventions. The researchers deliberately supported an image in the classification of nursing as a clinical discipline. This was a political decision, as several NIC team members noted in interviews, one said: "Nurses think that laying hands on patients is nursing. We would not have had the attention of the nursing community if we had not begun there."

Questions arose in the course of the project, however, about the distinction between direct and indirect care. For instance, if nurses must check resuscitation carts with every shift, and this is not included in NIC, then these activities will not be reimbursed when NIC is implemented. Time spent on this task will be invisible and thus fiscally wasteful. Over the course of the project, there has been increasing recognition of the importance of indirect interventions, and these were included in the second edition of the NIC classification system. The researchers have even adapted their initial definition of a nursing intervention to include indirect interventions. Nurses themselves are somewhat ambivalent about how to account for indirect care time. Statistical analyses based upon different validation studies reveal that

---

## Emergency Cart Checking                7660

**DEFINITION:** Systematic review of the contents of an emergency cart at established time intervals

---

**ACTIVITIES:**

Compare equipment on cart with list of designated equipment

Locate all designated equipment and supplies on cart

Ensure that equipment is operational

Clean equipment, as needed

Verify current expiration date on all supplies and medications

Replace missing or outdated supplies and equipment

Document cart check, per agency policy

Replace equipment, supplies, and medications as technology and guidelines are updated

Instruct new nursing staff on proper emergency cart checking procedures

---

**BACKGROUND READINGS:**

Copeland, W.M. (1990). Be prepared. Hospitals should develop methods to ensure emergency equipment is workable and available. Health Progress, 71(6), 80-81.

Shanaberger, C.J. (1988). Equipment failure is often human failure. Journal of Emergency Medical Services, 13(1), 124-125.

*Figure 7.4*
Emergency cart checking, NIC. An example of indirect care.
Source: *NIC*. second edition.

several of the indirect-care interventions are indeed considered in a different category by nurses responding to the surveys (see figure 7.4).

Administrative tasks as care are even more controversial. In interviews, one of the NIC collaborators whose main tasks are administrative expected that NIC would eventually also contain those kinds of interventions. "Nursing is very different in that when you make changes it involves many people, so the need for managers and supervision and coordination of planned change is so much more a part of nursing, there are so many more people that are a part of changing nursing. I think anything that reflects nursing, needs to reflect those kind of things." A majority of the design team and consulting group, however, was not sure whether administrative care was typical for nurses and thus whether it belonged in a nursing classification. "The administrators are not actually nursing. When they are not there, the nursing continues without them." Or in the words of Gloria Bulechek, "management science is a different discipline, all managers have to manage people and it is not unique to nursing." For the latter group, the need to make administrative care visible is not as urgent as the need to differentiate nursing as a hands-on clinical discipline. Although the nursing researchers are aware that the boundaries among direct, indirect, and administrative care are not firm, administrative care was not part of the first two editions of NIC. This dilemma about the encoding of administrative work points to a practical limit on the visibility-discretion tradeoffs. To fully abstract from the local, everything must be spelled out; to avoid resistance from nurses and nursing administrators, some specifications for work must be left implicit. What is left implicit becomes doubly invisible: it is the residue left over when other sorts of invisible work have been made visible (Strauss et al. 1985, Star 1991c; Star and Strauss 1999). Where claims are made for the completeness of an accounting system, that which is not accounted for may be twice overlooked. This is noted here as both a formal and a practical challenge for classification designers and users.

The tension between visibility and discretion became apparent when several group members noted that the classification is strong—perhaps too strong—within the nursing specialties of the system's developers such as the complex physiological domain. It is still underdeveloped in other nursing areas, however, such as community health and social-psychological nursing. Social-psychological care giving is one of the areas where the control-visibility dilemma is very difficult to grasp. As noted, NIC lists "humor" as one nursing intervention. How can one capture humor as a deliberate nursing intervention? Does sarcasm, irony, or laughter count as a nursing intervention? How to reimburse humor; how to measure this kind of care? No one would dispute its importance, but it is by its nature a situated and subjective action. Since NIC does not contain protocols and procedures for each intervention, a grey area of common sense remains for the individual staff nurse to define whether some of the nursing activities can be called nursing interventions or are worth charting. This same grey area also remains for more clinical interventions such as "cerebral edema management" or "acid-base monitoring."[40] But because the classification is modeled after a clinical model of nursing, the team felt it easier to define and include those more clinical interventions.

The borderland between professional control and the urge to make nursing visible is fraught with difficult choices and balances not only in the interventions themselves but also in the decisions underlying NIC. Team members recalled discussions where interventions were so singular and demarcated as to warrant inclusion, but they ended up not being included. For example, "leech therapy" was not accepted as an intervention in the first edition of the classification system, although there was enough research literature to support this intervention as typical nursing in many parts of the world. This strategic choice was a response to a prior history of nursing classifications not being taken seriously. It was feared that the mention of leech therapy, with its folk and medieval associations, would provide a red flag for critics (compare this situation with virus classifications discussed in chapter 2). It was introduced into the second edition, when the Iowa intervention team believed that they had sufficiently demonstrated their credentials. Also the advanced statistical analysis of the validation studies was located in what the design team members typified as "common sense." One could have a reliability coefficient of .73, but if it did not respond to a visible or controllable enough nursing reality, it became an outlier, a nonresult, or resulted in a residual category. As with all statistical analyses, a link with theory and practice must precede testing or the results are meaningless.

In other cases, the criteria for inclusion and control are themselves contested. One research member confided in an interview that her intervention was rejected because it was not supported with research evidence. Her plan was to first publish a paper about the intervention in a research journal and then resubmit the intervention for consideration with her own reference as research evidence.

In these examples, the goal of making as much visible as possible clashes with what should remain taken for granted. The nursing researchers temper their quest to make nursing visible with the image of what nursing is or should be about. Again, there is no final answer or algorithm but a complex balance of experience and rules. Common practice, contingency, and legitimacy temper visibility.

### Control, Discretion, and Reliability

There is a continuing tension within NIC between abstracting away from the local and rendering 'invisible work' visible. Nurses' work is often quintessentially invisible for a combination of good and bad reasons. Nurses have to ask mundane questions, rearrange bedcovers, move a patient's hand so that it is closer to a button, and sympathize about the suffering involved in illness (Olesen and Whittaker 1968).

***By the Book***

The movie *A Few Good Men* hinges on an anecdote about several soldiers who perform a "code red" on another soldier, during which he dies. A code red is an illegal informal punishment-harassment in the manner of a rough fraternity joke. The death of the soldier causes an investigation; the commanding officer is suspected of deliberately ordering the code. The harassing soldiers defend themselves by saying that they could not have been ordered to perform a code red because that was forbidden by the manual of conduct. The denouement of the film has the prosecuting attorney closely questioning one defendant, roughly as follows:
"Does he do everything by the book?"
"Yes."
"Does the book contain all knowledge about how to conduct oneself in military life?"
"Yes."
"Did he have breakfast this morning?"
"Yes."
"Does the manual specify how to get to the mess hall, or where it is located?"
"No."
"QED—the manual does not contain all knowledge."

Bringing this caring work out into the open and differentiating its components has encountered problems from the nurses themselves. In naming and differentiating someone's work, there is a fine line between being too obvious and being too vague once one has decided to take on naming as a central task. If the task that is brought under the glare of enlightenment science is too obvious and mundane, then some nurses who are testing the system find it insulting. To tell veteran nurses to shake down a thermometer after taking a temperature puts them in a childlike position. Some experienced nurses, encountering interventions they felt were too obvious, have called them NSS—"no shit, Sherlock" interventions—it does not take a Sherlock Holmes to realize that nurses have to do this! Creating difference by cutting up the continuum of duties that make up "looking after the patient's welfare" is thus sociologically as well as phenomenologically and philosophically difficult. One must be explicit enough for the novices, yet not insulting to the veterans. Reading the NIC minutes, one is frequently reminded of ethnomethodological texts: just how much common sense can be taken for granted is a perpetually open question, and to whom it is common sense is not always so obvious (for example, Sacks 1975).

But ethnomethodology alone will not solve the political and organizational controversies and dilemmas of discretion. We see a link here with all previous attempts to rationally reconstruct the workplace, especially those modeling work for information systems. As Schmidt and Bannon (1992) point out, the management of real-time contingencies ("articulation work") never goes away, but if ignored, will be costly in many ways.

One of the battlefields where comparability and control appear as opposing factors is in linking NIC to costing. NIC researchers assert that the classification of nursing interventions will allow a determination of the costs of services provided by nurses and planning for resources needed in nursing practice settings. Currently, nursing treatments "are lumped in with the room price." In interviews with team members, they noted that although nurses fill in for physical therapists during weekends, the nursing department is not always reimbursed for this service. Sometimes the money flows back to the hospital at large, to the physical therapy department, or these treatments are simply not reimbursed. According to the NIC researchers, NIC will allow hospital administrators to determine nursing costs and resource allocation and stop such apparent "freeloading." Until it is made explicit exactly what nurses do on a daily basis, administrators have trouble rationally allocating tasks. Similarly, NIC is used in the development of nursing health care systems and communication with the classification systems of other health care providers. This coordination provides a safety net and planning vehicle for untracked costs.

The horizon is not fully clear, however. Wagner (1993), Egger and Wagner (1993), Gray, Elkan, and Robinson (1991), Strong and Robinson (1990), and Bjerknes and Bratteteig (1987a and 1987b) have studied the implementation of similar measures in Europe.[41] While these measures have the effect of making nursing work visible and differentiated, nurses may also become a target for social control and surveillance. Visibility here works against control in the sense of discretionary judgment and common sense. Wagner (1993) states that while computerization of care plans in French and Austrian hospitals is partly designed to give nurses greater scope of responsibility and legitimize their care giving in some detail, it also has another side:

The idea of computerized care-plans, as put forward in nursing research, is to strengthen the focus on nurses' own preplanned nursing "projects." Like "the autonomous profession," nurses are seen as setting apart time for specialized activities, irrespective of ad-hoc-demands . . . the reality of computerized care plans—even when nurses themselves have a voice in their development—may lag far behind this idea, given the authority structures in hospitals. With management focusing on care plans as instruments that may help them with their legal and accreditation issues, and nurses having to continue documenting their work on the KARDEX and other forms as well, care plans cannot unfold their potential. (Wagner 1993, 12)

Once designed, a classification system is therefore not a black box before it becomes part of nursing practice. The designers' balancing act needs to continue on every ward of every hospital.

### Professionalization, Classification Systems, and Nursing Autonomy

Since the focus of the NIC is on making nursing visible, along with balancing out control and comparability, it is interesting to compare the strategies chosen by the NIC researchers to fully professionalize nursing to the range of strategies discussed by Abbott (1988, Hughes 1970) in *The System of Professions*. Abbott puts the struggle for jurisdiction in a central place, and his model of "the cultural machinery of jurisdiction" (Abbott 1988, 59) characterizes professional work in terms of diagnosis, treatment, inference, and academic work. The very words are drawn from the medical profession; staking out a jurisdictional claim within that profession is particularly difficult—what is specific in a "nursing diagnosis" that differentiates it from "medical diagnosis?" He does not describe any other case where a central tool has been the creation of a classification system. Yet within the medical system as a whole, having access to one's own classification has long been a control strategy. Kirk and Kutchins (1992), for example, discuss jurisdictional disputes between the ICD and the DSM, and they show convincingly that the DSM became a tool for a particular theory of psychiatry, empowering more physiologically based models at the expense of psychological models.

To gain equity with the medical profession (where they have often been seen as subordinate), nursing research is an important aspect of legitimization. In turn, classification of work is a cornerstone of this research. Nursing classification creates the possibility of equivalence on the research end. Because nursing had long been defined as the undifferentiated other (everything that doctors do not do in the treatment of patients), it was impossible to create precise arguments for professionalization based on research results.

But as nursing differentiates and becomes more autonomous, it too creates its own undifferentiated other. In what sense? As Abbott emphasized, professionalization depends upon the scope of the professions' jurisdiction. For NIC this implies that if nurses define a number of activities as specifically to do with nursing, they also claim only these activities. Although the researchers mean to include all the activities that nurses do, it is impossible to be totally inclusive, as we have demonstrated. Regional variations and those activities that cut across professional domains cannot be articulated in an interventions classification system. Some activities may be left in residual categories, or left for other health care groups such as licensed vocational nurses and technicians. Implicit in the physician's classification systems was the assumption that nurses would perform any unaccounted work that would support the fit between the doctors' prescription and the patient's health.[42] Now that nurses are creating their own classification system, they too might rely in a changed fashion on the invisible and unaccounted work of others.

The NIC group hopes that their classification system will sensitize the entire health care sphere to the contribution that nurses make to the well being of patients. But the road to such an outcome is a difficult (and potentially even dangerous) one for nurses as a group, as Wagner has shown for the European example. For instance, it is possible that NIC might be used against nursing professionalization in some computerization and surveillance scenarios. Imagine a hospital administrator who has implemented NIC and evaluates what the nurses are doing. To curtail costs and adequately allocate resources, the administrator might reallocate nursing activities in ways that are putatively more cost efficient. When asked about this issue, one of the principal investigators Joanne McCloskey emphasized that it is more important that nurses deal with those questions instead of leaving them tacit. "It may create some problems, but it forces nursing into the mainstream and forces nurses to be responsible, accountable, health care providers. Then, of course, you have to deal with the questions that physicians have had to deal with for a long time. And we ought to be able to deal with that and find a good new solution" (see also McCloskey and Bulechek 1994b).

A classification system is an important tool in the struggle for professional recognition. When the tensions among visibility, comparability, and control are skillfully managed in the construction of the classification system itself, the same processes need to be balanced at the level of users and policymakers. NIC's goal is to promote the work of nurses by communicating newly visible (in the sense of inscribed and legitimated) work practices and by leaving enough space for controllable action. But even if the designers succeed in creating equilibrium at the information system level, there are potential utilization problems in the political arena. Professionalization through visibility alone may have latent consequences: constant surveillance in the name of the panopticon of cost containment (Foucault 1979). In this era of information infrastructure shifts, the significance of this scenario is enormous.

### Conclusion

A classification of work becomes, then, a political actor in the attempts to establish power on broad institutional and historical levels. When such a classification system intends to promote a professional group, the challenges are geared toward their ability to enhance professionalization. In the best case, classification systems hold a memory of work that has been done (laboratory, organizational, epidemiological, sociological) and so permit the recommendation of a reasonable due process for future work (Gerson and Star 1986).

It is difficult to retrace these processes after the classification is black boxed. We have been fortunate to observe an effort to classify work in its early days, coordinated by a group of American nursing researchers, which is beginning to spread to other locations as well. Their work exemplifies a profoundly skilled balancing act revolving around managing the trade-offs outlined above. The NIC project team has a global strategy of balanced classification through a series of sophisticated moves of differentiation and dedifferentiation. This strategy assumes that the work of producing equivalence (making other things equal) will reduce the overall amount of effort: retraining when a nurse needs to move into a new situation, introducing the nurse to the medical information system in a new hospital, and so on. It is linked with the strategy of the creation of a single information infrastructure to facilitate hospital operation.

A favorite metaphor of NIC members to describe their task is to make the invisible work visible. As the layers of complexity involved in its architecture reveal, however, a light shining in the dark illuminates certain areas of nursing work but may cast shadows elsewhere: the whole picture is a very complex one. NIC is at once an attempt at a universal standardized tool with a common language; at the same time, its development and application is proceeding via managing and articulating the local and particular. It is in that sense a boundary object between communities of practice, with a delicate cooperative structure (Star and Griesemer 1989). At the same time, it is balanced in a given workflow and historical period that makes it a potential target for control. The fact that NIC researchers are carefully involving a huge web of nurses and nursing researchers and building slowly over time, with revisions, is key to this process. The conservation of work inscribed in the static list of concepts and activities that form a classification system will be inserted into a field of ongoing practices, negotiations, and professional autonomy disputes. These practices and the political field in which they occur form the architecture of intimacy, manageability, and standardization. The local and macro contexts of the classification system and its attendant practices determine in the final instance the extent of the displacement of nursing work. In classification systems, differentiation and dedifferentiation emerge as a continuous and negotiated accomplishment over time. The same lesson holds for the organization of nursing work through NIC as for the coordination of medical organizations of all kinds through the ICD as discussed in chapter 4: it is not a question of mapping a preexisting territory but of making the map and the territory converge.

# 8

# Organizational Forgetting, Nursing Knowledge, and Classification

*Introduction—Well do I remember . . .*

The last chapter looked at the ways in which NIC operated within multiple agendas through strategies for balanced tensions and strategically protecting ambiguities. Here we turn to the question of what happens when the system is used to encode and classify current and past knowledge and store it for the future.

Classification schemes always have the central task of providing access to the past. They are used to order archives, libraries, and the presentation of knowledge. Indeed, Auguste Comte argued that a good classification scheme could supplant the need for detailed history, since it could encode all valuable knowledge. Thus, classification schemes are used for various kinds of recall. Recall is in general a problematic concept, however, even when one can assume that people are trying to tell the truth about the past. Studies of people's intensely remembered "flashbulb memories" (What were you doing when Kennedy was assassinated?) have proved them to be often false (Brown and Kulik 1982). White House Counsel John Dean claimed fairly total recall at the time of Watergate. Ulric Neisser points out in his analysis of the tapes made in the Oval Office, however, that Dean remembered neither conversations nor even gists of conversations. Rather, Dean encoded an ideal set of possible conversations that embodied his perceived truth of the situation and his fantasies about his own role therein (Neisser 1982, see also the excellent critique in Edwards and Potter 1992[43]).

People cannot generally remember accurately how they felt in the past. They take the present as a benchmark and then work from a currently held belief about change or stability in their attitudes. Thus, when asked how they felt six months ago about, say, a TV series, their memories will necessarily be colored by what has happened since in that series (Linton 1982, Strauss 1959). It is hard to remember back past an act of infamy: my enemy today was always my enemy (I distrusted them when they appeared to be my friend, as I now recall). If on the other hand—as happens perhaps more often than inversely—my friend today was once an enemy, then I can tell you a conversion story that recasts their past, my past, or both.

If all history is in this sense history of the present, then one might surely think of memory as ineluctably a construction of the present. These studies from cognitive science and social psychology suggest that truth or falsity is not a simple concept when it comes to analyzing organizational memory in science or elsewhere (compare Hacking 1995: chapter 17 on the indeterminacy of the past). Thus Bannon and Kuutti (1996) stress that if "organizational memory" is at all a useful concept, it is so to the extent that it refers to active remembering that carries with it its own context. The memory comes in the form not of true or false facts but of multifaceted stories open to interpretation.

Neisser (1982), building on Tulving's distinction between episodic and semantic memory (remembering what versus remembering how) introduces a third kind of memory. *Repisodic* memory means remembering what was actually happening; by all accounts, including Neisser's own, this is an elusive positivist goal.

Against this increasing differentiation and specialization in the concept of memory, we find a single and undifferentiated definition of "forgetting." Forgetting is just "not remembering." Further, forgetting in all its guises has frequently been seen as necessarily a problem to be solved. Freud encouraged the recall of suppressed memories (see Hacking 1995 for a discussion of memory and veracity in Freud). Historians insist that we must learn the lessons of the past. Yrjö Engeström, in his paper on "organizational forgetting" (1990a) discusses problems raised by the ways in which doctors forget selectively and always linked with current exigencies. His activity theoretical perspective on the organic links between internal and external memory traces is particularly fruitful in that it provides a model for ethnographic studies of collective memory. But he still gives forgetting a negative spin. Bitner and Garfinkel (1967) are among the few to observe and describe a positive ecology of forgetting in their account of "good" organizational reasons for "bad" clinical records. Psychoanlysts do this as well to some degree—concepts such as repression and denial sound more negative than they are technically.

Total recall, in individuals or organizations, is neither desirable nor possible. There are indeed several good reasons for organizations to forget things about their own past. First, it might be the case that rediscovery is easier than remembering. This is especially so where the overhead of constructing a sufficiently precise archive, for a fine-grained situational memory, is high. For example, airline companies frequently do not retain a record of one's food or seating options. They process passengers anew each time they are encountered, which is easier from a data processing perspective.

Extending Chandler (1977), one can see the development of statistics as a filtering mechanism that allows a central office not to have to remember everything about a company's day-to-day running to make things run smoothly. The filtering works as proactive forgetting. Railroad companies do not need to know which particular piece of rolling stock is located where, but simply how many pieces of such and such a kind there are at any given location.

Another positive mode of forgetting occurs when an organization wants to change its identity. Here the argument that "we have always done things this way" stands in the way of breaking new ground. Hughes (1989 [1883]) described the change at Rugby school under Arnold in this light. He showed how Arnold imperceptibly changed the way things were done in this tradition-bound institution, such that group memory was never mobilized against the changes. Recent work in organizational theory has suggested that perhaps it is good on occasion to forget everything about the past to start over without being trapped in old routines (Wackers 1995). In general, if a set of archives indexed by a given classification scheme is being used as a tool of reification or projection, then it can have harmful consequences.

This chapter describes how organizations use classification schemes to selectively forget things about the past in the process of producing knowledge. We argue that there are two major kinds of organizational forgetting in the process of producing and then maintaining classification systems in the workplace:

- Clearance—the erection of a barrier in the past at a certain point so that no information or knowledge can leak through to the present.

- Erasure—the ongoing destruction of selective traces in the present.

Standardized classification systems may permit the organization to move from heterogeneous forms of memory operating within multiple frameworks to the privileging of a form of memory (potential memory) operating within a well-defined information infrastructure subtended by classification systems. In this process, the decision of whether to opt in to an infrastructure, with its attendant memory frames and modes of forgetting, or to stay out of it is of great political and ethical import. We first follow this set of arguments through with a case study of the development of NIC and then broaden the discussion out to more general considerations of classification and memory.

### Nursing Classifications and Organizational Forgetting

Nursing is particularly interesting with respect to forgetting. Nursing work has traditionally been invisible, and its traces removed at the earliest opportunity from the medical record. In general, the nursing profession has not been able as an institution to draw on an active archived memory. Rather, nursing has been seen as an intermediary profession that does not need to leave a trace; in accord with traditional gender expectations, nurses are "on call" (Star and Strauss 1999). As nursing informatician Castles notes, citing Huffman on medical records management, "the nursing records are the first to be purged from the patient records; there is thus no lasting documentation of nursing diagnoses or nursing interventions and no method of storage and retrieval of nursing data" (Castles 1981, 42).

There was a primal act of clearance in the very establishment of NIC. By clearance we mean a complete wiping away of the past of nursing theory in order to start with a clean slate (we draw here on Serres' (1993) work on clearance and origins in geometry). The nurses said that until now there had been no nursing science and therefore there was no nursing knowledge to preserve. One nursing informatician ruefully noted: "It is recognized that in nursing, overshadowed as it is by the rubrics of medicine and religion, no nurse since Nightingale has had the recognized authority to establish nomenclature or procedure by fiat. There are no universally accepted theories in nursing on which to base diagnoses, and, in fact, independent nursing functions have not yet gained universal acceptance by nurses or by members of other health professions" (Castles 1981, 40). Nursing, it was argued, had until now been a profession without form; nothing scientific *could* be preserved. There was no way of coding past knowledge and linking it to current practice. A conference was held to establish a standardized nursing minimum data set (information about nursing practice that would be collected from every care facility). It found that "the lists of

## Clearance and the Past

Grand historiographer Sima Qin (1994 [ca 100 B.C.]), writing of the burning of the books in 213 B.C., notes that the chief minister advised the emperor that "all who possess literature such as the Songs, the Documents, and the sayings of the hundred schools should get rid of it without penalty. If they have not got rid of it a full thirty days after the order has reached them, they should be branded and sent to do forced labor on the walls. There should be exemption for books concerned with medicine, pharmacy, divination by tortoise-shell and milfoil, the sowing of crops, and the planting of trees." In response to this, the emperor ordered the famous burning of the books. Citing Qin, "the First Emperor collected up and got rid of the Songs, the Documents, and the sayings of the hundred schools to make the people stupid and ensure that in all under heaven there should be no rejection of the present by using the past. The clarification of laws and regulations and the settling of statutes and ordinances all started with the first Emperor. He standardized documents."

interventions for any one condition are long partially because nursing has a brief history as a profession in the choosing of interventions and lacks information for decision making. As a profession, nursing has failed to set priorities among interventions; nurses are taught and believe they should do everything possible" (McCloskey and Bulechek 1992, 79).

In the face of this view of the nurse as the inglorious other—doing everything that no one else does—should all previous nursing knowledge be abandoned? William Cody, in an open letter to the Iowa Intervention Team who produced NIC (published in *Nursing Outlook* in 1995) charged that this was precisely what would follow from widespread adoption of NIC:

It would appear that the nursing theorists who gave nursing its first academic leg to stand on, as it were, are deliberately being frozen out. I would like to ask Drs. McCloskey and Bulechek, why is there no substantive discussion of nursing theory in your article? How can you advocate standardizing "the language of nursing" by adopting the language of only one paradigm? How do you envision the relationship between the "standardized" masses and those nurse scholars with differing views? (Cody 1995, 93)

The project team responded that indeed clearance was an issue. "The Iowa group contends that taxonomic development represents a radical shift in theory construction in which the grand conceptual models are not debated, but transcended. We believe that, as a scientific community, nursing has moved to the point of abandoning the conceptual models of nursing theorists as forming the science base of the discipline" (McCloskey, Bulechek, and Tripp-Reimer 1995, 95). It is not just at the level of nursing theory that this act of clearance is seen as unsettling. Practicing nurses implementing NIC at one of four test bed sites had several complaints. They stated that learning to use NIC together with the new computer system in which it was embedded was like going to a foreign country where you had to speak the language; to make matters worse, you had to go to a new country every day. More prosaically, they said that they felt they were going from being experts to novices.[44]

The argument was made that there has been no comparative work done in the past. "The discipline of nursing has not yet constructed a cohesive body of scientific knowledge" (Tripp-Reimer et al. 1996, 2). There is a complexity here, however, that often arises in connection with the strategy of clearance. One wants to be able to say that nurses now do something that is valuable and adaptable to scientific principles. At the same time, they maintain that nurses have not yet (until the development of the classification system) been able to develop any nursing theory and thence any systematic, scientific improvement in practice.

This difficulty is a general problem when new classification schemes are introduced. New schemes effectively invalidate much previous knowledge by creating new sets of categories. Yet, they seek to draw on the authority of the outdated knowledge while simultaneously supplanting it.

This same article, concerning the dimensional structure of nursing interventions, tackles this problem directly. Tripp-Reimer argues that there must be a cycle of forgetting in the development of the new classification scheme. The article begins with a quote from Chung Tzu:

The purpose of a fish trap is to catch fish. When the fish are caught, the trap is forgotten.
The purpose of a rabbit snare is to catch rabbits. When the rabbits are caught, the snare is forgotten.
The purpose of words is to convey ideas. When the ideas are grasped, the words are forgotten.
Seek those who have forgotten the words. (Tripp-Reimer et al. 1996, 2)

The authors argue here that the traditional grand theories had a "certain limited utility beyond their historical importance" in that they provided a structure for educational programs. In the field, however, expert nurses soon "forgot" these words and developed their own rubric to get at the deep structure of the nursing situation (there is indeed a reference to transformational grammar here). Using NIC categories as a research tool, one could uncover the three key dimensions of nursing work (the intensity, focus, and complexity of care) that experts already knew about without there having been a nursing science. Having passed through the purifying cycle of forgetting, one could finally "bring intuitive clinical decision making to a conscious level."

There is a double complexity to this cycle. The first is the fact that the first author, Toni Tripp-Reimer, is a cultural anthropologist turned nursing informatician well versed in Kuhn, Lakoff, and other philosophers of science and language. The organization that produces NIC has to be broadly enough construed, on occasion, to include the community of sociologists of science and linguists, even though this inclusion may never be represented overtly in the records of the classification scheme. In passing, these alliances can form a kind of organizational memory that becomes instead forgetting. It means storing information in locations once within the network of an organization but now outside of it; a variety of outsourcing gone sour. The alliances may be fragile, or historical circumstances may change. Thus, for example, the problem of using a centralized external memory source like the library at Alexandria. . . .

The second complexity is that de novo classifications reflect a bootstrapping between what practicing nurses already know and what the science of nursing will tell them. Thus, to get the category of culture brokerage in NIC (see figure 7.5), Tripp-Reimer had to get it into the research literature as something that was already being done by nurses (and indexed in databases!). The NIC team in general claims both that nursing is already a science and that it is a science that has not yet been formulated. They need both points for their project. That is, they need to maintain the former to justify the profession against current attacks and the latter to justify their classification system, which when in place will protect it from future attacks.

One is reminded of Piaget's (1969) assertion that our earliest intuitions are of the relativistic nature of time, and that we need to unlearn our school lessons both to access the latest science and to get back in touch with our childhood insights. The point here is to suggest that unlearning, like forgetting, may be a more pervasive feature of organizational and cognitive life than accounts of learning and of memory

---

## Culture Brokerage                                    7330

---

**DEFINITION:** Bridging, negotiating, or linking the orthodox health care system with a patient and family of a different culture

---

**ACTIVITIES:**

Determine the nature of the conceptual differences that the patient and nurse have of the illness

Discuss discrepancies openly and clarify conflicts

Negotiate, when conflicts cannot be resolved, an acceptable compromise of treatment based on biomedical knowledge, knowledge of the patient's point of view, and ethical standards

Allow the patient more than the usual time to process the information and work through a decision

Appear relaxed and unhurried in interactions with the patient

Allow more time for translation, discussion, and explanation

Use nontechnical language

Determine the "belief variability ratio"—the degree of distance the patient sees between self and cultural group

Use a language translator, if necessary (e.g., signing, or Spanish)

Include the family, when appropriate, in the plan for adherence with the prescribed regimen

Translate the patient's symptom terminology into health care language that other professionals can more easily understand

Provide information to the patient about the orthodox health care system

Provide information to the health care providers about the patient's culture

---

**BACKGROUND READINGS:**

Caudle, P. (1993). Providing culturally sensitive health care to Hispanic clients. Nurse Practitioner, 18(12), 40-51.

Jackson, L.E. (1993). Understanding, eliciting, and negotiating clients' multicultural health beliefs. Nurse Practitioner, 18(4), 36-42.

Rairdan, B., & Higgs, Z.R. (1992). When your patient is a Hmong refugee. American Journal of Nursing, 92(3), 52-55.

Sloat, A.R., & Matsuura, W. (1990). Intercultural communication. In M.J. Craft & J.A. Denehy (Eds.), Nursing Interventions for Infants and Children (pp. 166-180). Philadelphia: W.B. Saunders.

Tripp-Reimer, T., & Brink, P.J. (1985). Culture brokerage. In G.M. Bulechek & J.C. McCloskey (Eds.), Nursing Interventions: Treatments for Nursing Diagnoses (pp. 352-264). Philadelphia: W.B. Saunders.

*Figure 8.1*

Culture brokerage, NIC. An intervention requiring the nurse to mediate between medical belief systems.

Source: *NIC*, second edition.

---

### Erasure and the Present

Donald Crowhurst went quietly mad on a round-the-world yachting race and lay becalmed on the ocean developing a theory of the cosmic mind while at the same time completing and radioing in an immaculate official log that had him winning the race at a record pace. Crowhurst's double log surfaces within his madness as contemplation on the nature of time: "The Kingdom of God has an area measured in square hours. It is a kingdom with all the time in the world—we have used all the time available to us and must now seek an imaginary sort of time." (Tomalin and Hall 1970, 259).

---

might lead us to believe. The act of clearance is to take away useless theory; then ethnographic work will uncover the true science (always already there) that NIC can express. The act of clearance, therefore, is not one of simple denial of the past, though complex historical narratives need to be constructed to distinguish the two.

We do not accept the position that such clearance leads to the creation of some sort of truer science—the issue of the validity of nursing knowledge is entirely orthogonal to our purpose. We are producing an anatomy of what it has meant in the case of nursing work to create such a science. This is not an accidental feature of their work but can be seen as a core strategy over the centuries in the creation of sciences through the establishment of stable classification schemes. The strategy itself provides a way of managing a past that threatens to grow out of control. One can declare by fiat that the past is irrelevant to nursing science, while, in Tripp-Reimer's case, validating the past as embodied in current best practice. The development of a classification scheme will provide for a good ordering of memory in the future so that nothing henceforth deemed vital will be lost.

This claim that nothing vital will be lost is strategically important but largely unverifiable for two reasons. First, the classification scheme itself forms a relatively closed system with respect to the knowledge that it enfolds. Thus, in Latour's terms, it resists trials of strength. It becomes difficult to stand outside of it and demonstrate that something is being selectively deleted or overlooked from the archive it supports. Second, even if the classification scheme is in principle robust, it is by definition hard to remember what has been removed from the archive when the archive itself is basically the only memory repository at hand.

With the strategy of clearance, we saw the complete wiping clean of a historical slate. This made it possible for a single origin for nursing science to be created. From that point of origin nursing actions could be coded and remembered in an organizationally and scientifically useful fashion. A second mode of directed forgetting in organizations is erasure: the constant filtering out of information deemed not worthy of preserving for the organization's future purposes.

Historically, the selective erasure of nursing records within hospital information systems has been drastic. Nursing records are the first destroyed when a patient is released. The hospital administration does not require them (nursing is lumped in with the price of the room), doctors consider them irrelevant to medical research, and nursing theorists are not well enough entrenched to demand their collection. Huffman (1990: 319), in a standard textbook on medical records management writes:

As nurses' notes are primarily a means of communication between the physicians and nurses, they have served their most important function during the episode of care. Therefore, to reduce the bulk and make medical records less cumbersome to handle, some hospitals remove the nurses' notes from records of adult patients when medical record personnel assemble and check the medical record after discharge of the patient. The nurses' notes are then filed in chronological order in some place less accessible than the current files until the statute of limitations has expired and they are destroyed. (Huffman 1990, 319)

Traditionally nurses have been facilitated out of the equation: though they may not have an official trace of their own past, their duty is to remember for others. In one of those vague but useful generalizations that characterize information statistics, it was asserted, in a book on next-generation nursing information systems, that 24 percent of total hospital operating costs were devoted to information handling. Nursing "accounted for most of the information handling costs (28 percent to 34 percent of nurses' time);" and what is worse, "in recent years, external regulatory factors, plus increasing organizational and health care complexity, have augmented the central position of information in the health care environment" (Zielstorff et al. 1993, 5). The nursing profession acts as a distributed memory system for doctors and hospital administrators. Ironically, in so doing, it is denied its own official memory.

Even when the erasure is not mandated, it has been voluntary. One text on a nursing classification system cites as a motif of the profession an observation that "the subject of recordkeeping has probably never been discussed at a convention without some agitated nurse arising to ask if she is expected to neglect her patients to write down information about them . . ." (Martin and Scheet 1992, 21, echoing a 1917 source). And Joanne McCloskey, one of the two principal architects of NIC notes that "the most convincing argument against nursing service or Kardex care plans is the absence of them. Although written care plans are a requirement by the Joint Commission for Hospital Accreditation and a condition for participation in Medicare, few plans are, in fact, written" (McCloskey 1981, 120). In her study of the ICD, Ann Fagot-Largeault (1989) notes the same reluctance on the part of doctors to spend time accurately filling in a death certificate (itself a central tool for epidemiologists) when they might be helping live patients. Thus there is, in Engeström's (1990a) terms, a block between internal memory and external memory. Because representational work takes time, those filling out forms systematically erase the complex representations that they hold in their heads in favor of summary ones. In the case of the ICD, there are many complaints about the quality of data, due to the overuse of general disease terms or "other" categories. In the case of a computerized NIC, nurses are sometimes suspected by the NIC implementation team of using the choices that appear before them on a screen (which they can elect with a light pen) rather than searching through the system for the apt descriptor (IIP 6/8/95).

One of the main problems of nurses is that they are trying to situate their activity visibly within an informational world that has factored them out of the equation. It has furthermore maintained that they should be so excluded, since what nurses do can be defined precisely as that which is not measurable, finite, packaged, or accountable. In nursing theorist Jenkins' terms, "nurses have functioned in the post-World War II era as the humanistic counterbalance to an increasingly technology-driven medical profession" (Jenkins 1988, 92). Nursing informaticians face a formidable task. They have tried to define nursing as something that fits naturally into a world partly defined by the erasure of nursing and other modes of invisible and articulation work. This is parallel with technicians who seek new ways of writing scientific papers in a way that their work is acknowledged, and yet neither the nature of scientific truth nor its division of labor remains intact.

Sometimes the nurses are driven for these reasons by their own logic to impeach medical truth. At other times they challenge orthodoxy in organization science, or they seek to restructure nursing so that these challenges will not be necessary. In the end, there will be an information infrastructure for medical work that contains an account of nursing activity. The move to informational panopticons is overwhelming in this profession as in many others. With projects like NIC—which offer new classification systems to embed in databases, tools, and reports—we get to see what is at stake in making invisible work visible.

This section has explored two strategies, clearance and erasure. Some of the points made here in the context of organizational forgetting relate to arguments within the sociology of science about scientific representations of nature. These include the idea of deleting the work (Star 1991a, Shapin 1989) and the deletion of modalities in the development of scientific texts (Star 1983, Latour 1987). That is, as a scientific statement gets ever closer to being accepted as fact, historical contingencies get progressively stripped from its enunciation.

Why, then, talk at all about memory and forgetting when representation and its literature can do much of the same work? The concept of representation tends naturally to abstract away the ongoing work of individual or organizational agents (compare here Woolgar 1995, 163). It is difficult to express the fact that the representation can have different meanings at different times and places in the organization in a language that has been used rather to demonstrate the conjuring of a single articulation of "fact." The act of remembering a fact organizationally involves not only mobilizing a set of black-boxed allies (in Latour's terms) but also translating from the context of storage to the present situation (one might store a fact for reason x but recall it for reason y). This is a central problem for most classification systems operating as information storage tools. Within an organizational context, it is easier to explore the distribution of memory and forgetting than the distribution of representation.

Finally, there it is always a temptation when talking of representation to fall into a cognitivist trap of assuming the primacy of thinking. By concentrating on "following the actors," sociologists of science have as a rule produced a language that privileges the scientific "fact" and its circulation. They have in this put the infrastructure supporting that fact relatively into the background (what goes on inside the black box, or indeed what black boxes look like, is seen as irrelevant). From the perspective of organizational memory, a modality can be deleted in a number of different ways. It might be distributed (held in another part of the organization than in that which produces the text); it might be

---

### How to Forget

In a work reminiscent of Frances Yates' (1966), Fentress and Wickham argue that artificial memory systems went on the wane after Descartes. "Instead of a search for the perfectly proportioned image containing the 'soul' of the knowledge to be remembered, the emphasis was on the discovery of the right logical category. The memory of this system of logical categories and scientific causes would exempt the individual from the necessity of remembering everything in detail. . . . The problem of memorizing the world, characteristic of the sixteenth century, evolved into the problem of classifying it scientifically."

(James Fentress and Chris Wickham 1992, 13)

---

built into the infrastructure (the work environment is changed such that the modality is never encountered); or simply dismissed. Looking at ways of distributing memory and operating forgetting we can, therefore, look in more fine-grained detail at what happens as the representation moves into and out of circulation.

Clearance is a strategy employed internally within the profession of nursing as a tool for providing an origin for the science of nursing. Erasure is employed externally on the profession of nursing as a tool for rendering nursing a transparent distributed memory system. The logic of the relationship between clearance and erasure has been that the nurses are operating the clearance of their own past in order to combat the erasure of their present in the records of medical organizations. Medical information systems, they argue, should represent the profession of nursing as if it just began yesterday. Otherwise, they will copy the transparency of nursing activity from one representational space (the hospital floor and paper archives) to another (the electronic record). This poses, then, the question of what happens when a new ecology of attention (what can be forgotten and what should be remembered) is inaugurated with the development of a new information infrastructure.

Memory—individual and organizational—is in general filtered through classification systems. Such systems permit encoding of multiple bits of information about the environment into a single coherent framework (see Schachter 1996, 98–133). Edouard Clarapède (who performed the initial notorious experiment of having a stranger rush into the classroom, do something outrageous, and then have students describe what happened) noted as early as 1907 that "the past—even of a simple event—was less a record than a sort of taxonomy. Not perceptions, but categorization of familiar types was the major function of memory" (cited in Matsuda 1996, 109).

Any complex information infrastructure—paper or electronic, formal or informal—claims by its nature to contain all and only the information that is needed for the smooth running of that organization. Organizations frequently want to know everything relevant about some past action. For example, if there is a blackout along the West Coast due to a tree falling in Idaho, an awful amount of information needs to be recalled and synthesized for the connection to be made. Frequently, a prime function of recordkeeping in the organization is to keep track of what is going on so that, should anyone ever want to know (auditors, a commission of inquiry, and so forth), a complete reconstruction of the state of the organization at a particular moment can be made. For instance, Hutchins (1995, 20) talks about the role of the logs kept by navy ships of all their movements. "Aboard naval vessels . . . records are always kept—primarily for reasons of safety, but also for purposes of accountability. Should there be a problem, the crew will be able to show exactly where the ship was and what it was doing at the time of the mishap." For something to be remembered officially by an organization, however, it must be recorded on a form. Forms necessarily impose categories (Berg and Bowker 1997).

No reconstruction will cover literally everything that was going on at a particular moment. Rather, it will capture primarily objects that fit into the organization's accepted classification scheme of relevant events. The kind of memory that is encoded in an organization's files for the purposes of a possible future reconstruction could be called "potential memory." We are using the word potential to draw attention to the distributed, mediated nature of the record. No one person remembers everything about a medical intervention, and generally it can be processed through an organization without ever having been recalled. There is a possible need to recall any one intervention in huge detail, however, and the only way that the possible need can be met is through the construction of a classification system that allows for the efficient pigeonholing of facts.

Within the hospital, nursing work has been deemed irrelevant to any possible future reconstruction; it has been canonically invisible (Star 1991a, Star and Strauss 1999). The logic of NIC's advocates is that what has been excluded from the representational space of medical practice should be included.

Operating within the space of erasure that is at once home for them and a threat to their continued existence, the nurses in Iowa have thought long and hard about the politics and philosophy of classifying their activities so that they fit into the hospital's potential memory. They do not want to flip over from being completely invisible to being far too visible. They have decided to name their tasks, but not to name too much at too fine a grain of detail. To this end they have adopted their own practice of continuing partial erasure (where they limit the nature and scope of erasure) for three reasons:

• From within the exercise of the profession of nursing, to recognize local differences and protect local autonomy (so central to the nursing self-image) while providing the necessary degree of specification for entry into the world of potential memory. They have decided to specify only down to the level of interventions, but to leave the subcategories of activities as relatively fluid—several possibly contradictory activities are subsumed under a single intervention (see figure 7.1 for example).

• From within the hospitals of which nursing constitutes one administrative unit, to protect the nurses from too much scrutiny by accountants. It is harder to set off aspects of nursing duties and give them to lower paid adjuncts if that work is relatively opaque. The test sites that are implementing NIC have provided some degree of resistance here. They argue that activities should be specified so that, within a soft-decision support model, a given diagnosis can trigger a nursing intervention of a single, well-defined set of activities. As Marc Berg (1997b) has noted in his study of medical expert systems, such decision support can only work universally if local practices are rendered fully standard. A key professional strategy for nursing—particularly in the face of the ubiquitous process to reengineer—is realized by deliberate nonrepresentation in the information infrastructure. What is remembered in the formal information systems resulting is attuned to professional strategy and to the information requisites of the nurses' take on what nursing science is.

• From an information systems perspective, to ensure that information does in fact get recorded on the spot. There is a brick wall that such systems encounter when dealing with nurses on the hospital floor. If they overspecify an intervention (break it down into too many constituent parts), then it will be seen as an NSS classification—one that is too obvious. The project team sees the classification scheme as having to be very prolix at present; but when the practice of nursing itself is fully standardized, some of the words will be able to wither away. They point to intervention classifications used by doctors that are much less verbose—and can afford to be, they argue—because every doctor knows the standard form of treatment for, say, appendicitis. (Though they also argue that there are local variations in medical practice that have been picked up by good reporting procedures and that NIC will be able to provide such a service for nursing, leading to an improvement in the quality of practice). It is assumed that any reasonable education in nursing or medicine should lead to a common language wherein things do not need spelling out to any ultimate degree. The information space will be sufficiently well prestructured that some details can be assumed. Attention to the finer-grained details is delegated to the educational system where it is overdetermined.

These NIC erasure strategies—dealing with overspecification and the political drive to relative autonomy by dropping things out of the representational space—are essential for the development of a successful potential memory. Partial erasure of local context is needed to create the very infrastructure in which nursing can both become a science like any other and yet nursing as a profession can continue to develop as a rich, local practice. The ongoing erasure is guaranteed by the classification system. Only information about nursing practice recognized by NIC or by other classification schemes in use can be coded on the forms fed into a hospital's computers or stored in a file cabinet.

### Granularity and Politics

Nursing informaticians agree as a body that for proper health care to be given and for nursing to be recognized as a profession, hospitals should code for nursing within the framework of their memory systems. Nursing work should be classified and forms should be generated that utilize these classifications. There has been one notable disagreement, however, with respect to the best strategy for coding nursing work into memory systems.

To understand the difference that has emerged, recall one of those forms you have filled in that does not allow you to say what you think. You may, in a standard case, have been offered a choice of several racial origins, but may not believe in any such categorization. There is no room on the form to write an essay on race identity politics. So either you make an uncomfortable choice to get counted, and hope that enough of your complexity will be preserved by your set of answers to the form, or you do not answer the question and perhaps decide to devote some time to lobbying the producers of the offending form to reconsider their categorization of people. The NIC group has wrestled with the same strategic choice. It must fit its classification system into the Procrustean bed of all the other classification systems with which it must articulate in any given medical setting. Only thus may it come to form a part of a given organization's potential memory. The other choice is to reject the ways in which memory is structured in the organizations with which they are dealing.

Let us look first at the argument for including NIC within the potential memory framework of the hospital. The nursing team argues that NIC has to respond to multiple important agendas simultaneously. Consider the following litany of needs for a standard vocabulary of nursing practice:

It is essential to develop a standardized nomenclature of nursing diagnoses to name without ambiguity those conditions in clients that nurses identify and treat without prescription from other disciplines; such identification is not possible without agreement as to the meaning of terms. Professional standards review boards require discipline-specific accountability; some urgency in developing a discipline-specific nomenclature is provided by the impending National Health Insurance legislation, since demands for accountability are likely both to increase and become more stringent following passage of the legislation. Adoption of a standardized nomenclature of nursing diagnoses may also alleviate problems in communication between nurses and members of other disciplines, and improvement in interdisciplinary communication can only lead to improvement in patient care. Standardization of the nomenclature of nursing diagnoses will promote health care delivery by identifying, for legal and reimbursement purposes, the evaluation of the quality of care provided by nurses; facilitate the development of a taxonomy of nursing diagnoses; provide the element for storage and retrieval of nursing data; and facilitate the teaching of nursing by providing content areas that are discrete, inclusive, logical, and consistent. (Castles 1981, 38)

We have cited this passage at length since it incorporates most of the motivations for the development of NIC. The development of a new information infrastructure for nursing, heralded in this passage, will make nursing more "memorable." It will also lead to a clearance of past nursing knowledge—henceforth prescientific—from the textbooks, it will lead to changes in the practice of nursing (a redefinition of disciplinary boundaries), and to a shaping of nursing so that future practice converges on potential memory.

Many nurses and nursing informaticians are concerned that the profession itself may have to change too much to meet the requirements of the information infrastructure. In her study of nursing information systems in France, Ina Wagner speaks of the gamble of computerizing nursing records, "Nurses might gain greater recognition for their work and more control over the definition of patients' problems while finding out that their practice is increasingly shaped by the necessity to comply with regulators' and employers' definitions of 'billable categories'" (Wagner 1993, 7). Indeed, a specific feature of this "thought world" into which nurses are gradually socialized through the use of computer systems is the integration of management criteria into the practice of nursing. Wagner continues: "Working with a patient classification system with time units associated with each care activity enforces a specific time discipline on nurses. They learn to assess patients' needs in terms of working time."[45] This analytic perspective is shared by the Iowa nurses. They argue that documentation is centrally important; it not only provides a record of nursing activity but also structures the activity at the same time:

While nurses complain about paperwork, they structure their care so that the required forms get filled out. If the forms reflect a philosophy of the nurse as a dependent assistant to the doctor who delivers technical care in a functional manner, this is the way the nurse will act. If the forms reflect a philosophy of the nurse as a professional member of the health team with a unique independent function, the nurse will act accordingly. In the future, with the implementation of price-per-case reimbursement vis-à-vis diagnosis related groups, documentation will become more important than ever. (Bulechek and McCloskey 1985, 406)

As the NIC classification has developed, observes Joanne McCloskey, the traditional category of "nursing process" has been replaced by "clinical decision making plus knowledge classification." And in one representation of NIC that she produced, both the patient and the nurse had dropped entirely out of the picture (both were, she said, located within the "clinical decision making box" on her diagram) (IIP 6/8/95). A recent book about the next generation nursing information system argued that the new system:

cannot be assembled like a patchwork quilt, by piecing together components of existing technologies and software programs. Instead, the system must be rebuilt on a design different from that of most approaches used today: it must be a data-driven rather than a process-driven system. A dominant feature of the new system is its focus on the acquisition, management, processing, and presentation of "atomic-level" data that can be used across multiple settings for multiple purposes. The paradigm shift to a data-driven system represents a new generation of information technology; it provides strategic resources for clinical nursing practice rather than just support for various nursing tasks. (Zielstorff et al. 1993, 1).

This speaks to the progressive denial of process and continuity through the segmentation of nursing practice into activity units. Many argue that to "speak with" databases at a national and international level just such segmentation is needed. The fear is that unless nurses can describe their process this way (at the risk of losing the essence of that process in the description), then it will not be described at all. They can only have their own actions remembered at the price of having others forget, and possibly forgetting themselves precisely what it is that they do.

Some nursing informaticians have chosen rather to challenge the existing memory framework in the medical organizations they deal with. They have adopted a Batesonian strategy, responding to the threat of the new information infrastructure by moving the whole argument up one level of generality and trying to supplant data-driven categories with categories that recognize process on their own terms. Thus the Iowa team pointed to the fact that women physicians often spend longer with patients than male doctors; however, these physicians need to see patients less often as a result. The female physicians argue that just such a process-sensitive definition of productivity needs to be argued for and implemented in medical information systems so that nursing work be fairly represented (IIP 6/8/95). They draw from their tacit (because unrepresented) reservoir of knowledge about process to challenge the data-driven models from within.

Within this strategy the choice of allies is by no means obvious. Since with the development of NIC we are dealing with the creation of an information infrastructure, the whole question of how and what to challenge becomes very difficult. Scientists can only deal with data as presented to them by their information base just as historians of previous centuries must rely heavily on written traces. When creating a new information infrastructure for an old activity, questions have a habit of running away from one. A technical issue about how to code process can become a challenge to organizational theory and its database. A defense of process can become an attack on the scientific world-view. Susan Grobe, a nursing informatician, has made one of the chief attacks on the NIC scheme. She believes that rather than standardized nursing language, computer scientists should develop natural language processing tools so that nurse narratives can be interpreted. Grobe argues for the abandonment of any goal of producing "a single coherent account of the pattern of action and beliefs in science" (Grobe 1992, 92). She goes on to say that "philosophers of science have long acknowledged the value of a multiplicity of scientific views" (92). She excoriates Bulechek and McCloskey, architects of NIC, for having produced work "derived from the natural science view with its hierarchical structures and mutually exclusive and distinct categories" (93). She on the other hand is drawing from cognitive science, library science, and social science (94). Or again, a recent paper on conceptual considerations, decision criteria, and guidelines for the Nursing Minimum Data Set cited Fritjof Capra against reductionism, Steven Jay Gould on the social embeddedness of scientific truth, and praised Foucault for having developed a philosophical system to "grapple with this reality" (Kritek 1988, 24). Nurse scientists, it is argued, "have become quite reductionistic and mechanistic in their approach to knowledge generation at a time when numerous others, particularly physicists, are reversing that pattern" (ibid., 27). And nursing has to find allies among these physicists. "Nurses who deliver care engage in a process. It is actually the cyclic, continuous repetition of a complex process. It is difficult, therefore, to sketch the boundaries of a discrete nursing event, a unit of service, and, therefore, a unit of analysis. Time is clearly a central force in nursing care and nursing outcomes. Nurses have only begun to struggle with this factor. It has a centrality that eludes explication when placed in the context of quantum physics" (ibid., 28). The point here is not whether this argument is right or wrong. It is an interesting position. It can only be maintained, as can many of the other possible links that bristle through the nursing informatics literature, because the information infrastructure itself is in flux. When the infrastructure is not in place to provide a seemingly "natural" hierarchy of levels, then discourses can and do make strange connections among themselves.

To not be continually erased from the record, nursing informaticians are risking either modifying their own practice (making it more data driven) or waging a Quixotic war on database designers. The corresponding gain is great, however. If the infrastructure is designed in such a way that nursing information has to be present as an independent, well-defined category, then nursing itself as a profession will have a much better chance of surviving through rounds of business process reengineering and nursing science as a discipline will have a firm foundation. The infrastructure assumes the position of Bishop

Berkeley's God: as long as it pays attention to nurses, they will continue to exist. Having ensured that all nursing acts are potentially remembered by any medical organization, the NIC team will have gone a long way to ensuring the future of nursing.

### Classification Systems: Potential Memory and Forgetting

Three social institutions, more than any others, claim perfect memory: the institutions of science, the law, and religion. The legal and clerical professions claim perfect memory through an intricate set of reference works that can be consulted for precedence on any current case. The applicability of past to present is a matter of constant concern often argued in the law courts or in theological disputes. Scientific professionals, though, have often claimed that by its very nature science displays perfect memory. Furthermore they structure their recall primarily through a myriad of classification systems that give them a vast reserve of potential memory. Scientific articles are in principle— though never in practice—encoded in such a way that, hopefully, an experiment performed one day in Pesotum, Illinois, can be entirely replicated 100 years later in Saffron Walden, England.

The chapter now draws some more general conclusions about the ways in which classification systems structure memory within organizations, taking as a chief example the nature and operation of classification systems in science. There are two major reasons for choosing the institution of science for a wider discussion—the NIC development team claims to be rendering nursing scientific, and so these wider examples develop naturally out of that section above; and classification work has been more formalized in science than in other institutions.

It can readily be accepted that great discoveries were made but not recognized as such at the time (the cases of Kepler and Mendel are canonical). But it is not often held that discoveries were made, recognized, and then forgotten. Traditionally in science the discourse of perfect memory has not been that of the complete file folder, though notable publications have claimed to be archives for their respective disciplines. The more general claim to perfect memory is that this good recordkeeping is in the very nature of science. Take, for example, Henri Poincaré's *Science and Hypothesis* (1905). All scientific work, for Poincaré and many other positivists, went toward the construction of an eternal palace. Poincaré uses the metaphor of an army of scientists, foot soldiers, each adding a brick or so to the edifice of science. "The scientist must set in order. Science is built up with facts, as a house is with stones" (101). The thing about bricks is that they do not get forgotten: they are there in the nature of the edifice. Nobody need actively recall them: buildings do not remember. But each brick that is in a building is continuously present and is therefore ageless. In another metaphor, Poincaré sees the work of doing physics as similar to building a collection of books with the role of the theorist being to facilitate information retrieval, to catalogue:

Let us compare science to a library that ought to grow continually. The librarian has at his disposal for his purchases only insufficient funds. He ought to make an effort not to waste them.
It is experimental physics that is entrusted with the purchases. It alone, then, can enrich the library.
As for mathematical physics, its task will be to make out the catalogue. If the catalogue is well made, the library will not be any richer, but the reader will be helped to use its riches. (104)

The very nature of theory, then, is that it furnishes a classification system that can then be used to remember all (and only that) which is relevant to its associated practice.

All classification systems, however, face a bootstrapping problem. In a world of imperfect knowledge, any classificatory principle might be good, valid, useful: you will not know what makes a difference until you have built up a body of knowledge that relies, for its units of data, on the classification scheme that you have not yet developed. This is Spinoza's problem. Consider its form in the world of medical record-keeping—a world in which every trace might count for a patient's health or a clinical discovery. To maintain a good system of medical records, a state needs to classify a huge amount of information not only about its own citizens but about citizens of countries that it is in contact with (classification systems are necessarily imperialistic; witness the protests of African doctors to pressures from western AIDS researchers). As seen in chapter 2, the need for information and thence the burden of classificatory activity is effectively infinite.

In a world in which, as Ann Fagot-Largeault (1989, 6) has pointed out it is impossible to die of old age (the category of being "worn out" having been removed from the ICD) it appears that we are afloat in a sea of multiple, fractured causalities each demanding their own classification systems and their own apparatus of record collection. To deal with the plenum of information that all good organizations logically need, one can operate a distribution of memory in space (such and such a subgroup needs to hold such and such knowledge) and a distribution of memory in time (such and such a memory will only be recalled if a given occasion arises).

Classification systems provide both a warrant and a tool for forgetting at the same time as they operate this distribution. To take an overview of this process, consider the case of the classification of the sciences. Auguste Comte wrote about this in the first volume of his course of positive philosophy wherein he lays out a new classification of all the sciences in hierarchical order, each science having a statics and a dynamics. He argued that it was only at the current state of advancement of science that a true classification system could emerge since only now were the forces of religion and metaphysics sufficiently at bay that a true picture of the nature of knowledge could emerge.

At the same time, the sum total of scientific knowledge was sufficiently great that it was inconceivable to learn a science by mentally tracing its history. There were too many wrong turns, blind allies, or vagaries. (Just as one does not want to remember where one's keys are by tracing the series of actions that one has made in the past several hours). With the new classification system, knowledge could be arrayed logically and naturally. One would lose chronological order but gain coherence. Indeed, only in this way could science be logically taught; thus: ". . . the most important property of our encyclopedic formulation . . . is that it directly gives rise the true general plan of an entirely rational scientific education" (Comte 1975 [1830–45], 50).[46]

What is left in Comte's work is the positivist calendar, where certain great scientists have their days, just as the saints had theirs in the age of religion. Serres annotates this passage with the observation that the formation (training) of scientists covers up and hides the formation (production) of scientific knowledge (51).

Indeed, Comte sets in train a double motion. On the one hand, you will only learn science if you forget its history; on the other hand, you will only understand the history of science if you look at the entire history of humanity:

This vast chain is so real that often, to understand the effective generation of a scientific theory, the mind is led to consider the perfectioning of some art with which it has no rational link, or even some particular progress in social organization without which this discovery would not have taken place. . . . It follows therefore that one cannot know the true history of any science, that is to say the real formation of the discoveries it is composed of, without studying, in a general and direct manner, the history of humanity. (52)

On one side we have the complete history of humanity, where nothing can be forgotten because everything might be relevant. On the other, we have an efficient classification system that allows us to remember only what we need to remember about science. The classification system operates as a clearance: all that was religious and metaphysical is wiped away with a single gesture. It operates selective erasure in that even in the current scientific age the processes of the production of knowledge will have to be erased from the account of the knowledge itself. The classification system tells you what to forget and how to forget it. It operates a double distribution in space of scientific memory. First, the social story of science will be excluded from the organization of the sciences, and held outside of it (if at all) by historians. This is a form of erasure.

Second, it offers a natural hierarchy of the sciences, saying that a given discipline (say geology, statics) will need to remember all and only a given set of facts about the world. It also operates a distribution in time, saying that all scientific problems can be progressively unfolded so that at one point along the path in treating a problem you will need to draw on biology, then chemistry, then physics, then mathematics. Each type of memory that has been distributed in space will also be sequenced in time. The plenum is contained by the overarching organization constituted by the scientific community precisely through a controlled program of first clearance then continuing erasure. The work of conjuring the world into computable form (compare Hutchins 1995) has already begun by the setting up of a certain kind of formal memory system, for example, one in which facts can be stored in linear time and space.

In the history of science, we frequently encounter an apposition among clearance (the deliberate destruction of the past) and establishment of a classification system. When Lavoisier set out to found the new discipline of chemistry, he wrote a textbook that standardized the names of the elements (so that Ag became silver: not Diane's metal, a name that "remembered" the alchemical prehistory of the discipline). He also rewrote the history of chemistry so that his rivals, arguing the theory of affinities, no longer occupied a place in the textbooks: they were written out of the historical record. (Bensaude-Vincent 1989).

The strategy of clearance is a complete wiping clean of the slate so that one can start anew as if nothing had ever happened. As in the example of the burning of books, it is doubtful if clearance can ever work in the short term since people do remember things and institutional arrangements do bear traces of their past, for example, in the case of an outmoded classification system being reflected in the arrangements of artifacts in a museum. In the long term, however, by the time that the curricula have been redesigned, the manuals rewritten, and new nursing information systems produced, it can become a highly effective tool. Clearance is a pragmatic strategy. It may well be the case that a given organizational routine or piece of knowledge has roots in the distant past. At the same time, it may also be the case that in dealing with said routine or knowledge it is easier to act as if it had just arrived on the scene. For this reason the issue of truth or falsity of memory can be a red herring in treatments of organizational memory as well as analytically undecidable: a false memory, well constructed through a program of forgetting, can be of great use.

Erasure is a key dimension of classification work in all organizations. There is a famous passage in the Sherlock Holmes stories where Watson informs Holmes that the earth circles the sun; Holmes politely thanks Watson and then remarks that he will try to forget this fact as soon as possible, since it is a kind of fact that cannot possibly be relevant to the task that is at hand for him: the solution of a crime. In scientific organizations, things get deliberately forgotten in a variety of ways. Scientists classify away traces that they know to be relevant but which should not be officially recorded. For example, when looking at the early archives of the Schlumberger Company, Bowker was struck by a change in the written traces being left of company activity (1994). In the early days the boxes contained a series of highly detailed reports of daily activity sent by engineers in the field across the world to the company's center of calculation in Paris, to borrow Callon's felicitous phrase (Callon 1986). The theory, explicitly stated, was that the company needed the best possible records of what went on in the field to build up a sufficiently large database so as to construct scientific knowledge. It needed this as well to coordinate strategies for the insertion of the company into the oil field environment.

Then one day things changed. Detailed accounts in French of work practice became sketchy tables in English of numbers of oil wells logged. What had happened? The company had gotten involved in a legal suit with *Halliburton* and had come to realize that its own internal traces of activity were open to potential scrutiny by U.S. courts determining patent claims. There were two simultaneous realizations: first the records should be in English, since the French language could be read by a southern court as a foreign code. Second, the records should only contain kinds of facts that leant weight to the company's official presentation of itself: that is to say the cycle of accumulation of messy half-truths should be carried out elsewhere than in the organization's own potential memory system (Bowker 1994, chapter 3).

This strategy of distributed erasure is more punctillist than that of clearance: it involves the systematic and deliberate forgetting of some actions the better to remember others. In Adrienne Rich's words, this is an act of silence. "The technology of silence/The rituals, etiquette/the blurring of terms/silence not absence . . . Silence can be a plan/rigorously executed" (Rich 1978, 17).

Classification systems subtending information infrastructures operate as tools of forgetting (without representation in the medical informatics infrastructure, the profession of nursing is progressively erased from the annals both of history and of science). They also operate as tools for delegating attention (Latour 1996b has an extended discussion of this sense of delegation). Nurses do not want to have to carry around in their heads what drugs the patients on their wards need to be taking and when. They either use written traces or electronic means to hold the memory and perhaps automatically remind them (either directly by commanding attention through a beeping sound or routinely by constituting distributed traces that the nurse will encounter on their normal rounds, for example, the canonical chart at the foot of the patient's bed). The storage of information in a section of an organization's permanent record guarantees that attention (Weick and Roberts 1993) is paid to that information in either the production of organizational knowledge (formal accounts of how the organization works) or the organization's production of knowledge (how the hospital contributes to the production of nursing knowledge).

To produce nursing (and other) knowledge, then, various kinds of forgetting need to be operated on the permanent record held by organizations. This suggestion is fully complementary to the results from science studies and organization theory that many significant memories are held outside of formal information infrastructures. Ravetz (1971), Latour (1987), and many others have noted that one cannot do scientific work without being able to draw on information about specific local, organizational details of the operation of a given laboratory. And yet that information is nowhere systematically stored. In a series of studies of Xerox technicians, Julian Orr has shown that formal representations of fault diagnosis is often, on the spot, supplemented and indeed replaced by the swapping of war stories ("I had a machine that did something like that . . ." and so forth). We do not go into the preservation of nursing stories that Orr's work (Orr 1990, 1996) and others' assures us will be generated alongside of and as a complement to formal representations of nursing work. Further, new information infrastructures such as a hospital information system adopting NIC will in fact retain traces of organizational work and will despite themselves allow for the sharing of organizational memory. One might develop here the concept of organizational repression (by analogy to repressed memories). The argument comes down to asking not only what gets coded in but what gets read out of a given scheme; for example, who learns what from the fact that the coding book always falls open on a given page? (Compare Brown and Duguid 1994 on the importance of such peripheral clues.)

Just as oral history is a significant form of community memory, however, it is a different kind of memory (dates are far less important, stories migrate between characters, and so forth, see Vansina 1961) from that retained in the written record. This chapter has placed emphasis purely on the nature and articulation of what goes down in the continuing formal record that the organization preserves of its own past activity. This latter area is interesting in its own right because it is by using these memories that transportable formal accounts used in law, science, and management will be constructed.

### Conclusion

Information, in Bateson's famous definition, is about differences that make a difference. Designers of classification schemes constantly have to decide what really does make a difference; along the way they develop an economy of knowledge that articulates clearance and erasure and ensures that all and only relevant features of the object (a disease, a body, a nursing intervention) being classified are remembered. In this case, the classification system can be incorporated into an information infrastructure that is delegated the role of paying due attention. A corollary of the "if it moves, count it" theory is the proposition "if you can't see it moving, forget it." The nurses we looked at tried to guarantee that they would not be forgotten (wiped from the record) by insisting that the information infrastructure pay due attention to their activities.

This chapter has argued that here may indeed be good organizational reasons for forgetting. It has also argued that the ways in which things get forgotten are not merely images in a glass darkly of the way things get remembered; rather they are positive phenomena worthy of study in their own right. We have discussed two kinds of forgetting: clearance and erasure. From this emerged a consideration of forgetting and potential memory (mediated by classification systems).

The chapter has stressed that representation in the formal record is not the only way to be remembered: indeed, there is a complex ecology of memory practices within any one organization. The shift into long-term memory, however, that the infrastructure provides is significant, if fraught. The production of transportable knowledge used in other registers (scientific texts, the law) at present assumes that this knowledge can be stored and expressed in a quite restricted range of genres. As we saw with Poincaré, it can be argued that the work of much scientific theory is the storage of information as long-term memory. To prevent continuing erasure within hospital information systems, nurses have had to operate a clearance of their own past (recorded history begins today). The prize before their eyes is a science and a profession; the danger oblivion. (Either being definitively excluded from ongoing information practices and thus relegated to an adjunct role or being included but then distributed through reengineering.)

There is much to be done to understand the processes of commemoration, memory, history and recall in organizations. Organizational forgetting and organizational memory are useful concepts here because they allow us to move flexibly between the formal and the informal, the material and the conceptual. Designers of information superhighways need to take the occasional stroll down memory lane.

# IV

## The Theory and Practice of Classifications

This final part of the book attempts to weave the threads from each of the chapters into a broader theoretical fabric. Throughout the book we have demonstrated that categories are tied to the things that people do; to the worlds to which they belong. In large-scale systems those worlds often come into conflict. The conflicts are resolved in a variety of ways. Sometimes boundary objects are created that allow for cooperation across borders. At other times, such as in the case of apartheid, voices are stifled and violence obtains.

Chapter 9 discusses an abstract model of the several processes involved in both the development of boundary objects or any other alternatives. The key concept in this chapter is multiplicity both of people's memberships and of the ways in which objects are naturalized simultaneously in more than one world. People become members of many communities of practice. They do so at different rates and with different degrees of completeness. Some communities are all encompassing while others occupy very little of one's life space. Some things are shared quite locally; others become standardized across thousands of social worlds. While it is impossible, and will always be impossible, fully to map the myriad of relationships even a simple situation contains, it is possible to get at least a gestalt sense of the issues involved.

The chapter discusses the multiple trajectories of membership and naturalization. It discusses the consequences of some memberships being silenced, ignored, or devalued. It examines the notion of "cyborg" as a term for discussing the relationships between memberships and the naturalization of objects. The categorical exile of people and objects creates a monstrous landscape, such as those seen in chapters 5 and 6.

Chapter 9 concludes with recognition of the language that people often use in describing the complexities of people, things, and their relationships. We often turn to metaphors of texture to describe our sense of the moral, emotional, and organizational feelings of these relationships. A situation can be knotty and tangled; it can be handled smoothly and without ruffles. Following Lakoff and Johnson (1980), we take these "metaphors we live by" as more than poetic but imprecise means of expression. The metaphors are more systematic and more telling than that. Indeed, they are a key to understanding and perhaps to modeling some of the more complex phenomena facing us in what Ina Wagner (personal communication) has called "the classification society."

The book concludes in chapter 10 with a discussion of the ways in which research on classification and standards can inform a larger program of research into the building and maintaining of infrastructure—and its simultaneous material-cultural nature.

# 9

## Categorical Work and Boundary Infrastructures: Enriching Theories of Classification

Where do categories come from? How do they span the boundaries of the communities that use them? How can we see and analyze something so ubiquitous and infrastructural—something so "in between" a thing and an action? These questions have been at the heart of much of social science over the past 100 years. Sociology and history are both concerned with relationships—that are invisible except through indicators such as the actions people perform. One cannot directly *see* relations such as membership, learning, ignoring, or categorizing. They are names we give to patterns and indicators. If someone is comfortable with the things and language used by a group of others, we say that he or she is a member of that group. In this sense, categories—our own and those of others—come from action and in turn from relationships. They are, as sociologists like Aaron Cicourel (1964) remind us, continually remade and refreshed, with a lot of skilled work. The cases in this book are framed in dialogue with an extensive literature on language, group membership, and classification.

This chapter makes several aspects of that dialogue more explicit. Our goal here, however, is much more modest than a thoroughgoing analysis of categorization and language. We examine classification systems as historical and political artifacts very much as part of modern Western bureaucracy. Assigning things, people, or their actions to categories is a ubiquitous part of work in the modern, bureaucratic state. Categories in this sense arise from work and from other kinds of organized activity, including the conflicts over meaning that occur when multiple groups fight over the nature of a classification system and its categories.

This chapter picks up the theoretical strands of the cases in this volume to begin to develop a more general notion of these classification systems. By so doing, we take a step back and look at how the various kinds of classifcation we have discussed knit together to form the texture of a social space. We move from classifying and boundary objects to *categorical work and boundary infrastructures*, weaving along the way the many strands that our cases have presented. As noted in chapter 1, maintaining a vision that allows us to see the relationships among people, things, moral order, categories, and standards is difficult. It requires a good map and a working compass that we attempt to provide here.

The journey begins by clearing away some of the theoretical brush surrounding the very notions of categories and classification. Many scholars have seen categories as coming from an abstract sense of "mind," little anchored in the exigencies of work or politics. The work of attaching things to categories, and the ways in which those categories are ordered into systems, is often overlooked (except by theorists of language such as Harvey Sacks 1975, 1992).

We present classification systems in modern organizations as tools that are both material and symbolic. As information technologies used to communicate across the boundaries of disparate communities, they have some unique properties. Next, we present some basic propositions about large-scale information systems, examining how they are used to communicate across contexts. These systems are always heterogeneous. Their ecology encompasses the formal and the informal, and the arrangements that are made to meet the needs of heterogeneous communities—some cooperative and some coercive.

The third part of the journey involves understanding two sets of relationships: first, and analytically, between people and membership, and then between things and their naturalization by communities of practice.

The fourth step moves away from the analytical device of single person-single membership and single object—single naturalization—to describing a more complex set of multiple relationships. Everyone is part of multiple communities of practice. Things may be naturalized in more than one social world—sometimes differently, sometimes in the same fashion. Both people's memberships and the naturalization of objects are multiple, and these processes are, furthermore, intimately intertwined.

The fifth part of the chapter introduces the idea of categorical work—the work that people do to juggle both these multiple memberships and the multiple naturalizations of objects. In this work is the genius of what Sacks called "doing being ordinary" (1975) or what Strauss pointed to as "continual permutations of action" (1993). In the simplest seeming action, such as picking an article of clothing to wear, is embedded our complex knowledge of situations. (Where will I go today? What should I look like for the variety of activities in which I will participate?) These situations involve multiple memberships and how objects are used differently across communities. (Will this shirt "do" for a meeting with the dean, lunch with a prospective lover, and an appointment with the doctor at the end of the day?) Many of these choices become standardized and built into the environment around us; for example, the range of clothing we select is institutionalized by the retail stores to which we have access, traditions of costuming, and so forth. To think of this formally, the institutionalization of categorical work across multiple communities of practice, over time, produces the structures of our lives, from clothing to houses. The parts that are sunk into the built environment are called here boundary infrastructures— objects that cross larger levels of scale than boundary objects.

Finally, the chapter concludes with a discussion of future directions for research into classifications, standards, and their complex relationships with memberships in communities of practice. This includes ways we might visualize and model these intricate relationships.

The overall goal of the chapter is to trace theoretically what we have shown empirically and methodologically throughout the book: that categories are historically situated artifacts and, like all artifacts, are learned as part of membership in communities of practice. We want as well to talk about this insight beyond the individual "mind," task, or the small scale. Classifications as technologies are powerful artifacts that may link thousands of communities and span highly complex boundaries.

### What Sort of Thing Is a Category?

In so far as the coding scheme establishes an orientation toward the world, it constitutes a structure of intentionality whose proper locus is not the isolated, Cartesian mind, but a much larger organizational system, one that is characteristically mediated through mundane bureaucratic documents such as forms. *(Goodwin 1996, 65)*

Classification is a core topic within anthropology, especially cognitive anthropology, and within computer science. Recently, there has been a move to understand the practical, work-related aspects of classification as part of a larger project of revisioning cognition (e.g., Suchman 1988, Hutchins 1996, Keller and Keller 1996, Lave 1988).

## Revisioning Cognition

Within anthropology, psychology, and the sociology of science, the last two decades have seen a resurgence of the struggle to understand the material, social, and ecological aspects of cognition. The work in this book has been deeply informed by that intellectual movement. In brief, the research in this tradition seeks to ground activities previously seen as individual, mental, and nonsocial as situated, collective, and historically specific. On this view, for example, solving a mathematical problem is not a matter of mentally using an algorithm and coming up with the correct answer in a fashion that exists outside of time or culture, rather, it is a process of assembling materials close to hand and using them with others in specific contexts. Jean Lave, for example, studied mathematical problem solving in everyday life and contrasted it with formal testing situations (1988). She followed adults shopping for the best buy in a supermarket, people in Weight Watchers weighing cottage cheese to get the correct unit for the diet's specifications, and a variety of other mundane activities. She observed people performing highly abstract, creative mathematical problem solving in these circumstances. They were creating new units of analysis transposed against given ones in order to measure units, literally cutting up the cottage cheese, moving these material units around, or holding one can against another. These tasks were performed successfully by people who tested badly in a traditional math test. There was, she argued, no way to separate the material circumstances of the problem solving from the mathematical challenges. Those who appear to solve mathematics problems without such outside help are not working in a putative realm of pure number; rather, they and their observers have so naturalized the structures within which they are operating that they have become invisible. Lucy Suchman makes a similar argument for the process of planning as material resource, Ed Hutchins for navigation problems (1995), and Janet and Charles Keller (1996) for designing and measuring in doing iron blacksmithing work. In this book we join their effort at reforming the notion of categorizing and classifying so often seen as purely mental.

The problem of conceptualizing classifications is also akin to Cole's (1996) search for the nature of artifacts in mediated action. Cole notes,

"An artifact is an aspect of the material world that has been modified over the history of its incorporation into goal-directed human action. By virtue of the changes wrought in the process of their creation and use, artifacts are simultaneously *ideal* (conceptual) and *material*. They are ideal in that their material form has been shaped by their participation in the interactions of which they were previously a part and which they mediate in the present" (Cole 1996, 117). The materiality of categories, like that of other things associated with the purely cognitive, has been difficult to analyze. The Janus-faced conceptual-material notion of artifacts suggested by Cole combined with the attention to the use in practice of categories is a good way to begin. Classifications are both conceptual (in the sense of persistent patterns of change and action, resources for organizing abstractions) and material (in the sense of being inscribed, transported, and affixed to stuff).

Cole's intent is to emphasize the conceptual and symbolic sides of things often taken as only materials, tools, and other artifacts. It is similarly felicitous to emphasize the brute material force of that which has been considered ideal, such as categories.

### The Pragmatist Turn

The most radical turn taken by Pragmatist philosophers such as Dewey and Bentley, and closely followed by Chicago School sociologists such as Thomas and Hughes, is perhaps the least understood. It is related, both historically and conceptually, to the cognitive reforms detailed above. Consequences, asserted Dewey against a rising tide of analytic philosophy, are the thing to look at in any argument—not ideal logical antecedents. What matters about an argument is who, under what conditions, takes it to be true. Carried over into sociology, W. I. and Dorothy Thomas used it (as Howard Becker would some decades later) to argue against essentialism in examining so-called deviants or problem children (Thomas and Thomas 1917, Becker 1963). If social scientists do not understand people's definition of a situation, they do not understand it at all. That definition—whether it is the label of deviant or the performance of a religious ritual—is what people will shape their behavior toward.

This is a much more profound cut on social construction than the mere notion that people construct their own realities. It makes no comment on where the definition of the situation may come from—

human or nonhuman, structure or process, group or individual. It powerfully draws attention to the fact that the materiality of anything (action, idea, definition, hammer, gun, or school grade) is drawn from the consequences of its situation.

The Pragmatist turn, like the activity theoretical turn taken by Cole and others, emphasizes the ways in which things perceived as real may mediate action (Star 1996). If someone is taken to be a witch, and an elaborate technical apparatus with which to diagnose her or him as such is developed, then the reality of witchcraft obtains in the consequences—perhaps death at the stake. Classification systems are one form of technology, used in the sense Cole employs, linked together in elaborate informatic systems and enjoining deep consequences for those touched by them.

The following section discusses the problems of scaling up, from boundary objects and classifications systems on the one hand to a notion of boundary infrastructure. This analysis draws together the notions of multiplicity and the symbolic-material aspects of categories as artifacts discussed above.

### Information Systems across Contexts

At its most abstract, the design and use of information systems involves linking experience gained in one time and place with that gained in another, via representations of some sort. Even seemingly simple replication and transmission of information from one place to another involves encoding and decoding as time and place shift. Thus the context of information shifts in spite of its continuities; and this shift in context imparts heterogeneity to the information itself. Classifications are a very common sort of representation used for this purpose. Formal classification systems are, in part, an attempt to regularize the movement of information from one context to another; to provide a means of access to information across time and space. The ICD, for example, moves information across the globe, over decades, and across multiple conflicting medical belief and practice systems.

One of the interesting features of communication is that, broadly speaking, to be perceived, information *must* reside in more than one context. We know what something is by contrast with what it is not. Silence makes musical notes perceivable; conversation is understood as a contrast of contexts, speaker and hearer, words, breaks and breaths. In turn, in order to be meaningful, these contexts of information must be relinked through some sort of judgment of equivalence or comparability. This occurs at all levels of scale, and we all do it routinely as part of everyday life.

None of this is new in theories of information and communication: we have long had models of signals and targets, background, noise and filters, signals, and quality controls. We are moving this insight here to the level of social interaction. People often cannot see what they take for granted until they encounter someone who does not take it for granted.

A radical statement of this would be that information is only information when there are *multiple* interpretations. One person's noise may be another's signal or two people may agree to attend to something, but it is the tension between contexts that actually creates representation. What becomes problematic under these circumstances is the relationships among people and things, or objects, the relationships that create representations, not just noise. The ecological approach we have taken in this volume adds people as active interpreters of information who themselves inhabit multiple contexts of use and practice (Star 1991b). This multiplicity is primary, not accidental nor incidental.

Consider, for example, the design of a computer system to support collaborative writing. Eevi Beck (1995, 53) studied the evolution of one such system where "how two authors, who were in different places, wrote an academic publication together making use of computers. The work they were doing and the way in which they did it was inseparable from their immediate environment and the culture which it was part of." To make the whole system work, they had to juggle time zones, spouses' schedules, and sensitivities about parts of work practice such as finishing each other's sentences as well as manipulating the technical aspects of the writing software and hardware. They had to build a shared context in which to make sense of the information. Beck is arguing against a long tradition of decontextualized design where only the technical, or narrowly construed considerations about work hold sway.

We lack good relational language here. There is a permanent tension between the formal and the empirical, the local and the situated, and attempts to represent information across localities. It is this tension itself which is underexplored and undertheorized. It is not just a set of interesting metaphysical observations. It can also become a pragmatic unit of analysis. How can something be simultaneously concrete and abstract? The same and yet different? People are not (yet, we hope) used to thinking in this fashion in science or technology.[47] As information systems grow in scale and scope, however, the need for such complex analyses grows as well. In opposition to the old hierarchical databases, where relations between classes had to be decided once and for all at the time of original creation, many databases today incorporate object-oriented views of data whereby different attributes can be selected and combined on the fly for different purposes. The tailorability of software applications similarly becomes very important for customizing use in different settings (Trigg and Bødker 1994).

If we look at these activities in the context of practice, we see what Suchman and Trigg (1993) call the "artful integration" of local constraints, received standardized applications, and the re-representation of information. The tension between locales remains, and this tension it is not something to be avoided or deleted. When the sort of artful integration discussed by Suchman and Trigg becomes (a) an ongoing, stable relationship between different social worlds, and (b) shared objects are built across community boundaries, then boundary objects arise.

Boundary objects are one way that the tension between divergent viewpoints may be managed. There are of course many other ways. All of them involve accommodations, work-arounds, and in some sense, a higher level of artful integration. It too is managed by people's artful juggling, gestalt switching, and on the spot translating.

Too often, this sort of work remains invisible to traditional science and technology, or to rational analyses of process. This tension is itself collective, historical, and partially institutionalized. The medium of an information system is not just wires and plugs, bits and bytes, but also conventions of representation, information both formal and empirical. A system becomes a system in design and use, not the one without the other. The medium is the message, certainly, and it is also the case that both are political creations (Taylor and Van Every 1993). In Donna Haraway's words, "No layer of the onion of practice that is technoscience is outside the reach of technology of critical interpretation and critical inquiry about positioning and location; that is the condition of embodiment and mortality. The technical and the political are like the abstract and the concrete, the foreground and the background, the text and the context, the subject and the object" (Haraway 1997, 10). A fully developed method of multiplicity-heterogeneity for information systems must draw on many sources and make many unexpected alliances (Star 1989a, chapter 1, Star 1989b, Hewitt 1986, Goguen

1997). If both people and information objects inhabit multiple contexts and a central goal of information systems is to transmit information across contexts, then a representation is a kind of pathway that includes everything populating those contexts. This includes people, things-objects, previous representations, and information about its own structure. The major requirements for such an ecological understanding of the path of re-representation are thus:

1. How objects can inhabit multiple contexts at once, and have both local and shared meaning.

2. How people, who live in one community and draw their meanings from people and objects situated there, may communicate with those inhabiting another.

3. How relationships form between (1) and (2) above—how can we model the information ecology of people and things across multiple communities?

4. What range of solutions to these three questions is possible and what moral and political consequences attend each of them?

Standardization has been one of the common solutions to this class of problems.[48] If interfaces and formats are standard across contexts, then at least the first three questions become clear, and the fourth seems to become moot. But we know from a long and gory history of attempts to standardize information systems that standards do not remain standard for very long, and that one person's standard is another's confusion and mess (Gasser 1986, Star 1991b). We need a richer vocabulary than that of standardization or formalization with which to characterize the heterogeneity and the processual nature of information ecologies.

### Boundary Objects and Communities of Practice

The class of questions posed by the slippage between classifications and standards on the one hand, and the contingencies of practice on the other, form core problematics both in the sociology of science and in studies of use and design in information science. A rich body of work has grown up in both fields that documents the clever ways people organize and reorganize when the local circumstances of their activities do not match the prescribed categories or standards (see Gasser 1986, Kling and Scacchi 1982, Lave 1988, Sacks 1975, Star 1983). Making or using any kind of representation is a complex accomplishment, a balance of improvisation and accommodation to constraint.

People learn how to do this everyday, impossible action as they become members of what Lave and Wenger (1991) call *communities of practice,* or what Strauss (1978) calls *social worlds.* A community of practice (or social world) is a unit of analysis that cuts across formal organizations, institutions like family and church, and other forms of association such as social movements. It is, put simply, a set of relations among people doing things together (Becker 1986). The activities with their stuff, their routines, and exceptions are what constitute the community structure.[49] Newcomers to the community learn by becoming "sort of" members, through what Lave and Wenger (1991) call the process of "legitimate peripheral participation." They have investigated how this membership process unfolds and how it is constitutive of learning.

We are all in this sense members of various social worlds—communities of practice—that conduct activities together. Membership in such groups is a complex process, varying in speed and ease, with how optional it is and how permanent it may be. One is not born a violinist, but gradually becomes a member of the violin playing community of practice through a long period of lessons, shared conversations, technical exercises, and participation in a range of other related activities.

People live, with respect to a community of practice, along a trajectory (or continuum) of membership that has elements of both ambiguity and duration. They may move from *legitimate peripheral participation* to full membership in the community of practice, and it is extremely useful in many ways to conceive of learning this way.

### How Does this Include Categories?

Learning the ropes and rules of practice in any given community entails a series of encounters with the objects involved in the practice: tools, furniture, texts, and symbols, among others. It also means managing encounters with other people and with classes of action. Membership in a community of practice has as its sine qua non an increasing familiarity with the categories that apply to all of these. As the familiarity deepens, so does one's perception of the object as strange or of the category itself as something new and different.[50] Anthropologists call this the *naturalization* of categories or objects. The more at home you are in a community of practice, the more you forget the strange and contingent nature of its categories seen from the outside.

Illegitimacy, then, is seeing those objects as would a stranger—either as a naïf or by comparison with another frame of reference in which they exist. And this is not to be equated with an idealized notion of skill, but with membership. One does not have to be Isaac Stern to know fully and naturally what to do with a violin, where it belongs, and how to act around violins and violinists. But if you use a Stradivarius to swat a fly (but not as part of an artistic event!) you have clearly defined yourself as an outsider, in a way that a schoolchild practicing scales has not.

Membership can thus be described individually as the experience of encountering objects and increasingly being in a naturalized relationship with them. (Think of the experience of being at home, and how one settles down and relaxes when surrounded by utterly familiar objects; think of how demented one feels in the process of moving house.)

From the point of view of learning-as-membership and participation, then, the illegitimate stranger is a source of learning. Someone's illegitimacy appears as a series of interruptions to experience (Dewey 1916, 1929) or a lack of a naturalization trajectory. In a way, then, individual membership processes are about the resolution of interruptions (anomalies) posed by the tension between the ambiguous (outsider, naive, strange) and the naturalized (at home, taken-for-granted) categories for objects. Collectively, membership can be described as the processes of managing the tension between naturalized categories on the one hand and the degree of openness to immigration on the other. Harvey Sacks, in his extensive investigations into language and social life, notes that categories of membership form the basis of many of our judgments about ordinary action. "You can easily enough come to see that for any population of persons present there are available alternative sets of categories that can be used on them. That then poses for us an utterly central task in our descriptions; to have some way of providing which set of categories operate in some scene—in the reporting of that scene or in its treatment as it is occurring" (1992, vol. 1, 116). Sacks draws attention to the ways in which being ordinary are not pregiven but are in fact a kind of job—a job which asserts the nature of membership:

Whatever we may think about what it is to be an ordinary person in the world, an initial shift is not to think of an "ordinary person" as some person, but as somebody having as their job, as their constant preoccupation, doing "being ordinary." It's not that somebody *is* ordinary, it's perhaps that that's what their business is. And it takes work, as any other business does. And if you just extend the analogy of what you obviously think of as work—as whatever it is that takes analytic, intellectual, emotional energy—then you can come to see that all sorts of normalized things—personal characteristics and the like—are jobs which are done, which took some kind of effort, training, etc.. So I'm not going to be talking about an "ordinary person" as this or that person, or as some average, i.e., a nonexceptional person on some statistical basis, but as something that is the way somebody constitutes themselves, and, in effect, a job that they do on themselves. Fate and the people around may be coordinatively engaged in assuring that each of them are ordinary persons, and that can then be a job that they undertake together, to achieve that each of them, together, are ordinary persons. (1992, vol. 2, 216)

The performance of this job includes the ability to choose the proper categories under which to operate, to perform this ordinariness. The power of Sack's work, like that of John Dewey (e.g. 1929), is that he draws attention to the ways in which the ordinary—and the interruption to the expected experience—are delicate constructions made and remade every day.

### Boundary Objects

Science and technology are good places to study the rich mix of people and things brought to bear on complex problem-solving questions, although the points made here are more generally applicable as well. Categories and their boundaries are centrally important in science, and scientists are especially good at documenting and publicly arguing about the boundaries of categories. Thus, science is a good place to understand more about membership in communities. This point of departure has led us to try to understand people and things ecologically, both with respect to membership and to the things they live with, focusing on scientists (Star 1995a). One of the observations is that scientists routinely cooperate across many communities of practice. They thus bring different naturalized categories with them into these partnerships.

In studying scientific problem solving, we have been concerned for a number of years to understand how scientists could cooperate without agreeing about the classification of objects or actions. Scientific work is always composed of members of different communities of practice (we know of no science that is not interdisciplinary in this way, especially if—as we do—you include laboratory technicians and janitors). Thus, memberships (and divergent viewpoints or perspectives) present a pressing problem for modeling truth, the putative job of scientists. In developing models for this work, Star coined the term "boundary objects" to talk about how scientists balance different categories and meanings (Star and Griesemer 1989, Star 1989b). Again, the term is not exclusive to science, but science is an interesting place to study such objects because the push to make problem solving explicit gives one an unusually detailed amount of information about the arrangements.

Boundary objects are those objects that both inhabit several communities of practice and satisfy the informational requirements of each of them. Boundary objects are thus both plastic enough to adapt to local needs and constraints of the several parties employing them, yet robust enough to maintain a common identity across sites. They are weakly structured in common use and become strongly structured in individual-site use. These objects may be abstract or concrete. Star and Griesemer (1989) first noticed the phenomenon in studying a museum, where the specimens of dead birds had very different meanings to amateur bird watchers and professional biologists, but "the same" bird was used by each group. Such objects have different meanings in different social worlds but their structure is common enough to more than one world to make them recognizable, a means of translation. The creation and management of boundary objects is a key process in developing and maintaining coherence across intersecting communities.

Another way of talking about boundary objects is to consider them with respect to the processes of naturalization and categorization discussed above. Boundary objects arise over time from durable cooperation among communities of practice. They are working arrangements that resolve anomalies of naturalization without imposing a naturalization of categories from one community or from an outside source of standardization. (They are therefore most useful in analyzing cooperative and relatively equal situations; issues of imperialist imposition of standards, force, and deception have a somewhat different structure.) In this book, sets of boundary objects arise directly from the problematics created when two or more differently naturalized classification systems collide. Thus nursing administrators create classification systems that serve hospital administrators and nursing scientists; soil scientists create classifications of soil to satisfy geologists and botanists (Chatelin 1979). Other outcomes of these meetings are explored as well—the dominance of one over another or how claims of authority may be manipulated to higher claims of naturalness.

The processes by which communities of practice manage divergent and conflicting classification systems are complex, the more so as people are all members in fact of many communities of practice, with varying levels of commitment and consequence. Under those conditions a series of questions arise: How are boundary objects established and maintained? Does the concept scale up? What is the role of technical infrastructure? Is a standard ever a boundary object? How do classification systems, as artifacts, play a role?

### Membership and Naturalization: People and Things

As Engeström (1990b) and other activity theorists note so well, tools and material arrangements always mediate activity. People never act in a vacuum or some sort of hypothetical pure universe of doing but always with respect to arrangements, tools, and material objects. Strauss (1993) has recently made a similar point, emphasizing the continuity and permeability of such arrangements—action never really starts from scratch or from a tabula rasa. Both Engeström and Strauss go to great lengths to demonstrate that an idea, or something that has been learned, can also be considered as having material-objective force in its consequences and mediations.

"Object" includes all of this—stuff and things, tools, artifacts and techniques, and ideas, stories, and memories—objects that are treated as consequential by community members (Clarke and Fujimura 1992a, 1992b). They are used in the service of an action and mediate it in some way. Something actually *becomes* an object only in the context of action and use; it then becomes as well something that has force to mediate subsequent action. It is easier to see this from historical examples than it may be to look to contemporary ones. For instance, the category of hysteria was naturalized in medicine and in popular culture at the end of the nineteenth century. People used the diagnosis of hysteria for purposes of social control as well as for medical treatment. It became a category through which physicians, social theorists, and novelists discussed pain and anxiety and, arguably, the changing social status of women. The point is not who believed what when but rather that the category itself became an object existing in both communities. It was a medium of communication, whatever else it may also have been.

A community of practice is defined in large part according to the co-use of such objects since all practice is so mediated. The relationship of the newcomer to the community largely revolves around the nature of the relationship with the objects and not, counterintuitively, directly with the people. This sort of directness only exists hypothetically—there is always mediation by some sort of object. Acceptance or legitimacy derives from the familiarity of action mediated by member objects.

But familiarity is a fairly sloppy word. Here it is not meant instrumentally, as in proficiency, but relationally, as a measure of taken-for-grantedness. (An inept programmer can still be a member of the community of practice of computer specialists, albeit a low status one in that he or she takes for granted the objects to be used.) A better way to describe the trajectory of an object in a community is as one of naturalization. Naturalization means stripping away the contingencies of an object's creation and its situated nature. A naturalized object has lost its anthropological strangeness. It is in that narrow sense desituated —members have forgotten the local nature of the object's meaning or the actions that go into maintaining and recreating its meaning.[51] We no longer think much about the miracle of plugging a light into a socket and obtaining illumination, and we must make an effort of anthropological imagination to remind ourselves of contexts in which it is still not naturalized.

Objects become natural in a particular community of practice over a long period of time. (See Latour's (1987) arguments in *Science in Action* for a good discussion of this.) Objects exist, with respect to a community, along a trajectory of naturalization. This trajectory has elements of both ambiguity and duration. It is not predetermined whether an object will ever become naturalized, or how long it will remain so; rather, practice-activity is required to make it so and keep it so. The more naturalized an object becomes, the more unquestioning the relationship of the community to it; the more invisible the contingent and historical circumstances of its birth, the more it sinks into the community's routinely forgotten memory.[52] Light switches, for instance, are ordinary parts of modern life. Almost all people living in the industrialized world know about light bulbs and electricity, even if they live without it, and switches and plugs are naturalized objects in most communities of practice. People do not think twice about their nature, only about whether or not they can find them when needed. Commodity and infrastructural technologies are often naturalized in this way. In a sense they become a form of collective forgetting, or naturalization, of the contingent, messy work they replace. We wrote this chapter on Macintosh and IBM computers, for example, and cutting and pasting are no longer phenomenologically novel operations, although we can remember when they once were. We have naturalized the mouse, the operation of selecting text, and the anachronistic "cut and paste" metaphor.

## Multiplicity

This chapter, so far, has discussed analytically two sets of relationships: between people and membership on the one hand; and objects and naturalization on the other hand. In any given instance, *both membership and naturalization are relations along a trajectory.* In saying this, we do not want to recreate a great divide between people and objects, reifying an objectless human or wild child. Ironically, social science has spent incredible resources on precisely this sort of search. There is something compelling about the idea of a person without "a society," naked even of touch or language. The sad case of "Genie," a child kept captive by her parents for many years (Rymer 1993, Star 1995d), or the "wild child of Aveyron" who amazed eighteenth-century philosophers, are emblematic of this propensity. They have been seen as holding the key to language or in a way to what it is to be human.

Exactly the opposite, however, is true. People-and-things, which are the same as people-and-society, cannot be separated in any meaningful practical sense. At the same time, it is possible for analytical purposes to think of two trajectories traveling in tandem, membership, and naturalization. Just as it is not practically possible to separate a disease from a sick patient, yet it is possible to speak of the trajectories of disease and biography operating and pulling at one another, as seen in chapter 5 in the case of tuberculosis.

## Residual Categories, Marginal People, and Monsters

People often see multiplicity and heterogeneity as accidents or exceptions. The marginal person, who is for example of mixed race, is portrayed as the troubled outsider; just as the thing that does not fit into one bin or another gets put into a "residual" category. This habit of purity has old and complicated origins in western scientific and political culture (e.g. as explicated by Dewey 1916). The habit perpetuates a cruel pluralistic ignorance. No one is pure. No one is even average. And all things inhabit someone's residual category in some category system. The myriad of classifications and standards that surround and support the modern world, however, often blind people to the importance of the "other" category as constitutive of the whole social architecture (Derrida 1980).

Communities vary in their tastes for openness, and in their tolerance for this ambiguity. Cults, for example, are one sort of collective that is low on the openness dimension and correspondingly high on the naturalization-positivism dimension—us versus them.

In recent years social theorists have been working toward enriching an understanding of multiplicity and misfit, decentering the idea of an unproblematic mainstream. The schools of thought grappling with this include feminist research (e.g., Haraway 1997), multicultural or race-critical theory (e.g., Ferguson et al. 1990), symbolic interactionism, and activity theory (e.g., Cole 1996, Wertsch 1991, 1998). During the same period, such issues have become increasingly of concern to some information scientists. As the information systems of the world expand and flow into each other, and more kinds of people use them for more different things, it becomes harder to hold to pure or universal ideas about representation or information.

Some of these problems are taken up in the intellectual common territory sometimes called "cyborg." Cyborg, as used for example by Donna Haraway (1991) and Adele Clarke (1998), means the intermingling of people, things (including information technologies), representations, and politics in a way that challenges both the romance of essentialism and the hype about what is technologically possible. It acknowledges the interdependence of people and things, and it shows just how blurry the boundaries between them have become. The notion of cyborg has clearly touched a nerve across a broad spectrum of intellectual endeavors. The American Anthropological Association has hosted sessions on cyborg anthropology for the past several years; the weighty *Cyborg Handbook* was published a few years ago (Gray 1995).

Through looking at ubiquitous classification systems and standards, it is possible to move toward an understanding of the stuff that makes up the networks that shape much of modern daily life in cyborg fashion. We draw attention here to the places where the work gets done of assuring that these networks will stick together: to the places where human and nonhuman are constructed to be operationally and analytically equivalent. By so doing, we explore the political and ethical dimensions of classification theory.

Why should computer scientists read African-American poets? What does information science have to do with race-critical or feminist methods and metaphysics? The collective wisdom in those domains is one of the richest places from which to understand these core problems in information systems design: how to preserve the integrity of information without a priori standardization and its often attendant violence. In turn, if those lessons can be taken seriously within the emerging cyberworld, there may yet be a chance to strengthen its democratic ethical aspects. It is easy to be ethnocentric in virtual space; more difficult to avoid stereotypes. The lessons of those who have lived with such stereotypes are important, perhaps now more than ever.

### Borderlands and Monsters

People who belong to more than one central community are also important sources for understanding more about the links between moral order and categorization. Such "marginal" people have long been of interest to social scientists and novelists alike. Marginality as a technical term in sociology refers to human membership in more than one community of practice.[53] Here we emphasize those people who belong to communities that are different in key, life-absorbing ways, such as racial groups (see our discussion in chapter 6). A good example of a marginal person is someone who belongs to more than one race, for example, half white and half Asian. Again, we are not using marginality here in the sense of center-margin or center-periphery (e.g., not "in the margins"), but rather in the old-fashioned sense of Robert Park's marginal man, the one who has a double vision by virtue of having more than one identity to negotiate (Park 1952, Stonequist 1937, Simmel 1950 [1908], Schütz 1944). Strangers are those who come and stay a while, long enough so that membership becomes a troublesome issue—they are not just nomads passing through, but people who sort of belong and sort of do not.

Marginality is an interesting paradoxical concept for people and things. On the one hand, membership means the naturalization of objects that mediate action. On the other, everyone is a member of multiple communities of practice. Yet since different communities generally have differently naturalized objects in their ecology, how can someone maintain multiple membership without becoming simply schizophrenic? How can they naturalize the same object differently, since naturalization by definition demands forgetting about other worlds?

There are also some well-known processes in social psychology for managing these tensions and conflicts: passing, or making one community the shadow for the other; splitting, or having some form of multiple personality; fragmenting or segmenting the self into compartments; becoming a nomad, intellectually and spiritually if not geographically (Larsen (1986) covers many of these issues in her exquisite fiction).

One dissatisfaction we have with these descriptions is that they all paint each community of practice as ethnocentric, as endlessly hungry and unwilling collectively to accommodate internal contradictions. There is also an implicit idea of a sort of imperialist über-social world (the mainstream) that is pressing processes of assimilation on the individual (e.g., Americanization processes in the early twentieth century). Communities vary along this dimension of open-closedness, and it is equally important to find successful examples of the nurturing of marginality (although it is possible that by definition they exist anarchically and not institutionally-bureaucratically). Here again, feminism has some important lessons. An important theme in recent feminist theory is resistance to such imperializing rhetoric and the development of alternative visions of coherence without unconscious assumption of privilege. Much of it emphasizes a kind of double vision, such as that taken up in the notion of borderlands by Anzaldúa (1987), or the qualities of partiality and modesty of Haraway's cyborg (1991).

Charlotte Linde's book on the processes of coherence in someone's life stories also provides some important clues. She especially emphasizes accidents and contingency in the weaving together of a coherent narrative (Linde 1993). The narratives she analyses are in one sense meant to reconcile the heterogeneity of multiply naturalized object relations in the person, where the objects in question are stories-depictions of life events. Linden (1993) and Strauss (1959) have made similar arguments about the uncertainty, plasticity, and collectivity of life narratives.

In traditional sociology this model might have overtones of functionalism, in its emphasis on insiders-outsiders and their relations. But functionalists never considered the nature of objects or of multiple legitimate memberships. If we think in terms of a complex cluster of multiple trajectories simultaneously of both memberships and naturalizations, it is possible to think of a many-to-many relational mapping.

The mapping suggested here pushes us further into the analysis of the cyborg. On the one hand, cyborgs as an image are somehow grotesque. Imagining the relationships between people and things such that they are truly interpenetrated means rethinking human nature itself. It is reminiscent somehow of bad science fiction. Yet analytically, it is a crucial notion for understanding technoscience and classifications as artifacts.

How can we think of cyborgs in the analysis presented in this chapter? The mapping among things, people, and membership provides a way in. Anzaldúa's work on borderlands rejects any notion of purity based on membership in a single, pristine racial, sexual, or even religious group (1987). Haraway's work pushes this analysis a bit further. In speaking of borderlands, both those concerning race and those concerning the boundaries between humans and things, she employs the term "monsters."

A monster occurs when an object refuses to be naturalized (Haraway 1992). A borderland occurs when two communities of practice coexist in one person (Anzaldúa 1987). Borderlands are the naturalized home of those monsters known as cyborgs. If we read monsters as persistent resisters of transparency-naturalization within some community of practice, then the experience of encountering an anomaly (such as that routinely encountered by a newcomer to science, for instance most women or men of color) may be keyed back into membership. A person realizes that they do not belong when what appears like an anomaly to them seems natural for everyone else. Over time, collectively, such outsider experiences (the quintessential stranger) can become monstrous in the collective imagination. History and literature are full of the demonizing of the stranger. Here is what Haraway (1992) has called "the promise of monsters" and one of the reasons that for years they have captured the feminist imagination.[54] Frankenstein peering in the warmly lit living room window; Godzilla captured and shaking the bars of his cage are intuitions of exile and madness, and served as symbols of how women's resistance and wildness have been imprisoned and reviled, kept just outside.

In a more formal sense, monsters and freaks are also ways of speaking about the constraints of the classifying and (often) dichotomizing imagination. Ritvo (1997) writes of the proliferation of monsters in the eighteenth and nineteenth century, linking it to a simultaneous increase in public awareness of scientific classification and hunger for the exotic. As classification schemes proliferated, so did monsters:

Monsters were understood, in the first instance, as exceptions to or violations of natural law. The deviations that characterized monsters, however, were both so various and in some cases, so subtle as significantly to complicate this stock account . . . . As a group, therefore, monsters were united not so much by physical deformity or eccentricity as by their common inability to fit or be fitted into the category of the ordinary—a category that was particularly liable to cultural and moral construction. (Ritvo 1997, 133–134)

In a practical sense, this is a way to talk about what happens to any outsider. For example, it could refer to experience in the science classroom when someone comes in with no experience of formal science, or to the transgendered person who does not fit cultural gender dichotomies (Stone 1991). It is not simply a matter of the strangeness, but of the politics of the mapping between the anomalies and the forms of strangeness-marginality.

In accepting and understanding the monsters and the borderlands there may be an intuition of healing and power, as Gloria Anzaldúa  (1987) shows us in her brilliant and compassionate writing. In her essay, "La conscientia de la mestiza," the doubleness and the ambiguity of the male-female, straight-gay, Mexican-American borderland becomes the cauldron for a creative approach to surviving, a rejection of simplistic purity and of essentialist categories (1987). At the same time, she constantly remembers the physical and political suffering involved in these borderlands, refusing a romanticized version of marginality that often plagued the early sociological writers on the topic.

The path traced by Anzaldúa is not an easy healing and certainly not a magic bullet but a complex and collective twisted journey, a challenge to easy categories and simple solutions. It is, in fact, a politics of ambiguity and multiplicity—this is the real possibility of the cyborg. For scholars, this is necessarily an exploration that exists in interdisciplinary borderlands and crosses the traditional divisions between people, things, and technologies of representation.

### Engineered versus Organic Boundary Objects

Would it be possible to design boundary objects? To engineer them in the service of creating a better society? On the surface, this idea is tempting. In some sense, this has been the goal of progressive education, multiculturalism in the universities, and the goal of the design of information systems that may be accessed by people with very different points of view.

Most schools now are lousy places to grow boundary objects because they both strip away the ambiguity of the objects of learning and impose or ignore membership categories (except artificial hierarchically assigned ones).[55] In mass schooling and standardized testing, an attempt is made to insist on an engineered community of practice, where the practices are dictated and the naturalization process is monitored and regulated while ignoring borderlands. They are virtual factories for monsters. In the 1970s and 1980s many attempts were made to include other communities in the formula via affirmative action and multicultural initiatives. But where these lacked the relational base between borderlands and the naturalization of objects, they ran aground on the idea of measuring progress in learning. This is partly a political problem and partly a representational one. As feminists learned so painfully over the years, a politics of identity based on essences can only perpetuate vicious dualisms. If a white male science teacher were to bring in an African-American woman as a (Platonic) representative of African-American-ness and/or woman-ness, for example, then attempt to match her essential identity to the objects in the science classroom (without attending much to how they are fully naturalized objects in another community of practice), costly and painful mismatches are inevitable. The teacher risks causing serious damage to her self-articulation (especially where she is alone) and her ability to survive (a look at the dismal retention statistics of women and minority men in many sciences and branches of engineering will underscore this point). Any mismatch becomes her personal failure, since the measurement yardstick remains unchanged although the membership criteria appear to have been stretched. Again, both borderlands and anomalous objects have been deleted. Kal Alston (1993), writing of her experience as an African-American Jewish feminist, has referred to herself as a unicorn—a being at once mythical and unknowable, straddling multiple worlds.

But all people belong to multiple communities of practice—it is just that in the case of the African-American woman in science, the visibility and pressure is higher, and her experience is especially rich, dense in the skill of surviving multiplicity. Thus Patricia Hill Collins' title, "Learning from the Outsider Within," has many layers and many directions to be explored as we all struggle for rich ways of mapping that honor this experience and survival (1986). Karla Danette Scott (1995) has recently written about the interwoven languages of black women going to college, and how language becomes a resource for this lived complexity. They "talk black" and "talk white" in a seamless, context-driven web, articulating the tensions between those worlds as a collective identity. This is not just code switching but braided identity—a borderland.

## Wildness

Things and people are always multiple, although that multiplicity may be obfuscated by standardized inscriptions. In this sense, with the right angle of vision, things can be seen as heralds of other worlds and of a wildness that can offset our naturalizations in liberatory ways. Holding firmly to a relational vision of people-things-technologies in an ethical political framework, there is a chance to step off the infinite regress of measuring the consumption of an object naturalized in one centered world, such as the objects of Western science, against an infinitely expanded set of essentially-defined members as consumers.

By relational here we argue against misplaced concretism or a scramble unthinkingly to assimilate the experiences of things to pregiven categories. We affirm the importance of process and ethical orientations. We also mean to take seriously the power of membership, its continual nature (i.e., we are never *not* members of some community of practice), and the inherent ambiguity of things. Boundary objects, however, are not just about this ambiguity, they are not just temporary solutions to disagreements about anomalies. Rather, they are durable arrangements among communities of practice. Boundary objects are the canonical forms of all objects in our built and natural environments. Forgetting this, as people routinely do, means empowering the self-proclaimed objective voice of purity that creates the suffering of monsters in borderlands. Due attention to boundary objects entails embracing the gentle and generous vision of *mestiza* consciousness offered by Anzaldúa.

## Casual versus Committed Membership

Another dimension to acknowledge here is the degree to which membership demands articulation at the higher level. Being a woman and African-American and disabled are three sorts of membership that are nonoptional, diffused throughout life, and embedded in almost every sort of practice and interaction.[56] So it is not equitable to talk about being a woman in the same breath as being a scuba diver—although there are ways in which both can be seen under the rubric community of practice (Lagache 1995). But if we go to the framework presented above, there is a way to talk about it. Where the joint objects are both multiply naturalized in conflicting ways and diffused through practices that belong to many communities, they will defy casual treatment. So for scuba diving—it is primarily naturalized in a leisure world and not

especially central to any others. Its practice is restricted and membership contained, neither contagious nor diffuse. On the other hand,
learning mathematics is multiply naturalized across several powerful
communities of practice (mathematics and science teachers and practitioners). At the same time it is both strange and central to others
(central in the sense of a barrier to further progress). It is also diffused
through many kinds of practices, in various classrooms, disciplines,
and workplaces (Hall and Stevens 1995). Some communities of practice expect it to be fully naturalized—a background tool or a substrate-
infrastructure—to get on with the business of being, for example, a
scientist (Lave 1988). There is no map or sense of the strangeness of
the object, however, across other memberships. Here, too, information
technologies are both diffused and strange with rising expectations of
literacy across worlds.

These relations define a space against which and into which information technologies of all sorts enter. These technologies of representation are entering into all sorts of communities of practice on a
global scale, in design and in use. They are a medium of communication and broadcast as well as of standardization. The toughest problems in information systems design are increasingly those concerned
with modeling cooperation across heterogeneous worlds, of modeling
articulation work and multiplicity. If we do not learn to do so, we face
the risk of a franchised, dully standardized infrastructure ("500 chan-
nels and nothing on," in the words of Mitch Kapor from the Electronic
Frontier Foundation) or of an Orwellian nightmare of surveillance.

Feminism and race-critical theory offer traditions of reflective de-
naturalization, of a politics of simultaneity and contradiction intuited
by the term cyborg. Long ago feminists began with the maxim that the
personal is political and that each woman's experience has a primacy
we must all learn to afford. Feminism went from reductionist identity
politics to cyborg politics in less than twenty years. Much of this was
due to the hard work and suffering of communities of practice that
had been made monstrous or invisible, especially women of color and
their articulation of the layered politics of insider-outsider and borderlands. One part of the methodological lesson from feminism read
in this way is that experience-experiment incorporates an ethics of
ambiguity with both modesty and anger. This means that how we hear
each other is a matter of listening forth from silence. Listening is active,
not passive; it means stretching to affiliate with multiplicity. In Nell
Morton's words, this is "hearing to speech":

- Not only a new speech but a new hearing.
- Hearing to speech is political.
- Hearing to speech is never one-sided. Once a person is heard to speech she becomes a hearing person.
- Speaking first to be heard is power over. Hearing to bring forth speech is empowering. (Morton 1985, 210).

Part of the moral vision of this book concerns how we may, through challenge and analysis of infrastructure, better hear each other to speech.

## Multiple Marginality, Multiple Naturalizations: Categorical Work

The model proposed here takes the form of a many-to-many relational mapping, between multiple marginality of people (borderlands and monsters) and multiple naturalizations of objects (boundary objects and standards). Over time, the mapping is between the means by which individuals and collectives have managed the work of creating coherent selves in the borderlands on the one hand and creating durable boundary objects on the other.

It is also not just many-to-many relational, but meta-relational. By this we mean that the map must point simultaneously to the articulation of selves and the naturalization of objects. One of the things that is important here is honoring the work involved in borderlands and boundary objects. This work is almost necessarily invisible from the point of view of any single community of practice. As Collins (1986) asks, what white really sees the work of self-articulation of the black who is juggling multiple demands-audiences-contingencies? It is not just willful blindness (although it can be that), but much more akin to the blindness between different Kuhnian paradigms, a revolutionary difference. Yet the juggling is both tremendously costly and brilliantly artful. Every community of practice has its overhead: *"paying your dues, being regular, hangin', being cool, being professional, people like us, conduct becoming, getting it, catching on."* And the more communities of practice one participates in, the higher the overhead not just in a straightforwardly additive sense, but interactively. Triple jeopardy (i.e., being old, black, and female) is not just three demographic variables or conditions added together, but a tremendously challenging situation of marginality requiring genius for survival. *The overheads interact.*

### From Articulation Work to Categorical Work

What is the name for this work of managing the overheads and anomalies caused by multiple memberships on the one hand and multiply naturalized objects on the other? Certainly, it is invisible. Most certainly, it is methodological, in the sense of reflecting on differences between methods and techniques. At first glance, it resembles articulation work, that is, work done in real time to manage contingencies; work that gets things back on track in the face of the unexpected, that modifies action to accommodate unanticipated contingencies. Within both symbolic interactionism and the field of computer-supported cooperative work, the term articulation work has been used to talk about some forms of this invisible juggling work (Schmidt and Bannon 1992, Gerson and Star 1986).

Articulation work is richly found for instance in the work of head nurses, secretaries, homeless people, parents, and air traffic controllers, although of course all of us do articulation work to keep our work going. Modeling articulation work is one of the key challenges in the design of cooperative and complex computers and information systems. This is because real-time contingencies, or in Suchman's (1987) terms, situated actions, always change the use of any technology (for example, when the host of a talk forgets to order a computer projector, can one quickly print out and assemble a handout?)

Other aspects of cooperative work concern novelty and the ways in which one person's routine may be another's emergency or anomaly (Hughes 1970), or in the words of Schmidt and Simone (1996) both the consequences and the division of labor of cooperative work. The act of cooperation is the interleaving of distributed tasks; articulation work manages the *consequences* of this distributed aspect of the work.[57] Schmidt and Simone note the highly complex dynamic and recursive relationship between the two—managing articulation work can itself become articulation work and vice versa, ad infinitum.

The consequences of the distribution of work, and its different meanings in different communities, must be managed for cooperation to occur. The juggling of meanings (memberships and naturalizations), is what we term *categorical work*. For example, what happens when one clerk, User A, entering data into a large database does not think of abortion as a medical matter, but as a crime; while another, User B, thinks of it as a routine medical procedure? User A's definition excludes abortion from the medical database, User B's includes it. The resulting data will be, at the least, incomparable, but in ways that may be completely invisible to User C, compiling statistics for a court case arguing for the legalization of abortion based on prevalence. When this aspect of the coordination of work is deleted and made invisible in this fashion, then voices are suppressed and we see the formation of master narratives and the myth of the mainstream universal (Star and Strauss 1999).

Thus, we can see categorical work as partly about managing the mismatches between memberships and naturalization. One way to think about this is through the management of anomalies as a tracer. Anomalies or interruptions, the cause of contingency, come when some person or object from outside the world at hand interrupts the flow of expectations. One reason that glass-box technology or pure transparency is impossible is that anomalies always arise when multiple communities of practice come together, and useful technologies cannot be designed in all communities at once. Monsters arise when the legitimacy of that multiplicity is denied. Our residual categories in that case become clogged and bloated.

Transparency is in theory the endpoint of the trajectory of naturalization, as complete legitimacy or centrality is the endpoint of the trajectory of membership in a community of practice. Due to the multiplicity of membership of all people, however, and the persistence of newcomers and strangers as well as the multiplicity of naturalization of objects, this is inherently nonexistent in the real world. For those brief historical moments where it appears to be the case, it is unstable.

In place of transparency—and it is a good enough counterfeit to work most of the time as transparency—one encounters convergence: the mutual constitution of a person or object and their representation. People get put into categories and learn from those categories how to behave. Thus there is the ironic observation that East Enders in London learn cockney (and how to be cockney) through watching the soap opera East Enders on television. "I am an East Ender therefore I must talk like this; and I must drink such and such a brand of beer." Aided by bureaucratic institutions, such cultural features take on a real social weight. If official documents force an Anglo-Australian to choose one identity or the other—and if friends and colleagues encourage that person, for the convenience of small talk, to make a choice—then they are likely to become ever more Australian, suffering alongside his or her now fellow countrypeople if new immigration measures are introduced in America or if "we" lose a cricket test. The same process occurs with objects—once a film has been thrown into the x-rated bin, then there is a strong incentive for the director to make it really x-rated; once a house has been posted as condemned, then people will feel free to trash it.

Where the difference lies between transparency and convergence is that where transparency ideally just produces a reflection of the way things really are (and so, in Jullien's (1995) beautiful phrase captures the "propensity of things" in any situation); convergence can radically break down—over time or across geographical borders. When categories do break down in this fashion they leave no continuous trace back to the previous regime. So, for instance, when the category of "hysteric" became medically unfashionable, then people with (what used to be called) hysteria were distributed into multiple widely scattered categories. At that juncture, there was no point in their seeing the same doctors, or learning from each other what hysteria was.

### Scaling Up: Generalization and Standards

Similarity is an institution.
—*Mary Douglas (1986, 55)*

In this whole complicated coconstruction process, what are the things that make objects and statuses seem given, durable, and real? For, as Desrosières (1990) reminds us, partly through classification work, large-scale bureaucracies are very good at making objects, people, and institutions hold together. Some objects are naturalized in more than one world. They are not then boundary objects, but rather they become standards within and across the multiple worlds in which they are naturalized. Much of mathematics and, in the West, much of medicine and physiology fits this bill. In the Middle Ages a lot of Christian doctrine fit this, too. The hegemony of patriarchy rose from the naturalization of objects across a variety of communities of practice, with the exclusion of women from membership and the denial of their alternative interpretations of objects (Kramarae 1988, Merchant 1980, Croissant and Restivo 1995).

When an object becomes naturalized in more than one community of practice, its naturalization gains enormous power to the extent that a basis is formed for dissent to be viewed as madness or heresy. It is also where ideas like "laws of nature" get their power because we are always looking to other communities of practice as sources of validity, and if as far as we look we find naturalization, then the invisibility layers up and becomes doubly, triply invisible. Sherry Ortner's (1974) classic essay on "man: culture-woman: nature" shows that this has held for the subjugation of women even where specific cultural circumstances vary widely, and her model of the phenomenon rests on the persistent misunderstanding of borderlands and ambiguity in many cultures. Before her, Simone de Beauvoir (1948) wrote of the ethics of ambiguity, showing the powerful negative consequences of settling for one naturalized mode of interaction. We need an ethics of ambiguity, still more urgently with the pressure to globalize, and the integration of systems of representation through information technologies worldwide.

We have presented here a model of memberships, naturalizations, and the work we do in managing their multiplicity. Further analysis is needed to examine different types of categorical work and how they emerge under different circumstances. The next section continues with a discussion of boundary infrastructures.

### Boundary Infrastructure

Any working infrastructure serves multiple communities of practice simultaneously be these within a single organization or distributed across multiple organizations. A hospital information system, for example, has to respond to the separate as well as the combined agendas of nurses, records clerks, government agencies, doctors, epidemiologists, patients, and so forth. To do so, it must bring into play stable regimes of boundary objects such that any given community of practice can interface with the information system and pull out the kinds of information objects it needs.

Clearly boundary infrastructures are not perfect constructions. The chimera of a totally unified and universally applicable information system (still regrettably favored by many) should not be replaced by the chimera of a distributed, boundary-object driven information system fully respectful of the needs of the variety of communities it serves. To the contrary, as we saw in the case of NIC, nurses have needed to make a series of serious concessions about the nature and quality of their data before hoping to gain any kind of entry into hospital information systems. These difficulties generalize, though they are to some extent counterposed by processes of convergence.

Boundary infrastructures by and large do the work that is required to keep things moving along. Because they deal in regimes and networks of boundary objects (and not of unitary, well-defined objects), boundary infrastructures have sufficient play to allow for local variation together with sufficient consistent structure to allow for the full array of bureaucratic tools (forms, statistics, and so forth) to be applied. Even the most regimented infrastructure is ineluctably also local: if work-arounds are needed, they will be put into place. The ICD, for example, is frequently used to code cultural expectations (such as low heart attack rates in Japan) even though these are nowhere explicitly part of the classification system.

What we gain with the concept of boundary infrastructure over the more traditional unitary vision of infrastructures is the explicit recognition of the differing constitution of information objects within the diverse communities of practice that share a given infrastructure.

### Future Directions: Texture and Modeling of Categorical Work and Boundary Infrastructures

If you could say it, you would not need metaphor. If you could conceptualize it, it would not be metaphor. If you could explain it, you would not use metaphor.
*(Morton 1985, 210)*

So far this chapter has given a series of analytic categories that we hope will prove useful in the analysis and design of information infrastructures. At the limit, as Nelle Morton points out, we arrive at the sets of metaphors that people use to describe information networks of all kinds. These metaphors we live by are powerful means of organizing work and intellectual practice. We will now look at one cluster of metaphors—centered on the concept of filiations—which we believe, offers promise for future analytical work.

### How Are Categories Tied to People?

The frequency with which metaphors of weaving, threads, ropes and the like appear in conjunction with contextual approaches to human thinking is quite striking.
*(Cole 1996, 135)*

Categories touch people in a variety of ways—they are assigned, they become self-chosen labels, they may be statistical artifacts. They may be visible or invisible to any other group or individual. We use the term filiation here—related via Latin to the French "fil" for thread—as a thread that goes from a category to a person. This metaphor allows

---

*filiation (fIlI'eIS&schwa.n). Also 6 filiacion.*

1. Theol. The process of becoming, or the condition of being, a son.
   Many Dicts. have a sense 'adoption as a son,' illustrated by the first of our quotes from Donne. The sense is etymologically justifiable, and may probably exist; but quot. 1628 seems to show that it was not intended by Donne.

2. The designating (of a person) as a son; ascription of sonship.

3. The fact of being the child of a specified parent. Also, a person's parentage; "whose son one is."

4. The fact of being descended or derived, or of originating from; descent, transmission from.

5. The relation of one thing to another from which it may be said to be descended or derived; position in a genealogical classification.

6. Formation of branches or offshoots; chiefly concr., a branch or off-shoot of a society or language.

7. = Affiliation 3. lit. and fig. (*Oxford English Dictionary*, 2)

---

a rich examination of the architecture of the multiple categories that touch people's lives. Threads carry a variety of textural qualities that are often applied to human interactions: tension, knottiness or smoothness, bundling, proximity, and thickness. We select a small number here to focus on.

### Loosely Coupled—Tightly Coupled

A category (or system of categories) may be loosely or tightly coupled with a person. Gender and age are very tightly coupled with a person as categories. One of the interesting aspects of the investigation of virtual identities in Multi User Dungeons (MUDs) and elsewhere on line is the loosening of these traditionally tightly coupled threads under highly constrained circumstances (e.g., Turkle 1995). Loosely coupled categories may be those that are transient, such as the color one is wearing on a given day or one's position in a waiting line. Somewhere in the middle are hair color, which may shift slowly over a lifetime or change in an afternoon, or marital status.

### Scope

Categories' filiations have variable scope. Some are durable threads that cover many aspects of someone's identity and are accepted as such on a very wide or even global scale. (Noting for the record that none are absolute, none cover all aspects of someone's identity, and there is no category that is completely globally accepted.) The category alive or dead is quite thick and nearly global. So we can think of two dimensions of scope: thickness and scale. How thick is the individual strand—gossamer or thickest rope? With how many others is it shared?

### What Is Its Ecology?

Classifications have habitats. That is, the filiations between person and category may be characterized as inhabiting a space or terrain with some of the properties of any habitat. It may be crowded or sparse, peaceful or at war, fertile or arid. In order not to mix too many metaphors. Important questions about filiations and their ecology that may be visualized in thread-like terms are: How many ties are there? That is, how many other categories are tied to this person, and in what density? Do these threads contradict or complement (torque versus boundary object of cooperation)? That is, are the threads tangled, or smoothly falling together?

### Who Controls the Filiation?

The question of who controls any given filiation is vital to an ethical and political understanding of information systems whose categories attach to individuals. A first crude characterization concerns whether the filiation was chosen or imposed (an echo of the sociological standard, achieved or ascribed); whether it may be removed and by whom; and under whose control and access is the apparatus to do so. Questions of privacy are important here, as with medical information classifying someone with a social stigmatized condition. The nature of the measure for the filiation here is important loci of control as well. For example, an IQ test may be an important way to classify people. People at some remove from those who take the test developed it. The measure, IQ, is controlled from afar. On these grounds, past criticisms of IQ tests charge that this control is racially biased and biased by gender.

### Is It Reversible or Irreversible?

Finally, there is the important question of whether the filiation is reversible. The metaphor of branding someone is not accidental in this regard, branding meaning that a label is burned into the skin and completely irreversible. Some forms of filiation have this finality for the individual, regardless of how the judgment was later regarded (e.g., a charge of guilt for murder may mean permanent public guilt regardless of a jury's verdict. Many are somewhere between, but knowing how reversible is the filiation is important for understanding its impact.

The metaphor of filiation presented here could be used to characterize a texture of information systems where categories touch either individuals or things. The aesthetics of the weave and the degree to which the individual is bound up or supported by it are among the types of characterizations that could be made. There are brute renderings, such as having two thick, irreversible threads tying one person to conflicting categories. More subtly, it is possible to think of something like Granovetter's strength of weak ties and characterize the thousand and one classifications that weakly tie people to information systems as binding or torquing in another way.

The metaphor of filiation is useful to the extent that it can be used to ask questions of working infrastructures in new and interesting ways. Two questions that rise directly out of our treatment of the metaphor for any individual or group filiation are: What will be the ecology and distribution of suffering? Who controls the ambiguity and visibility of categories?

## Conclusion

This chapter has argued that there is more to be done in the analysis of classification systems than deconstructing universal master narratives. Certainly, such narratives should be challenged. We have attempted to show, however, that there are ways of scaling up from the local to the social, via the concept of boundary infrastructures, and that we can in the process recognize our own hybrid natures without losing our individuality. The value of this approach is that it allows us to intervene in the construction of infrastructures—which surely exist and are powerful—as not only critics but also as designers.

# 10
## Why Classifications Matter

At the beginning of this book we told the story of the homicidal maniac who needed the insight of a psychic to understand his murderous urges as such. "Don't you get it, son? You're a homicidal maniac." End of explanation. The story is powerful and funny because it reminds us, ironically, that a classification is not of itself an explanation. All we understand at the end of the scene is that the maniac now has a label that others, and he himself, can apply to his behavior. Although the classification does not provide psychological depth, it does tie the person into an infrastructure—into a set of work practices, beliefs, narratives, and organizational routines around the notion of "serial killer." Classification does indeed have its consequences—perceived as real, it has real effect.

Classifications are powerful technologies. Embedded in working infrastructures they become relatively invisible without losing any of that power. In this book we demonstrate that classifications should be recognized as the significant site of political and ethical work that they are. They should be reclassified.

In the past 100 years, people in all lines of work have jointly constructed an incredible, interlocking set of categories, standards, and means for interoperating infrastructural technologies. We hardly know what we have built. No one is in control of infrastructure; no one has the power centrally to change it. To the extent that we live in, on, and around this new infrastructure, it helps form the shape of our moral, scientific, and esthetic choices. Infrastructure is now the great inner space.

Ethnomethodologists and phenomenologists have shown us that what is often the most invisible is right under our noses. Everyday categories are precisely those that have disappeared into infrastructure, into habit, into the taken for granted. These everyday categories are seamlessly interwoven with formal, technical categories and specifications. As Cicourel notes:

The decision procedures for characterizing social phenomena are buried in implicit common sense assumptions about the actor, concrete persons, and the observer's own views about everyday life. The procedures seem intuitively "right" or "reasonable" because they are rooted in everyday life. The researcher often begins his classifications with only broad dichotomies, which he expects his data to "fit," and then elaborates on these categories if apparently warranted by his "data." (1964, 21)

The hermeneutic circle is indeed all around us.

There is no simple unraveling of the built information landscape, or, *pace* Zen practice, of unsettling our habits at every waking moment. Black boxes are necessary, and not necessarily evil. The moral questions arise when the categories of the powerful become the taken for granted; when policy decisions are layered into inaccessible technological structures; when one group's visibility comes at the expense of another's suffering.

There are as well basic research questions implied by this navigation into infrastructural space. Information technology operates through a series of displacements, from action to representation, from the politics of conflict to the invisible politics of forms and bureaucracy. Decades ago, Max Weber wrote of the iron cage of bureaucracy. Modern humans, he posited, are constrained at every juncture from true freedom of action by a set of rules of our own making. Some of these rules are formal, most are not. Information infrastructure adds another level of depth to the iron cage. In its layers, and in its complex interdependencies, it is a gossamer web with iron at its core.

We have looked at several sets of classification schemes—the classifications of diseases, viruses, tuberculosis, race, and of nursing work. These are all examples of working classification systems: they are or have been maintained by organizations, governments, and individuals. We have observed several dances between classifier and classified, but have nowhere seen either unambiguous entities waiting to be classified or unified agencies seeking to classify them. The act of classification is of its nature infrastructural, which means to say that it is both organizational and informational, always embedded in practice (Keller and Keller 1996).

In our interviews of public health officials, nurses, or scientists, we have found that they recognize this about their own classification systems. At the same time, there is little inducement to share problems across domains. Because of the invisible work involved in local struggles with formal classification systems and standards, a great deal of what sociologists would call "pluralistic ignorance" obtains. There is the feeling that "I am the only one." People often have a picture that somehow their problems are unique: they believe that other "real" sciences do not have the same set of makeshift compromises and work-arounds.

It is important in the development and implementation of classifications (and many related fields such as the development and deployment of standards or archives) that we get out of the loop of trying to emulate a distant perfection that on closer analysis turns out to be just as messy as our own efforts. The importance lies in a fundamental rethinking of the nature of information systems. We need to recognize that all information systems are necessarily suffused with ethical and political values, modulated by local administrative procedures. These systems are active creators of categories in the world as well as simulators of existing categories. Remembering this, we keep open and can explore spaces for change and flexibility that are otherwise lost forever.

Such politics are common to most systems employing formal representations. Rogers Hall, in his studies of algebra problem solving by both children and professional math teachers, talks about the shame that children feel about their unorthodox methods for arriving at solutions (1990). Often using innovative techniques such as imaginary devices, but not traditional formulaic means, they achieved the right answer the wrong way. One child called this "the dirt way." A grown-up version of the dirt way is related by the example given earlier of the "good organizational reasons for bad organizational records" (Bitner and Garfinkel 1967). There are good organizational reasons for working around formal systems; these adaptations are necessarily local. What is global is the need for them.

In this book we have attempted to develop tools for maintaining these open spaces. Michel Serres has best expressed the fundamental ethical and political importance of this task. He has argued that the sciences are very good at what they do: the task of the philosopher is to keep open and explore the spaces that otherwise would be left dark and unvisited because of their very success, since new forms of knowledge might arise out of these spaces. Similarly, we need to consistently explore what is left dark by our current classifications ("other" categories) and design classification systems that do not foreclose on rearrangements suggested by new forms of social and natural knowledge.

There are many barriers to this exploration. Not least among them is the barrier of boredom. Delving into someone else's infrastructure has about the entertainment value of reading the yellow pages of the phone book. One does not encounter the dramatic stories of battle and victory, of mystery and discovery that make for a good read.

In an introductory chapter we laid the theoretical framework for the discussion of classification as an infrastructural practice, stressing the political and ethical texturing of classification schemes. In part I we examined the International Classification of Diseases (ICD) as a large-scale, long-term system ingrained in the work practices of multiple organizations and states. We argued that their organizational roots and operational exercise texture such systems. Such texture is an inescapable, appropriate feature of their constitution, and it is a feature that merits extended consideration in a discussion of the politics of infrastructure. In part II, we looked at the intersection between classification and individual biography in the case of the classification of tuberculosis and of race classification under apartheid in South Africa. Generalizing the arguments made in these chapters, we maintained that individuals in the modern state operate within multiple classification systems, from the small-scale, seminegotiated system—as with the informal classification of tuberculosis patients negotiated with doctors—up to enforced universal systems such as race classification. We drew attention to the torquing of individual biographies as people encounter these reified classifications. Finally, we examined classification and work practice, taking the example of the classification of nursing work. We argued that multiple tensions between representation and autonomy, disability and discretion, forgetting the past and learning its lessons, make such classifications a key site of political and professionalization work. We are all called upon to justify our productivity when we are embedded in complex modern organizations. The dilemma faced by nurses in accounting for their work is omnipresent in the modern organization. Even children are not exempt.

We have seen throughout this book that people (and the information systems they build) routinely conflate formal and informal, prototypical and Aristotelian aspects of classification. There is no such thing as an unambiguous, uniform classification system. (Indeed, the deeper one goes into the spaces of classification expertise—for example, librarianship or botanical systematics—the more perfervid one finds the debates between rival classificatory schools.) This in turn means that there is room in the constitution of any classification system with organizational and political consequences—and few schemes if any are without such dimensions—for technical decisions about the scheme to systematically reflect given organizational and political positions. Since we are dealing, then, with an agonistic field, there will be no pure reflection of a single position but rather dynamic tensions among multiple positions. And finally, since the classification system is not a pure reflection of such positions (an impossible aim in its own right—no classification system can reflect either the social or the natural world fully accurately) but also integrally a tool for exploring the real world, there is no simple prediction from how a given set of alliances or tensions leads invariably to a given classification used in a given way.

As sets of classification systems coalesce into working infrastructures they become integrated into information systems of all sorts. Thus we have argued throughout this book that information systems design should be informed by organizational and political analysis at this level. We are not offering this as an ex cathedra design principle. Rather, we have—along with many researchers in the field of social informatics—demonstrated empirically that invisible organizational structures influence the design and use of systems: the question is not whether or not this occurs but rather how to recognize, learn from, and plan for the ineluctable presence of such features in working infrastructures. We have suggested one design aid here—long-term and detailed ethnographic and historical studies of information systems in use—so that we can build up an analytic vocabulary appropriate to the task.

Working infrastructures contain multiple classification systems that are both invisible, in the senses above, and ubiquitous. The invisibility of infrastructure makes visualization or description difficult. The metaphors we reach for to describe infrastructure are ironic and somehow childish. We speak of "way down in the underwear," "underneath the system," or use up-down metaphors such as "runs under," or "runs on top of." Lakoff and Johnson (1980) write of metaphors we live by. Our infrastructural metaphors show how baffled we often are by these systems. They are like undergarments or tunnel dwellers.

Another set of metaphors often used in organizations speaks indirectly to the experience of infrastructure. These are the metaphors of texture omnipresent in human relationships. Texture metaphors speak to the densely patterned interaction of infrastructures and the experience of living in the "classification society." Texture speaks to the way that classifications and standards link the individual with larger processes and structures. These links generate both enabling-constraining patterns over a set of systems (texture) and developmental patterns for an individual operating within a given set (trajectory).

Thus we have used the metaphor of the texture of a classification system to explore the fact that any given classification provides surfaces of resistances (where the real resists its definition), blocks against certain agendas, and smooth roads for others. Within this metaphorical landscape, the individual's trajectory—often, for all that, perceived as continuous and self-consistent—is at each moment twisted and torqued by classifications and vice versa.

Therefore we have, through our analysis of various classification systems, attempted to provide a first approximation to an analytic language that recognizes that the architecture of classification schemes is simultaneously a moral and an informatic one. This book has brought to light as crucial to the design process the reading of classification schemes as political and cultural productions. We have stressed that any classification scheme can be read in this fashion. We initially deliberately eschewed cases like DSM-IV, where categories have often already become explicit objects of political contention, such as "homosexual" or "premenstrual tension." In the psychiatric case, there can in this sense often be a more direct read-off from political exigencies to disease categories. Although such readings are of course highly valuable in their own right (see Kirk and Kutchins 1992, Kutchins and Kirk 1997, and Figert 1996), we first took the more muted cases posed by the ICD where the politics were quieter. This we hoped would show the generalizability of the thesis that all category systems are moral and political entities. This was balanced later in the book with an analysis of the much more obviously politically laden categories generated by the proapartheid government and its scientific apologists.

This book has implications for both designers and users (and we are all increasingly both) of complex information spaces. It provides intellectual and methodological tools for recognizing and working with the ethical and political dimensions of classification systems. In particular we have underlined several design exigencies that speak both to the architecture of information systems encoding classification systems and to their development and change:

• *Recognizing the balancing act of classifying.* Classification schemes always represent multiple constituencies. They can do so most effectively through the incorporation of ambiguity—leaving certain terms open for multiple definitions across different social worlds: they are in this sense boundary objects. Designers must recognize these zones of ambiguity, protecting them where necessary to leave free play for the schemes to do their organizational work.

• *Rendering voice retrievable.* As classification systems get ever more deeply embedded into working infrastructures, they risk getting black boxed and thence made both potent and invisible. By keeping the voices of classifiers and their constituents present, the system can retain maximum political flexibility. This includes the key ability to be able to change with changing natural, organizational, and political imperatives. A caveat here, drawn from chapter 7's lesson about the invisibility of nursing work: we are not simply celebrating visibility or naively proposing a populist agenda for the empire of classification. Visibility is not an unmitigated good. Rather, by retrievability, we are suggesting that under many circumstances, the "rule by no one" or the "iron cage of bureaucracy" is strengthened by its absence. When classification systems and standards acquire inertia because they are part of invisible infrastructure, the public is de facto excluded from policy participation.

• *Being sensitive to exclusions.* We have in particular drawn attention here to the distribution of residual categories (who gets to determine what is "other"). Classification systems always have other categories, to which actants (entities or people) who remain effectively invisible to the scheme are assigned. A detailed analysis of these others throws into relief the organizational structure of any scheme (Derrida 1998). Residual categories have their own texture that operates like the silences in a symphony to pattern the visible categories and their boundaries.

Stewart Brand's (1994) wonderful book, *How Buildings Learn,* gives many examples of how buildings get designed as they are used as much as on the architect's drawing board. Thus a house with a balcony and numerous curlicues around the roof will become a battened-down square fortress block under the influence of a generation of storms from the northeast. Big single-family mansions become apartment buildings as a neighborhood's finances change. These criteria generalize to classification systems. Through these three design criteria we are drawing attention to the fact that architecture becomes archaeology over time. This in turn may become a cycle.

Overall, we have argued that classifications are a key part of the standardization processes that are themselves the cornerstones of working infrastructures. People have always navigated sets of classification spaces. Mary Douglas (1984), among others, has drawn attention to this feature of all societies from the indigenous and tribal to the most industrialized.[58] Today, with the emergence of new information infrastructures, these classification systems are becoming ever more densely interconnected. This integration began roughly in the 1850s, coming to maturity in the late nineteenth century with the flourishing of systems of standardization for international trade and epidemiology. Local classification schemes (of diseases, nursing work, viruses) are now increasingly giving way to these standardized international schemes that themselves are being aligned with other large-scale information systems. In this process, it is becoming easier for the individual to act and perceive him or her self as a completely naturalized part of the "classification society," since this thicket of classifications is both operative (defining the possibilities for action) and descriptive.[59] As we are socialized to become that which can be measured by our increasingly sophisticated measurement tools, the classifications increasingly naturalize across wider scope. On a pessimistic view, we are taking a series of increasingly irreversible steps toward a given set of highly limited and problematic descriptions of what the world is and how we are in the world.

For these reasons, we have argued in this book that it is politically and ethically crucial to recognize the vital role of infrastructure in the "built moral environment." Seemingly purely technical issues like how to name things and how to store data in fact constitute much of human interaction and much of what we come to know as natural. We have argued that a key for the future is to produce flexible classifications whose users are aware of their political and organizational dimensions and which explicitly retain traces of their construction. In the best of all possible worlds, at any given moment, the past could be reordered to better reflect multiple constituencies now and then. Only then we will be able to fully learn the lessons of the past. In this same optimal world, we could tune our classifications to reflect new insitutional arrangements or personal trajectories—reconfigure the world on the fly. The only good classification is a living classification.

# Notes

1. Two notable exceptions are Lucy Suchman and Sanford Berman. Suchman's article challenging the categories implicit in a popular software system was entitled "Do Categories Have Politics?" (Suchman 1994). This article/critique has helped open up the discussion of values and categories in the field of computer-supported cooperative work (CSCW). It is, importantly, a gloss on an earlier article by Langdon Winner (1986), "Do Artifacts Have Politics?" which similarly drew attention to the moral values inscribed in aspects of the built environment. Berman (1984, 1993) has done invaluable work in the library community with his critiques of the politics of catologuing. See also *Library Trends* special issue on classification, edited by Geoffrey Bowker and Susan Leigh Star (1998).

2. As authors, we recognize that "we" is problematic here and throughout this work. At the same time, it would be awkward to qualify each of these sentences by saying Western, academic, middle-class people. We the authors recognize that not everyone—Western or not—holds individualist, rational choice moral models. Where possible, we have tried to qualify the voice assumed throughout this book. Furthermore, the book's entire argument is directed at subverting any sense of an overriding master voice. We are grateful to Kathy Addleson for bringing the question of voice to our attention.

3. As Holmes explained to Watson when he uncovered the chain of deductions (each link so simple) that allowed him to produce a thrust of "magical" insight. See Star and Strauss (1999).

4. O'Connell (1993) gives a fine analysis of the development of electrical standards. The study of standards has been an exciting strand in recent science studies—as witness recent work in *Social Studies of Science* devoted to the topic: Alder (1998), Curtis (1998), Mallard (1998), and Timmermans and Berg (1997).

5. The Journal of Online Nursing at http://www.nursingworld.org/ojin/tpc7/intro.htm presents a good introduction to issues of classification in nursing.

6. Chapter title shamelessly stolen from Howard Becker's *Tricks of the Trade* (Chicago: University of Chicago Press 1998).

7. This is formally similar to Hewitt's open systems properties (1986); see also Star (1989).

8. We refer to Latour (1993 and 1996a, b) in seeing knowledge production and political work as twin outcomes of a single set of processes.

9. This useful term means administrative procedures, things, and technologies that are combined to produce a given effect. The punishment of a prisoner in jail, for instance, is a dispositif technique combining walls and bars, prison procedures and routines, judicial rules, and computerized crime records. In this, the term is close to that of Kling and Scacchi's notion of the "web of computing"—workable computer systems mean that hardware, software, and organizational and cultural mores are working together (1982).

10. We borrow this phrase from Hacking (1995); it is also explored in C. Becker (1967).

11. Social informatics is the study of the design, use, and impact of information technology considered from the point of view of social organization.

12. One finds similar complaints today about the World Wide Web, to the point where a special electronic journal has been founded: Journal of Internet Cataloging: The International Quarterly of Digital Organization, Classification, and Access. Its URL is: http://jic.libraries.psu.edu/. See also Marcia Bates' (in press) excellent article on incomparability between Web search engines.

13. Under external causes of morbidity and mortality, contact with venomous snakes and lizards is X20. There is a list of eight snakes and one Gila monster to be included, but these are not broken down in the actual coding. So the rural inhabitant could not distinguish the density of sidewinders versus rattlesnakes, as they may well want to do for safety purposes.

14. Rodney King was stopped and beaten by several police officers; this was captured on videotape and led to a celebrated trial involving issues of due force.

15. Ironically, the slogan, "nobody dies of old age" was an anti-ageist aphorism first popular in the 1980s and used by groups such as the Grey Panthers. It was meant to imply that the social invisibility of old people led to them being medically invisible or overlooked as well. It is an interesting example of the inversion of the prototypical and Aristotelian aspects of death!

16. As Everett Hughes (1970) was fond of saying about sociological analysis: "It might have been otherwise."

17. The general principle is: "when more than one condition is entered on the certificate, the condition entered alone on the lowest used line of part I should be selected only if could have given rise to all the conditions entered above it" (ICD-10, 3: 34).

18. *Pace* cybernetics (Bowker 1993).

19. A famous example of such bootstrapping from the history of science is the story of Newton's prism used for his optical experiments. Italian researchers got different results using different prisms; and Newton only succeeded in establishing the veracity of his experiments once he had succeeded in imposing his prism as the standard, and he could therefore ascribe failures to replicate his experiment to defective prisms. The only way of choosing between Italian and English prisms, however, was whether or not they gave suitable results to Newton's experiments (Schaffer 1989). This interpretation has been questioned by Shapiro (1996); our thanks to an anonymous reviewer for pointing this out.

20. Again, similar to the story H. Becker (1982) tells of the intertwining of aesthetics with materials and conventions in his classic *Art Worlds*.

21. DRGs are used for medical accounting and rely on rearrangements of medical classifications and procedures.

22. This is essentially the same as what organizational theorists call the garbage can approach to decision making. Since the garbage category has a specific meaning here, we have maintained that terminology.

23. The original scientific aphorism, attributed to the medieval philosopher William of Occam, was "thou shalt not multiply entities without necessity." It is often interpreted as a value of parsimony in scientific explanation; equally, here, it applies to the design of forms!

24. AIDS presents a similar challenge as a condition, not per se a disease, and equally protean in expression.

25. Roth (1963) makes an eloquent analysis of how this image has come to be a powerful one in the medical literature; he argues it is in fact statistically quite rare.

26. In the European context here, "sister" means "nurse."

27. After the 1847 Dumas novel, *La Belle Dame aux Camélias*.

28. At this point, following Dubow (1995), we stop putting quotation marks around words such as race or coloured. We trust the reader to recognize that the entire argument here opposes any essentialist or simplistic interpretation of these terms, or acceptance of racist constructions! Except in direct quotes, we conform to the MIT style of using lowercase for 'black', 'white', and 'colored'. The South African usage was not standardized.

29. "They will count us. My friend was checked yesterday. Count us, count us!" (authors' translation from the Afrikaans).

30. An early antiapartheid organization noted for greeting officials at airports and the like wearing black sashes of protest.

31. Although DeKlerk in 1962 attributes this to purely technical reasons, "to use descent as a test it would have meant digging far back into the past for

proof, and the moment one has to start digging into the past one becomes lost in a labyrinth . . . in earlier years there was no reliable record of many of these facts. For example, in some of the provinces facts of this kind this have only been noted since 1915 in registrations of births and deaths. In other words one cannot trace the origin and the race of the person" (De Klerk 1962, 9).

32. "Kaffir" is a rude word equivalent to "nigger."

33. Not surprisingly, the school officials were of little help. The mother superior of a convent explained, "If I could have had my way I would have admitted the little girl. But we depend on public goodwill, and as I see it we would only have trouble if we admitted her. We have to consider the feelings of our parents and children" (*Ebony* June 1968, 88).

34. "This classification provides a minimum standard for maintaining, collecting, and presenting data on race and ethnicity for all federal reporting purposes. The categories in this classification are social-political constructs and should not be interpreted as being scientific or anthropological in nature. They are not to be used as determinants of eligibility for participation in any federal program. The standards have been developed to provide a common language for uniformity and comparability in the collection and use of data on race and ethnicity by federal agencies.

The standards have five categories for data on race: American Indian or Alaska Native, Asian, Black or African American, Native Hawaiian or other Pacific Islander, and white. There are two categories for data on ethnicity: "Hispanic or Latino," and "not Hispanic or Latino."

1. Categories and definitions
The minimum categories for data on race and ethnicity for federal statistics, program administrative reporting, and civil rights compliance reporting are defined as follows:

• American Indian or Alaska Native. A person having origins in any of the original peoples of North and South America (including Central America), and who maintains tribal affiliation or community attachment.

• Asian. A person having origins in any of the original peoples of the Far East, Southeast Asia, or the Indian subcontinent, including, for example, Cambodia, China, India, Japan, Korea, Malaysia, Pakistan, the Philippine Islands, Thailand, and Vietnam.

• Black or African American. A person having origins in any of the black racial groups of Africa. Terms such as "Haitian" or "negro" can be used in addition to "black or African American."

• Hispanic or Latino. A person of Cuban, Mexican, Puerto Rican, Cuban, South or Central American, or other Spanish culture or origin, regardless of race. The term, "Spanish origin," can be used in addition to "Hispanic or Latino."

• Native Hawaiian or other Pacific Islander. A person having origins in any of the original peoples of Hawaii, Guam, Samoa, or other Pacific Islands.

• White. A person having origins in any of the original peoples of Europe, the Middle East, or North Africa.

Respondents shall be offered the option of selecting one or more racial designations. Recommended forms for the instruction accompanying the multiple response question are "mark one or more" and "select one or more."

2. Data Formats
The standards provide two formats that may be used for data on race and ethnicity. Self-reporting or self-identification using two separate questions is the preferred method for collecting data on race and ethnicity. In situations where self-reporting is not practicable or feasible, the combined format may be used.

In no case shall the provisions of the standards be construed to limit the collection of data to the categories described above. The collection of greater detail is encouraged; however, any collection that uses more detail shall be organized in such a way that the additional categories can be aggregated into these minimum categories for data on race and ethnicity.

With respect to tabulation, the procedures used by federal agencies shall result in the production of as much detailed information on race and ethnicity as possible. Federal agencies shall not present data on detailed categories, however, if doing so would compromise data quality or confidentiality standards.

a. Two-question format
To provide flexibility and ensure data quality, separate questions shall be used wherever feasible for reporting race and ethnicity. When race and ethnicity are collected separately, ethnicity shall be collected first. If race and ethnicity are collected separately, the minimum designations are:

Race:

• American Indian or Alaska Native

• Asian

• Black or African American

• Native Hawaiian or Other Pacific Islander

• White

Ethnicity:

• Hispanic or Latino

• Not Hispanic or Latino

When data on race and ethnicity are collected separately, provision shall be made to report the number of respondents in each racial category who are Hispanic or Latino." See http://www.ameasite.org/omb15v97.html.

35. Our colleague Stefan Timmermans provided valuable assistance on earlier drafts of the argument in this chapter. We gratefully acknowledge his help. (See Timmermans, Bowker, and Star 1998.)

36. Although it may appear at first sight that comparability and standardization are the same thing, we see an important difference between the two concepts. Two things can be comparable but not standardized. You can compare an education at Harvard with an education at the local community college, for example, because you know that in general a lot more resources are pumped into Harvard and outcomes tend to be different because of the homogeneity of backgrounds. In this case, one would be high on the comparability side of standardization but low on the standardization side: no exact metric exists for the differences. If you then subject all students to a single standardized test, you have to measure comparability to provide standardization (and in the case of comparing educational systems, this is both politically and organizationally complex and fraught).

37. The NIC principal investigators maintain that at present there is effectively no scientific nursing knowledge: it is only with the creation and maintenance of a stable classification system that the groundwork will have been done to make such knowledge attainable.

38. Personal communication.

39. From a talk given at the Program for Cultural Values and Ethics, University of Illinois, December 1993.

40. See Michael Lynch's (1984) work on turning up signs in neurological diagnosis for an example of the inexhaustible discretion and improvisation in every human activity—the study of which has been a major contribution of ethnomethodology and phenomenology.

41. See Wagner (1993), Egger and Wagner (1993), Gray et al. (1991), and Strong and Robinson (1990).

42. Strauss et al. (1985) call this activity articulation work.

43. One of their arguments is that the truth of a memory is constructed in discourse in social settings and so is never fixed for all time.

44. Notes taken at Iowa intervention project meeting of 8 June 1995. (Hereinafter IIP 6/8/95.)

45. This is clearly a reference to Thompson's classic (1967) "Time, Work Discipline and Industrial Capitalism." It is questionable of course whether all nursing has ever been thought of as process, just as industrial work has often had its rhythmic side (the cycles of boom and bust in the eighteenth and nineteenth centuries, for example).

46. The translations from Comte's French are Bowker's.

47. Such a way of thinking is common in art, myth, and literature—especially in surrealist art and multivocal fiction and film—and in aspects of feminist and race-critical theory.

48. The two other types are (1) formal or axiomatic approaches and (2) encyclopedic listings with flattened or standardized nomenclatures. Both present other sorts of equally interesting political problems (Star 1989).

49. The term community of practice is interchangeable with the term social world (Strauss 1978, Clarke 1991, 1990) although they have different historical origins.

50. Clearly questions of language are central here as well, and we do not mean to exclude them by emphasizing things. Language considered as situated tool, in relationship with other tools and things, is part of this model.

51. The work of Schütz (1944) and subsequent ethnomethodologists such as Cicourel, Sacks, and Schegeloff, among many others, investigates this naturalization process through language.

52. Deconstructing this invisibility is one of the major shared projects of ethnomethodology, symbolic interactionist studies of science and of gender, and the Annalist school of historiography.

53. Things, strictly speaking, do not analytically have membership, in the sense of negotiated identity.

54. Thanks to Peter Garrett for insightful discussions of this topic.

55. We borrow the phrase from Howard Becker's classic, "A School is a Lousy Place to Learn Anything In," an essay that covers related ground (1972).

56. One of the intriguing features of electronic interaction is that it makes disclosure of these memberships voluntary, or at least problematic, where participants do not know each other in real life.

57. This distinction is in line with Strauss' original distinction between production work and articulation work (1988, Strauss et al. 1985).

58. The classification of societies, ranging from "primitive" to "developed" is of course a particularly tendentious one with its own complex political history. For a direct criticism from the library vantagepoint, see Berman (1993, 1984) and Dodge and DeSirey (1995).

59. We are grateful to Ina Wagner (personal communication July 8, 1998) for coining this term.

# References

## Archival Material

CH/experts stat/78. Dr. E. Roesle, "The International Recommendations for Determining the Causes of Death Drawn up by the Health Section of the League of Nations in 1925, and their Applicability as Regards the Reform of the German Statistics of Causes of Death."

CH/experts stat/34. 2 December 1927. "Report of the Committee of the Vital Statistics Section of the American Public Health Association on the Accuracy of Certified Causes of Death and its Relation to Mortality Statistics and the International List."

C. H./experts stat/46. Doc. 43806, Doss. 22685. 22 December, 1927. "Report of the Committee on Definition of Stillbirth." League of Nations Health Organization, Commission of Expert Statisticians.

CH/experts stat/80. "Government Commission on Morbidity and Mortality Statistics in Austria," with reference to the fourth revision.

CH/experts stat/43. 20 December 1927. Dr. P. I. Kurkin, "Note sur la nomenclature des maladies et des causes de décès en Russie."

CH/experts stat/88. 19 March, 1929. Dr. Teleky, "La Statistique de Morbidité des Caisses d'Assurance-Maladie en Allemagne."

CH/experts stat/87. Registrar General of England and Wales, "Observations upon Dr. Roesle's Memorandum upon the Comparative Study of Morbidity."

Commission Internationale Nomenclature Internationale des Maladies. 1910. Procès Verbaux.

La Réunion du Conseil de la Société des Nations. 1923. "Saint Sebastian et l'Organisation Internationale d'Hygiène Publique."

Société des Nations. 1921. Box R822. File12458. Session of the OIHP (Paris 1921).

Société des Nations, Organisation d'Hygiène, Commission d'Experts Statisticiens, CH/experts stat/1–43. 1927. "Communication du Chef de Service de la Statistique Médicale au Ministère Polonnais de l'Intèrieure."

WHO Archives. 453–1–4. 1946. E. J. Pampana. "Malaria as a problem for the WHO."

WHO archives. 455–3–3. 1947. "Collaborating with International Institute of Statistics, 29/1/47, Huber."

WHO archives. 455–3–4. 31/3/48. Expert Committee for the Preparation of the sixth Decennial Revision of the International List of Diseases and Causes of Death, "Assignment of Causes of Death."

*Printed Material*

Abbate, J., and B. Kahin (eds.). 1995. *Standards Policy for Information Infrastructure*. Cambridge, MA: MIT Press.

Abbott, Andrew. 1988. *The System of Professions: An Essay on the Division of Expert Labor*. Chicago: University of Chicago Press.

Addelson, Kathryn Pyne. 1994. *Moral Passages: Toward a Collectivist Moral Theory*. New York: Routledge.

Akrich, Madeleine, and Bruno Latour. 1992. "A Summary of a Convenient Vocabulary for the Semiotics of Human and Nonhuman Assemblies." In W. E. Bijker and J. Law (eds.). *Shaping Technology/Building Society: Studies in Sociotechnical Change*. Cambridge, MA: MIT Press, 259–264.

Akrich, Madeleine. 1992. "The De-Scription of Technical Objects." In W. E. Bijker and J. Law (eds.). *Shaping Technology/Building Society: Studies in Sociotechnical Change*. Cambridge, MA: MIT Press, 205–224.

Alder, Ken. 1998. "Making Things the Same: Representation, Tolerance, and the End of the Ancien Regime in France." In *Social Studies of Science* 28: 499–546.

Allegre, Claude J. 1992. *From Stone to Star: A View of Modern Geology*. Cambridge, MA: Harvard University Press.

Alston, Kal. 1993. "A Unicorn's Memoirs: Solitude and the Life of Teaching and Learning." In Delear Wear (ed.). *The Center of the Web: Women and Solitude*. New York: SUNY, 95–107.

Anderson, Robert E. 1984. "The Autopsy as an Instrument of Quality Assessment: Classification of Premortem and Postmortem Diagnostic Discrepancies." In *Archives of Pathology and Laboratory Medicine* 108: 490–493.

Anderson, Warwick. 1996. "Race, Medicine, and Empire." In *Bulletin of the History of Medicine* 70: 62–75.

Anzaldúa, Gloria. 1987. *Borderlands = La Frontera: The New Mestiza*. San Francisco: Spinsters/Aunt Lute.

Arendt, Hannah. 1963. *Eichmann in Jerusalem: A Report on the Banality of Evil*. New York: Viking Press.

Atran, Scott. 1990. *Cognitive Foundations of Natural History: Towards an Anthropology of Science*. Cambridge, UK: Cambridge University Press.

Bamford, B. R. 1967. "Race Reclassification." *In South African Law Journal* 84 (February): 37–42.

Bannon, Liam, and K. Kuutti. 1996. "Shifting Perspectives on Organizational Memory: From Storage to Active Remembering." In *Proceedings of the 29th HICSS*, Vol.III, *Information Systems—Collaboration Systems and Technology*. Washington, DC: IEEE Computer Society Press, 156–167.

Bardet, Jean-Pierre, Patrice Bourdelais, Pierre Guillaume, François Lebrun, and Claude Quêtel, eds. 1988. *Peurs et Terreurs Face à la Contagion*. Paris: Fayard.

Barnett, G. Octo. 1975. *COSTAR Computer-Stored Ambulatory Record*. Boston, MA: Massachusetts General Hospital.

Barnes, Barry, David Bloor, and John Henry. 1996. *Scientific Knowledge: A Sociological Analysis*. London: Athlone.

Bates, Barbara. 1992. *Bargaining for Life: A Social History of Tuberculosis, 1876–1938*. Philadelphia: University of Pennsylvania.

Bates, Marcia. In press. "Indexing and Access for Digital Libraries and the Internet: Human, Database, and Domain Factors." Journal of the American Society for Information Science.

Baudrillard, Jean. 1990. *Cool Memories*. New York: Verso.

Beauvoir, Simone de. 1948. *The Ethics of Ambiguity*: translated by Bernard Frechtman. Secaucus, N.J.: Citadel Press.

Beck, Eevi. 1995. "Changing Documents/Documenting Changes: Using Computers for Collaborative Writing over Distance." In Susan Leigh Star (ed.) *The Cultures of Computing*. Oxford: Blackwell, 53–68.

Becker, Carl L. 1967. *Detachment and the Writing of History: Essays and Letters of Carl L. Becker*. Phil L. Snyder (ed.). Ithaca, NY: Cornell University Press.

Becker, Howard S. 1953–54. "Becoming a Marihuana User." *American Journal of Sociology* 59: 235–42.

Becker, Howard S. 1963. *Outsiders; Studies in the Sociology of Deviance*. London: Free Press of Glencoe.

Becker, Howard S. 1972. "A School is a Lousy Place to Learn Anything In." In *American Behavioral Scientist* 16: 85–105.

Becker, Howard S. 1982. *Art Worlds.* Berkeley, CA: University of California Press.

Becker, Howard S. 1986. *Doing Things Together: Selected Papers.* Evanston, IL: Northwestern University Press.

Becker, Howard S. 1996. "The Epistemology of Qualitative Research." In Richard Jessor, Anne Colby, and Richard A. Shweder (eds.). *Ethnography and Human Development; Context and Meaning in Social Inquiry.* Chicago, University of Chicago: 53–71.

Becker, Howard S. 1998. *Tricks of the Trade.* Chicago: University of Chicago Press.

Beghtol, Claire. 1995. "'Facets' as Interdisciplinary Undiscovered Public Knowledge: S. R. Ranganathan in India and L. Guttman in Israel." *Journal of Documentation* 51: 194–224.

Beniger, James R. 1986. *The Control Revolution: Technological and Economic Origins of the Information Society.* Cambridge, MA: Harvard University Press.

Bensaude-Vincent, Bernadette. 1989. "Lavoisier: Une Révolution Scientifique." In Michel Serres (ed.). Eléments d'Histoire des Sciences. Paris: Bordas, 363–386.

Berg, Marc. 1992. "The Construction of Medical Disposals: Medical Sociology and Medical Problem Solving in Clinical Practice." In *Sociology of Health and Illness.* 14: 151–180.

Berg, Marc. 1997a. "Formal Tools and Medical Practices: Getting Computer-Based Decision Techniques to Work." In Geoffrey C. Bowker, Les Gasser, Susan Leigh Star, and William Turner (eds.). *Social Science, Technical Systems and Cooperative Work.* Princeton, NJ: L. Erlbaum Associates.

Berg, Marc. 1997b. *Rationalizing Medical Work—Decision Support Techniques and Medical Problems.* Cambridge, MA: MIT Press.

Berg, Marc. 1998. "Order(s) and Disorder(s): Of Protocols and Medical Practices." In Marc Berg and Annemarie Mol (eds.). *Differences in Medicine: Unraveling Practices, Techniques, and Bodies.* Durham, NC: Duke University Press.

Berg, Marc, and Geoffrey C. Bowker 1997. "The Multiple Bodies of the Medical Record–Toward a Sociology of an Artefact." In *The Sociological Quarterly* 38: 513–37.

Berman, Sanford (ed.). 1984. *Subject Cataloging: Critiques and Innovations.* NY: Haworth Press.

Berman, Sanford. 1993. *Prejudices and Antipathies: A Tract on the LC Subject Heads Concerning People*. Jefferson, N.C.: McFarland & Co.

Bertillon, Jacques. 1890. *Sur une Nomenclature Uniforme des Causes de Décès*. Kristiania [Oslo]: Th. Steen

Bertillon, Jacques. 1887. *Rapport sur les Travaux de l'Institut de Statistique* (Session de Rome, Avril 1887) *et sur l'Organisation de la Direction Générale de Statistique en Italie*. Paris: Société d'Editions Scientifiques.

Bertillon, Jacques. 1892. *De la Fréquence des Principales Causes de Décès en Paris pendant la Seconde Moitié du XIXème Siècle et Notamment pendant la Période* 1886–1905. Paris: Société d'Editions Scientifiques.

Bertillon, Jacques. 1895. *Cours Elémentaire de Statistique Administrative. Elaboration des Statistiques*. Organisation des Bureaux de Statistique. Elements de Démographie. Ouvrage Conforme au Programme Arrêté par le Conseil Supérieur de Statistique pour l'Examen d'Admission dans Diverses Administrations Publiques. Paris: Société d'Editions Scientifiques.

Bertillon, Jacques. 1900. *Sur une Nomenclature Uniforme des Causes de Décès*. Kristiania [Oslo]: Th. Steen.

Bigart, Homer. 1960. "Apartheid Drive Studies Ancestry." In *The New York Times*. Sunday, May 1, 14–15.

Bijker, Wiebe, and John Law. 1992. "General Introduction." In Wiebe Bijker and John Law (eds.). *Shaping Technology/Building Society: Studies in Sociotechnical Change*, Cambridge, MA: MIT Press, 1–19.

Biraben, Jean-Noel. 1988. "La Tuberculose et la Dissimulation des Causes de Décès." In Bardet, Jean-Pierre, Patrice Bourdelais, Pierre Guillaume, François Lebrun, Claude Quêtel (eds.). *Peurs et Terreurs Face à la Contagion*. Paris: Fayard, 184–198.

Bitner, Egon, and Harold Garfinkel. 1967. "'Good' Organizational Reasons for 'Bad' Clinical Records." In Harold Garfinkel. *Studies in Ethnomethodology*. Englewood Cliffs, NJ: Prentice Hall.

Bjerknes, Gro, and Tone Bratteteig. 1987a. "Perspectives on Description Tools and Techniques in System Development." In P. Docherty (ed.). *System Design for Human Development and Productivity: Participation and Beyond*. Amsterdam: Elsevier-North Holland, 319–330.

Bjerknes, Gro, and Tone Bratteteig. 1987b. "Florence in Wonderland: System Development with Nurses." In Gro Bjerknes, Pelle Ehn, and Morten Kyng (eds.). *Computers and Democracy: A Scandinavian Challenge*. Avebury, UK: Aldershot, 281–295.

Black Sash. 1971. Memorandum on the Application of the Pass Laws and Influx Control. Johannesburg, South Africa.

Blois, Marsden S. 1984. *Information and Medicine: The Nature of Medical Descriptions*. Berkeley, CA: University of California Press.

Bloor, David. "Durkheim and Mauss Revisited: Classification and the Sociology of Knowledge." In *Studies in the History and Philosophy of Science* 13 (4): 267–292.

Bloor, M. 1991. "A Minor Office: The Variable and Socially Constructed Character of Death Certification in a Scottish City." In *Journal of Health and Social Behavior* 32: 273–87.

Boland, Richard, and W. Day. 1989. "The Experience of System Design: A Hermeneutic of Organizational Action," In *Scandinavian Journal of Management* 5: 87–104.

Boland, Richard, and Rudy Hirscheim (eds.). 1987. *Critical Issues in Information Systems*. London: John Wiley.

Boltanski, Luc, and Laurent Thévenot. 1991. *De la Justification: les Economies de la Grandeur*. Paris: Gallimard.

Bolter, J. David. 1991. *Writing Space: The Computer, Hypertext, and the History of Writing*. Hillsdale, NJ.: L. Erlbaum Associates.

Boronstein, Richard. 1988. "L'Invention des 'Colors'." In Claude Meillassoux (ed.). *Verouillage Ethnique en Afrique du Sud*. (Etude préparé pour la Division des droits de l'homme et de la paix de l'UNESCO.) Paris: UNESCO/OU: 51–62.

Bowker, Geoffrey C. 1993. "How to be Universal: Some Cybernetic Strategies," In *Social Studies of Science* 23: 107–127.

Bowker, Geoffrey C. 1994. *Science on the Run: Information Management and Industrial Geophysics at Schlumberger, 1920–1940*. Cambridge, MA: MIT Press.

Bowker, Geoffrey C. 1998. "Modest Reviewer Goes on a Virtual Voyage: Some Recent Literature of Cyberspace." In *Technology and Culture* 39(3): 499–511.

Bowker, Geoffrey C. In press. "The Game of the Name: Nomenclatural Instability in Botany." To appear in Robert V. Williams (ed.), *The History of Information Science*.

Bowker, Geoffrey C. and Susan Laigh Star (ed.). 1998. "How Classifications Work: Problems and Challenges in an Electronic Age." In *Library Trends* 47(2).

Brand, Stewart. 1994. *How Buildings Learn: What Happens after They're Built*. New York: Viking.

Brock, Thomas D. 1988. *Robert Koch: A Life in Medicine and Bacteriology*. Berlin: Springer-Verlag.

Brookes, Edgar H. 1968. *Apartheid: A Documentary Study of Modern South Africa*. New York: Barnes and Noble.

Brown, John Seely, and Paul Duguid. 1994. "Borderline Issues: Social and Material Aspects of Design." In *Human-Computer Interaction* 9: 3–36.

Brown, R., and J. Kulik. 1982. "Flashbulb Memories." In Ulric Neisser (ed.). *Memory Observed: Remembering in Natural Contexts*. San Francisco, CA: W.H. Freeman and Company, 23–40.

Bud-Frierman, Lisa (ed.). 1994. *Information Acumen: The Understanding and Use of Knowledge in Modern Business*. London: Routledge.

Bulechek, Gloria M., and JoAnne C. McCloskey. 1985. "Future Directions." In Gloria M. Bulechek and Joanne C. McCloskey. *Nursing Interventions: Treatments for Nursing Diagnoses*. Philadelphia, PA: Saunders, 401–408.

Bulechek, Gloria M., and JoAnne C. McCloskey.1989. "Nursing Interventions: Treatments for Potential Diagnoses." In *Proceedings of the Eighth NANDA Conference*. R. M. Carroll-Johnson (ed.). Philadelphia: J.B. Lippincott, 23–30.

Bulechek, Gloria M., and JoAnne C. McCloskey. 1993. "Response to Grobe." In Canadian Nurses Association, *Papers from the Nursing Minimum Data Set Conference*. Edmonton, Canada: Canadian Nurses Association, 158–160.

Busch, Lawrence. 1995. "The Moral Economy of Grades and Standards," Invited paper presented at a conference on Agrarian Questions, Wageningen, Netherlands, May.

Callon, Michel. 1986. "Some Elements of a Sociology of Translation." In John Law (ed.). *Power, Action, and Belief: A New Sociology of Knowledge?* London: Routledge and Kegan Paul, 196–233.

Cambrosio, Alberto, and Peter Keating. *Exquisite Specificity: The Monoclonal Antibody Revolution*. New York: Oxford Universiy Press, 1995.

Campbell, E.J. M., J.G. Scadding, and R.S. Roberts. 1979. "The Concept of Disease." *In British Medical Journal* 2: 757–762.

Campbell-Kelly, Martin. 1989. *ICL: A Business and Technical History*. Oxford: Clarendon Press

Campbell-Kelly, Martin. 1994. "The Railway Clearing House and Victorian Data Processing." In Lisa Bud-Frierman (ed.). *Information Acumen: The Understanding and Use of Knowledge in Modern Business*. London: Routledge, 51–74

Cameron, H.M., and E. McGoogan. 1981. "A Prospective Study of 1152 Hospital Autopsies: 1. Inaccuracies in Death Certification." *In Journal of Pathology* 133: 273–283.

Carruthers, Mary. 1992. *Book of Memory: A Study of Memory in Mediaeval Culture*. 2nd. edition. Cambridge: Cambridge University Press.

Carter, John R. 1980. "The Problematic Death Certificate." *In New England Journal of Medicine* 313 (20): 1284–1286.

Casper, Monica. 1994a. "At the Margins of Humanity: Fetal Positions in Science and Medicine." In *Science, Technology and Human Values* 19:307–323.

Casper, Monica. 1994b. "Reframing and Grounding Nonhuman Agency: What Makes a Fetus an Agent." *In American Behavioral Scientist* 37: 839–856.

Casper, Monica. 1995. "Fetal Cyborgs and Technomoms on the Reproductive Frontier, or Which Way to the Carnival?" In Chris Hables Gray, Heidi Figueroa-Sarriera, and Steven Mentor (eds.). *The Cyborg Handbook*. New York: Routledge.

Casper, Monica. 1998. *The Making of the Unborn Patient: A Social Anatomy of Fetal Surgery*. New Brunswick, NJ: Rutgers University Press.

Castles, M.R. 1981. "Nursing Diagnosis: Standardization of Nomenclature." In Harriet H. Werley and Margaret R. Grier (eds.). *Nursing Information Systems*. New York: Springer, 36–44.

Cell, John. 1982. *The Highest Stage of White Supremacy: The Origins of Segregation in South Africa and the American South*. Cambridge: Cambridge University Press.

Chandler, Alfred D. 1977. *The Visible Hand: The Managerial Revolution in American Business*. Cambridge, MA: Belknap Press.

Charmaz, Kathleen. 1991. *Good Days, Bad Days: The Self in Chronic Illness and Time*. New Brunswick, NJ: Rutgers.

Chatelin, Yvon. 1979. *Une Epistmologie des Sciences du Sol*. Paris: ORSTOM.

Chronic Fatigue Syndrome Newsletter. 1997. Date: Thursday, 20 Feb 1997 22:31:18–0500. From: CFS-NEWS Electronic Newsletter. Subject: #64. "Change the Name Survey /Comments on Royal Colleges Report."

Cicourel, A. 1964. *Method and Measurement in Sociology*. New York: Free Press of Glencoe.

Cimino, James J., Geogre Hripcsak, Stephen B. Johnson, and Paul D. Clayton, Center for Medical Informatics, Columbia University, Columbia-Presbyterian Medical Center. 1989. "Designing an Introspective, Multipurpose, Controlled Medical Vocabulary." In Lawrence C. Kingsland III (ed.). *Proceedings of the Thirteenth Annual Symposium on Computer Applications in Medical Care, November 5–8, 1989*. Washington DC: IEEE Computer Society Press, 513–523.

Clarke, Adele. 1984. "Subtle Sterilization Abuse: A Reproductive Rights Perspective." In Rita Arditti, Renata Duelli Klein, and Shelly Minden (eds.). *Test Tube Women: What Future for Motherhood?* Boston: Pandora/Routledge, 188–212.

Clarke, Adele. 1990. "Controversy and the Development of Reproductive Sciences." In *Social Problems* 37: 18–37.

Clarke, Adele. 1991. "Social Worlds/Arenas Theory as Organizational Theory." In David Maines (ed.). *Social Organization and Social Process: Essays in Honor of Anselm Strauss*. Hawthorne, NY: Aldine de Gruyter, 119–158.

Clarke, Adele. 1998. *Disciplining Reproduction : Modernity, American Life Sciences, and the Problem of Sex.* Berkley, CA: University of CA Press

Clarke, Adele, and Monica Casper. 1992. From Simple Technology to Complex Arena: Classification of Pap Smears,1917–1990. Working Paper, Dept. of Sociology, University of California, San Francisco.

Clarke, Adele, and Joan H. Fujimura (eds.). 1992a. *The Right Tools For The Job: At Work in Twentieth-Century Life Sciences.* Princeton, NJ: Princeton University Press.

Clarke, Adele, and Joan H. Fujimura. 1992b. Introduction. In *The Right Tools For The Job: At Work in Twentieth-Century Life Sciences.* Princeton, NJ: Princeton University Press, 3–44.

Cody, William K. 1995. "Letter from William K. Cody." In Nursing Outlook. 43 (2): 93–94.

Cole, Michael. 1996. *Cultural Psychology: A Once and Future Discipline.* Cambridge, MA: Belknap Press of Harvard University Press.

Coleman, Linda, and Paul Kay. 1981. "Prototype Semantics: The English Word Lie," *In Language* 57: 26–44.

Collins, Patricia Hill. 1986. "Learning from the Outsider Within: The Sociological Significance of Black Feminist Thought." In *Social Problems* 33: 14–32.

Comstock, George W., and Robert E. Markush. 1986. "Further Comments on Problems in Death Certification." *In American Journal of Epidemiology* 124: 188–181.

Comte, August. 1975 [1830–1845]. Philosophie Première; Cours de Philosophie Positive, Leçons 1 à 45. Paris: Hermann.

Condon, Mark Casey, and David Schweingruber. 1994. "The Morality of Time and the Organization of a Men's Emergency Shelter." Unpublished Manuscript, Department of Sociology, University of Illinois, Urbana-Champaign.

Corbin, Juliet, and Anselm Strauss. 1988. *Unending Work and Care: Managing Chronic Illness at Home.* San Francisco: Jossey-Bass.

Corbin, Juliet, and Anselm Strauss. 1991. "Comeback: The Process of Overcoming Disability." In G. Albrecht and J. Levy (eds.). *Advances in Medical Sociology*, vol. 2. Greenwich, CT: JAI Press, 137–158.

Cornell, Margaret. 1960. "The Statutory Background of Apartheid: A Chronological Survey of South African Legislation." In *The World Today* 16: 181–194.

Cowan, Ruth Schwartz. 1985. "How the Refrigerator Got its Hum." In Judy Wacjman and Donald MacKenzie (eds.). *The Social Shaping of Technology: How the Refrigerator Got its Hum.* Milton Keynes: Open University Press.

Croissant, Jennifer, and Sal Restivo. 1995. "Science As a Social Problem." In Susan Leigh Star (ed.). *Ecologies of Knowledge*. Albany, NY: SUNY Press.

Curtis, Bruce. 1998. "From the Moral Thermometer to Money: Metrological Reform in Pre-Confederation Canada." *In Social Studies of Science* 28: 547–570.

Dagognet, Franois. 1970. *Le Catalogue de la Vie; Etude Méthodologique sur la Taxanomie*. Paris: Presses Universitaires de France.

David, Paul, and Geoffrey S. Rothwell. 1994. *Standardization, Diversity and Learning: Strategies for the Coevolution of Technology and Industrial Capacity*. Stanford, CA: Center for Economic Policy Research, Stanford University.

Davis, Fred. 1963. *Passage Through Crisis: Polio Victims and their Families*. Indianapolis: Bobbs-Merrill.

Davis, F. James. 1991. *Who Is Black? One Nation's Definition*. University Park, PA: Pennsylvania State University Press.

Dean, John. 1979. "Controversy over Classification: A Case Study from the History of Botany." In Barry Barnes and Steven Shapin (eds.), *Natural Order: Historical Studies of Scientific Culture*. London: Sage, 211–230.

de Klerk, J. 1962. "Know Ye Not their Colour?" In *Forum*. Pretoria, May: 8–11.

Derrida, Jacques. 1980. *La Carte Postale: de Socrate à Freud et Au-dela*. Paris: Flammarion.

Derrida, Jacques. 1998. *Of Grammatology*. Translated by Gayatri Chakravorty Spivak. Corrected ed. Baltimore: Johns Hopkins University Press.

Desrosières, Alain. 1990. "How to Make Things Which Hold Together: Social Science, Statistics and the State." In P. Wagner, B. Wittrock, and R. Whitley (eds.). *Discourses on Society: the Shaping of the Social Science Disciplines*. London: Kluwer.

Desrosières, Alain. 1993. *La Politique des Grands Nombres: Histoire de la Raison Statistique*. Paris: Editions La Découverte.

Desrosières, Alain, and Laurent Thévenot.1988. *Les Catégories Socio-professionnelles*. Paris: Découverte.

Dewey, John. 1929. The *Quest for Certainty: A Study of the Relation of Knowledge and Action*. New York: Minton, Balch.

Dewey, John. 1916. *Essays in Experimental Logic*. Chicago, IL: The University of Chicago Press.

*Diagnostic Standards and Classification of Tuberculosis*. 1955 edition. New York: National Tuberculosis Association.

*Diagnostic Standards and Classification of Tuberculosis*. 1961 edition. New York: National Tuberculosis Association.

Dodge, Chris, and Jan DeSirey (eds.). 1995. *Everything You Always Wanted to Know About Sandy Berman but Were Afraid to Ask*. Jefferson, NC: McFarland.

Dodier, Nicolas. 1994. "Expert Medical Decisions in Occupational Medicine: A Sociological Analysis of Medical Judgment." In *Sociology of Health and Illness* 16: 489–514.

Doman, J. 1975. "The In-betweeners." In *Optima* 25: 130–151.

Douglas, Mary. 1984. *Purity and Danger: An Analysis of the Concepts of Pollution and Taboo*. London: Routledge and Kegan Paul.

Douglas, Mary. 1986. *How Institutions Think*. Syracuse, NY: Syracuse University Press.

Douglas, Mary, and David L. Hull. 1992. *How Classification Works: Nelson Goodman among the Social Sciences*. Edinburgh: Edinburgh University Press.

Dubow, Saul. 1995. *Scientific Racism in Modern South Africa*. Cambridge: Cambridge University Press.

Dumas, Alexandre. 1858. *La Dame aux Camélias*; Préface de Jules Janin. Ed. illustré par Gavarni. Paris: G. Havard.

Duncan, Thomas, and Tod F. Stuessy. 1984. *Cladistics: Perspectives on the Reconstruction of Evolutionary History*. New York: Columbia University Press.

Durkheim, Emile. 1982. *The Rules of Sociological Method*. Edited with an Introduction by Steven Lukes. Translated by W.D. Halls. New York: Free Press.

Durkheim, Émile, and Marcel Mauss. 1969. "De Quelques Formes Primitives de Classification: Contribution à l'Etude des Représentations Collectives." In Marcel Mauss, *Oeuvres. 2. Représentations Collectivés et Diversité des Civilisations*. Paris: Les Editions de Minuit, 9–105.

*Ebony*. 1968. "A Shade of Shame?" In vol. 23 (June): 85–90.

Edwards, D., and J. Potter. 1992. *Discursive Psychology*. London: Sage.

Egger, Eddeltrud, and Ina Wagner. 1993. "Negotiating Temporal Orders: The Case of Collaborative Time Management in a Surgery Clinic." In *Computer Supported Cooperative Work (CSCW)* 1: 255–275.

Ehrenreich, Barbara, and Deirdre English. 1973. *Complaints and Disorders: the Sexual Politics of Sickness*. Old Westbury, NY: The Feminist Press.

Eisenstein, Elizabeth L. 1979. *The Printing Press as an Agent of Change: Communications and Cultural Transformations in Early Modern Europe*. Cambridge: Cambridge University Press.

Engestrom, Yrjo. 1990a. "Organizational Forgetting: an Activity-Theoretical Perspective." In Yrjo Engestrom, *Learning, Working and Imagining: Twelve*

*Studies in Activity Theory.* Jyvaskylassa: Painettu Kirjapaino at Oma Kyssa, 196–226.

Engestrom, Yrjo. 1990b. *Learning, Working and Imagining.* Helsinki: Orienta Konsultit Oy.

Epstein, Steven. 1996. *Impure Science: AIDS, Activism, and the Politics of Knowledge.* Berkeley: University of California Press.

Ewald, Francois. 1986. *L'Etat Providence.* Paris: B. Grasset.

Fagot-Largeault, Anne. 1989. *Causes de la Mort: Histoire Naturelle et Facteurs de Risque.* Paris : Librairie Philosophique J. Vrin.

Farr, William. 1885. *Vital Statistics: A Memorial Volume of Selections from the Reports and Writings of William Farr, M. D., D. C. L., C. B., F. R. S.* London: Offices of the Sanitary Institute.

Fentress, James, and Chris Wickham. 1992. *Social Memory: New Perspectives on the Past.* Oxford: Blackwell.

Ferguson, Russell, Martha Gever, Trinh T. Minh-ha, and Cornel West (eds.). 1990. *Out There: Marginalization and Contemporary Cultures.* Cambridge, MA: MIT Press.

Figert, Ann. 1996. *Women and the Ownership of PMS: The Structuring of a Psychiatric Disorder.* New York: Aldine de Gruyter.

Forsythe, Diana. 1993. "Engineering Knowledge: The Construction of Knowledge in Artificial Intelligence." In *Social Studies of Science* 23: 445–477.

Foucault, Michel. 1970. *The Order of Things: An Archaeology of the Human Sciences.* London: Tavistock Publications.

Foucault, Michel. 1975. *Surveiller et Punir: Naissance de la Prison.* Paris: Gallimard.

Foucault, Michel. 1979. *Discipline and Punish: The Birth of the Prison.* Translated by Alan Sheridan. New York: Vintage Books.

Foucault, Michel. 1982. *The Archaeology of Knowledge.* Translated from the French by A. M. Sheridan Smith. New York: Pantheon Books.

Foucault, Michel. 1991. "Governmentality." In Graham Burchill, Colin Gordon and Peter Miller (eds.). *The Foucault Effect: Studies in Governmentality.* Chicago, IL: University of Chicago

Frankel, Philip. 1979. "The Politics of Passes: Control and Change in South Africa." In *The Journal of Modern African Studies* 17: 199–217.

Freeman, Jo. 1972. "The Tyranny of Structurelessness." In *Berkeley Journal of Sociology* 17: 151–164.

Friedlander, Amy. 1995. *Emerging Infrastructure: The Growth of Railroads.* Reston, VA: Corporation for National Research Initiatives.

Frisby, Michael K. 1995–96. "Black, White or Other." In *Emerge.* Reprinted at: http://www.usis.usemb.se/sft/142/sf142//.htm.

Froom, Jack. 1975. "International Classification of Health Problems in Primary Care." Guest editorial in *Journal of the American Medical Association* 234(12): 1257.

Fujimura, Joan. 1987. "Constructing 'Do-able' Problems in Cancer Research: Articulating Alignment." In *Social Studies of Science* 17: 257–93.

Furet, François. 1978. *Penser la Revolution Française.* Paris: Gallimard.

Galtier, J. 1986. "Taxonomic Problems Due to Preservation: Comparing Compression and Permineralized Taxa." In Robert A. Spicer and Barry A. Thomas (eds.). *Systematic and Taxonomic Approaches in Palaeobotany.* The Systematics Association Special Volume No.31. Oxford: Clarendon Press, 1–16.

Gasser, Les. 1986. "The Integration of Computing and Routine Work." In *ACM Transactions on Office Information Systems* 4: 205–225.

Gebbie, K. M., and M. A. Lavin. 1975. *Classification of Nursing Diagnoses. Proceedings of the First National Conference.* St. Louis: Mosby Co.

Geist, P., and M. Hardesty. 1992. *Negotiating the Crisis: DRGs and the Transition of Hospitals.* Hillsdale, NJ: Lawrence Erlbaum.

Gerson, Elihu, and Susan Leigh Star. 1986. "Analyzing Due Process in the Workplace." In *ACM Transactions on Office Information Systems* 4: 257–270.

Glaser, Barney, and Anselm Strauss. 1965. *Awareness of Dying.* Chicago: Aldine.

Goguen, Joseph. 1997. "A Social, Ethical Theory of Information." In Geoffrey C. Bowker, Les Gasser, Susan Leigh Star and William Turner (eds.). *Social Science, Technical Systems and Cooperative Work.* Princeton, NJ: L. Erlbaum Associates.

Goffman, Erving. 1959. *The Presentation of Self in Everyday Life.* Garden City, NY: Doubleday.

Goldstein, Jan. 1987. *Console and Classify: The French Psychiatric Profession in the Nineteenth Century.* Cambridge, UK: Cambridge University Press.

Goodwin, Charles. 1996. "Practices of Color Classification. Ninchi Kagaku." In *Cognitive Studies: Bulletin of the Japanese Cognitive Science Society* 3(2): 62–82.

Goody, Jack. 1971. *The Domestication of the Savage Mind.* Cambridge: Cambridge University Press.

Goody, Jack. 1987. *The Interface Between the Written and the Oral.* Cambridge: Cambridge University Press.

Granovetter, M. "The Strength of Weak Ties." *American Sociological Review* 70: 1360–1380.

Graunt, John. 1662. *Natural and Political Observations Mentioned in a Following Index and Made upon the Bills of Mortality.* London: Roycroft, (reprinted 1975, New York: Arno Press).

Gray, A., R. Elkan, and J. Robinson. 1991. *Policy Issues in Nursing.* Milton Keynes: Open University Press.

Gray, Chris Hables, with the assistance of Heidi Figueroa-Sarriera and Steven Mentor (eds.). 1995. *The Cyborg Handbook.* New York: Routledge.

Grmek, Marcel. 1990. *History of AIDS: Emergence and Origin of a Modern Pandemic.* Princeton, NJ: Princeton University Press.

Grobe, Susan. 1992. "Response to J. C. McCloskey's and G. M. Bulechek's Paper on Nursing Intervention Scheme." In The Canadian Nurses Association. *Papers from the Nursing Minimum Data Set Conference, October 27–29, 1992.* Edmonton, Alberta: The Canadian Nurses Association.

Hacking, Ian. 1986. "Making Up People." In T. C. Heller et al. (eds.). *Reconstructing Individualism.* Stanford: Stanford University Press, 222–236.

Hacking, Ian. 1990. *The Taming of Chance.* Cambridge: Cambridge University Press.

Hacking, Ian. 1992. "World Making by Kind Making: Child Abuse for Example." In Douglas, Mary and David L. Hull. *How Classification Works: Nelson Goodman among the Social Sciences.* Edinburgh: Edinburgh University Press, 180–238.

Hacking, Ian. 1995. *Rewriting the Soul: Multiple Personality and the Sciences of Memory.* Princeton, NJ: Princeton University Press.

Hall, Rogers. 1990. "Making Mathematics on Paper: Constructing Representations of Stories about Related Linear Functions." Ph. D. Dissertation, Department of Information and Computer Science, University of California, Irvine.

Hall, Rogers, and Reed Stevens. 1995. "Making Space: A Comparison of Mathematical Work in School and Professional Design Practices." In Susan Leigh Star (ed.). *The Cultures of Computing.* Oxford: Blackwell.

Haraway, Donna J. 1991. *Simians, Cyborgs, and Women: The Reinvention of Nature.* New York: Routledge.

Haraway, Donna. 1992. "The Promises of Monsters: A Regenerative Politics for Inappropriate/d Others." In Paula Treichler, Cary Nelson, and Larry Grossberg (eds.). *Cultural Studies Now and in the Future.* New York: Routledge, 295–337.

Haraway, Donna. 1997. *Modest-Witness@Second-Millennium. FemaleMan-Meets-OncoMouse™: Feminism and Technoscience*. New York: Routledge.

Harding, Sandra (ed.). 1993. *The "Racial" Economy of Science: Toward a Democratic Future*. Bloomington : Indiana University Press.

Harvey, Francis. 1997. "Quality Needs More than Standards." Paper presented at the first Cassini International Workshop on Data Quality in Geographical Information: From Error to Uncertainty. Paris, 21–23 April.

Hawking, Stephen W. 1980. *Is the End in Sight for Theoretical Physics? An Inaugural Lecture*. Cambridge: Cambridge University Press.

Heidenstrom, P. N. 1985. Research unit, Accident Compensation Corporation, Wellington, New Zealand. "Accident Statistics, Coding Systems, and the New Zealand Experience." In R. A. Coté, D. J. Protti, and J. R. Scherer (eds.), *Role of Informatics in Health Data Coding and Classification Systems*. Amsterdam: Elsevier, 69–80.

Heller, Joseph. 1961. *Catch-22*. New York: Dell.

Hewitt, Carl. 1985. "The Challenge of Open Systems." In *BYTE* 10: 223–42.

Hewitt, Carl. 1986. "Offices are Open Systems." In *ACM Transactions on Office Information Systems* 4: 271–87.

Hirschauer, Stefan. 1991. "The Manufacture of Bodies in Surgery." In *Social Studies of Science* 21(2): 279–319.

Hope, Christopher. 1980. *A Separate Development*. Johannesburg: Ravan Press.

Horrell, Muriel. 1958. *Race Classification in South Africa: Its Effects on Human Beings, no. 2*. Johannesburg: South African Institute of Race Relations.

Horrell. Muriel. 1960. *The "Pass-Laws." South African Institute of Race Relations, no. 7*. Johannesburg: South African Institute of Race Relations.

Horrell, Muriel. 1969. *A Survey of Race Relations*. Johannesburg: South African Institute of Race Relations.

Huffman, E. 1990. *Medical Record Management*. Berwyn, IL: Physicians' Record Company.

Hughes, Everett C. 1970. "Mistakes at Work." In his *The Sociological Eye*. Chicago: Aldine.

Hughes, Thomas. 1989 [1883]. *Tom Brown's Schooldays*. Oxford: Oxford University Press.

Hughes, Thomas P. 1987. "The Evolution of Large Technological Systems." In W. E. Bijker, T. P. Hughes, and T. J. Pinch (eds.). *The Social Construction of Technological Systems: New Directions in the Sociology and History of Technology*. Cambridge, MA: MIT Press.

Hughes, Thomas P. 1983. *Networks of Power: Electrification in Western Society, 1880–1930.* Baltimore: Johns Hopkins University Press.

Hunn, Eugene. 1982. "The Utilitarian Factor in Folk Biological Classification." In *American Anthropologist* 84: 830–847.

Hutchins, E. 1995. *Cognition in the Wild.* Cambridge, MA: MIT Press.

ICD-9CM. 1996. *ICD-9-CM Fifth Edition; The International Classification of Diseases, 9th Revision, Clinical Modification, vol. 1; Diseases: Tabular List, vol. 2 Diseases: Alphabetic Index.* New York: McGraw-Hill, Inc.

ICD-10. 1992a. *ICD-10. International Statistical Classification of Diseases and Related Health Problems, Tenth Revision, vol. 1.* Geneva: World Health Organization.

ICD-10. 1992b. *International Statistical Classification of Diseases and Related Health Problems. Tenth Revision, vol. 2. Instruction Manual.* Geneva: World Health Organization.

Ignacio, Emily. 1998. *The Quest for a Filipino Identity: The Construction of Ethnic Identity within a Transnational Location.* Ph. D. dissertation, Department of Sociology, University of Illinois, Urbana-Champaign.

Iowa Intervention Project. 1993. "The NIC Taxonomy Structure." In *IMAGE: Journal of Nursing Scholarship* 25: 187–192.

Israel, Robert A., Harry M. Rosenberg, and Lester R. Curton. "Analytical Potential for Multiple Cause-of-Death Data." In *American Journal of Epidemiology* 124: 161–179.

Jenkins, T. 1988. "New Roles for Nursing Professionals." In M. J. Ball, K. J. Hannah, U. Gerdin Jelger, H. Peterson (eds.). *Nursing Informatics: Where Caring and Technology Meet.* NY: Springer, 88–95.

John, Richard R. 1994. "American Historians and the Concept of the Communications Revolution." In Lisa Bud-Frierman (ed.). *Information Acumen: The Understanding and Use of Knowledge in Modern Business.* London: Routledge 98–112.

The Journal of Online Nursing. Special issue on naming—at *http://www.nursingworld.org/ojin/tpc7/intro.htm.*

Jucovy, Peter M. 1982. "Developing a Critical Model for Diagnostic Language." In IEEE Computer Society. *Proceedings MEDCOMP '82; First IEEE Computer Society International Conference on Medical Computer Science/Computational Medicine.* Washington, DC: IEEE Computer Society Press, 465–469.

Jullien, François. 1995. *The Propensity of Things: Toward a History of Efficacy in China.* Translated by Janet Lloyd. New York: Zone Books.

Kahn, E. J., Jr. 1966. *The Separated People: A Look at Contemporary South Africa.* New York: Norton.

Karnik, Niranjan. 1994. *Western and Traditional Medicine in India.* Working report, Illinois Research Group on Classification.

Keller, Charles M., and Janet Dixon Keller. 1996. *Cognition and Tool Use : The Blacksmith at Work.* Cambridge: Cambridge University Press.

Kindleberger, Charles. 1983. "Standards as Public, Collective and Private Goods." In *Kyklos* 36 (Fasc. 3): 377–396.

King, Lester. 1982. *Medical Thinking: A Historical Preface.* Princeton, NJ: Princeton University Press.

King, John, and Susan Leigh Star. 1990. "Conceptual Foundations for the Development of Organizational Decision Support Systems." In *Proceedings of the 23rd Hawaiian International Conference on Systems Sciences.* Washington, DC: IEEE Computer Society Press 3: 143–151.

Kingsland, Lawrence C. III (ed.). 1989. *Proceedings of the Thirteenth Annual Symposium on Computer Applications in Medical Care. November 5–8, 1989.* Washington DC: IEEE Computer Society Press.

Kirk, Stuart A., and Herb Kutchins. 1992. *The Selling of the DSM: The Rhetoric of Science in Psychiatry.* New York: Aldine de Gruyter.

Kling, Rob, Spencer Olin, and Mark Poster (eds). 1991. *Postsuburban California: The Transformation of Orange County since World War II.* Berkeley: University of California Press.

Kraemer, K., S. Dickhoven, S. F. Tierney, S. F., and J. L. King. 1987. *Datawars: The Politics of Modeling in Federal Policymaking.* New York: Columbia University Press.

Kramarae, Cheris (ed). 1988. *Technology and Women's Voices: Keeping in Touch.* New York: Routledge.

Kritek, P. B. 1988. "Conceptual Considerations, Decision Criteria, and Guidelines for the Nursing Minimum Data Set from a Practice Perspective." In Harriet H. Werley and Norma M. Lang (eds.). *Identification of the Nursing Minimum Data Set.* New York: Springer, 22–33.

Kupka, Karel. 1978. "International Classification of Diseases: Ninth Revision." In *WHO Chronicle* 32: 219–225.

Kutchins, Herb, and Stuart A. Kirk. 1997. *Making Us Crazy: DSM: The Psychiatric Bible and the Creation of Mental Disorders.* New York: Free Press.

Kwasnik, Barbara. 1988. "Factors Affecting the Naming of Documents in an Office." In Borgman, Christine L.; Pai, Edward Y. H. (eds.). *ASIS "88: Proceedings of the American Society for Information Science (ASIS) 51st Annual Meeting,* volume 25; 1988 October 23–28; Atlanta, GA. Medford, NJ: Learned Information, Inc. for the American Society for Information Science, 100–106.

Kwasnik, Barbara. 1991. "The Importance of Factors that Are Not Document Attributes in the Organisation of Personal Documents," *Journal of Documentation* 47: 389–398.

Lagache, Edouard. 1995. Diving into Communities of Learning: Existential Perspectives on Communities of Practice at Zone of Proximal Development. Ph. D. Dissertation. School of Education, University of California, Berkeley.

Lakatos, Imre. 1976. *Proofs and Refutations: The Logic of Mathematical Discovery.* Cambridge: Cambridge University Press.

Lakoff, George. 1987. *Women, Fire, and Dangerous Things: What Categories Reveal about the Mind.* Chicago: University of Chicago Press.

Lakoff, George, and Mark Johnson. 1980. *Metaphors We Live By.* Chicago: University of Chicago Press.

Landis, Elizabeth. 1961. "South African Apartheid Legislation: Fundamental Structure." *Yale Law Journal* 1: 4–16.

Larsen, Nella. 1986. [1928, 1929] *Quicksand* and *Passing.* Deborah E. McDowell (ed.). New Brunswick, NJ: Rutgers University Press.

Latour, Bruno. 1987. *Science in Action: How to Follow Scientists and Engineers Through Society.* Milton Keynes: Open University Press.

Latour, Bruno. 1988. *The Pasteurization of France.* Cambridge, MA: Harvard University Press.

Latour, Bruno. 1993. *We Have Never Been Modern.* Cambridge, MA: Harvard University Press.

Latour, Bruno 1996a. *Aramis or the Love of Technology.* Cambridge, MA: Harvard University Press.

Latour, Bruno. 1996b. *Petite Réflexion sur le Culte Moderne des Dieux Faitiches.* Paris: Les Empecheurs de Penser en Rond.

Latour, Bruno. forthcoming. "Did Ramses II Die of Tuberculosis? On the Partial Existence of Existing and Nonexisting Objects." Typescript from author.

Latour, Bruno, and Steve Woolgar. 1979. *Laboratory Life: The Construction of Scientific Facts.* Thousand Oaks, CA: SAGE.

Lave, Jean. 1988. *Cognition in Practice: Mind, Mathematics, and Culture in Everyday Life.* Cambridge: Cambridge University Press.

Lave, Jean and Etienne Wenger. 1991. *Situated Learning: Legitimate Peripheral Participation.* Cambridge: Cambridge University Press.

League of Nations. 1927. CH/experts stat/34 of 2 December 1927 (held in the League of Nations archives at the United Nations Library in Geneva). *Report of the Committee of the Vital Statistics Section of the American Public Health Association on the Accuracy of Certified Causes of Death and its Relation to Mortality Statistics and the International List.*

League of Nations, 1938. "International Lists of Causes of Death adopted by the Fifth International Conference for Revision, Paris, October 3-7, 1938." In *Bulletin of the Health Organisation* VII (6): 944–987

Lelyveld, Joseph. 1985. *Move Your Shadow: South Africa, Black and White*. New York: Times Books.

Lemke, Jay. 1995. *Textual Politics : Discourse and Social Dynamics*. London; Bristol, PA : Taylor & Francis.

Leroi-Gourhan, André. 1965. *Le Geste et La Parole: La Mémoire et les Rythmes*. Paris: Albin Michel.

Levy, David. 1994. "Fixed or Fluid? Document Stability and New Media." In *ACM European Conference on Hypermedia Technology 1994 Proceedings ECHT 94. UK-Edinburgh, 18–23 September 1994*. New York: ACM, 24–33.

Linde, Charlotte. 1993. *Life Stories: The Creation of Coherence*. New York: Oxford University Press.

Linton, M. 1982. "Transformations of Memory in Everyday Life." In Ulric Neisser (ed.). *Memory Observed: Remembering in Natural Contexts*. San Francisco, CA: W. H. Freeman and Company, 77–91.

López, Ian F. Haney. 1996. *White by Law*. New York: New York University Press.

Lund, Johan. 1985. "Nordic Classification for Accident Monitoring, and its Implementation in Norway." In R. A. Côté, D. J. Protti and J. R. Scherrer (eds.). *Role of Informatics in Health Data Coding and Classification Systems*. Amsterdam: Elsevier, 81–87.

Lynch, Michael. 1984. "Turning Up Signs in Neurobehavioral Diagnosis." In *Symbolic Interaction* 7: 67–86.

Lynch, Michael. 1995. "Laboratory Spaces and the Technological Complex: An Investigation of Topical Contextures." In Susan Leigh Star (ed.). *Ecologies of Knowledge*. Albany, NY: SUNY Press.

MacKenzie, Donald A. 1990. *Inventing Accuracy: An Historical Sociology of Nuclear Missile Guidance*. Cambridge, MA: MIT Press.

Mallard, Alexandre. 1998. "Compare, Standardize, and Settle Agreement: On Some Usual Metrological Problems." In *Social Studies of Science* 28: 571–602.

Mann, Thomas. 1992 [1929]. *The Magic Mountain*. Trans. H-T Lowe-Porter. New York: Modern Library.

March, James G., and Herbert Simon. 1958. *Organizations*. New York: Wiley.

Martin, K. S., and N. Scheet. 1992. *The Omaha System: Applications for Community Health Nursing.* Philadelphia: W. B. Saunders.

Mathabane, Mark. 1986. *Kaffir Boy: The True Story of a Black Youth's Coming of Age in Apartheid South Africa.* New York: Macmillan.

Matsuda, Matt. 1996. *The Memory of the Modern.* New York: Oxford University Press.

Matthews, R. E. F. 1983. *A Critical Appraisal of Viral Taxonomy.* Boca Raton, Rl: CRC Press.

McCloskey, JoAnne C. 1981. "Nursing Care Plans and Problem-Oriented Health Records." In Harriet H. Werley and Margaret R. Grier (eds.), *Nursing Information Systems.* New York: Springer, 120–142.

McCloskey, JoAnne C., and Gloria M. Bulechek. 1992. "Nursing Intervention Schemes." In *Papers from the Nursing Minimum Data Set Conference.* Edmonton, Alberta: The Canadian Nurses Association, 77–91.

McCloskey, JoAnne C., and Gloria M. Bulechek. 1993. *Nursing Interventions Classification.* St. Louis, MO: Mosby Year Book.

McCloskey, JoAnne C., and Gloria M. Bulechek. 1994a. "Standardizing the Language for Nursing Treatments: An Overview of the Issues." In *Nursing Outlook* 42 (2): 56–63.

McCloskey, JoAnne C., and Gloria M. Bulechek. 1994b. "Response to Edward Halloran." In *IMAGE: Journal of Nursing Scholarship* 26 (2): 93.

McCloskey, JoAnne C., and Gloria Bulechek. 1995. "Validation and Coding of the NIC Taxonomy Structure." In *IMAGE: Journal of Nursing Scholarship* 27(1): 43–49.

McCloskey, JoAnne C., and Gloria M. Bulechek. 1996. *Iowa Intervention Project—Nursing Interventions Classification (NIC),* second edition. St Louis, MO: Mosby.

McCloskey, JoAnne C., and Gloria M. Bulechek, and Toni Tripp-Reimer. 1995. "Response to Cody." In *Nursing Outlook* 43(2): 95.

McKeown, Thomas. 1976. *The Modern Rise of Population.* London: Edward Arnold.

McKeown, Thomas. 1983. "A Basis for Health Strategies: A Classification of Disease." In *British Medical Journal* 287, 27 August: 594–596.

Merchant, Carolyn. 1980. *The Death of Nature: Women, Ecology, and the Scientific Revolution.* San Francisco: Harper & Row.

Metropolis, M., J. Howlett, and Gian-Carlo Rota (eds.). 1980. *International Research Conference on the History of Computing (1976: Los Alamos Scientific*

*Laboratory) A History of Computing in the Twentieth Century: A Collection of Essays.* New York: Academic Press.

Miller, George A., Eugene Galanter, and Karl H. Pribram. 1960. *Plans and the Structure of Behavior.* New York: Holt.

Monteiro, Eric, Ole Hanseth, and Morten Hatling. "Developing Information Infrastructure: The Tension Between Standardization and Flexibility." In *Science, Technology and Human Values* 21: 407–426.

Morgan, Gareth. 1986. *Images of Organization.* Beverly Hills: Sage.

Morton, Nelle. 1985. *The Journey is Home.* Boston: Beacon Press.

Munson, Eve S. 1997. *Adobe Walls and Black-Tinted Windshields: The Press and Community Identity in Santa Fe, NM.* Ph. D. dissertation, Institute of Communications Research, University of Illinois, Urbana-Champaign.

Murphy, F. A., C. M. Fauquet,D. H. L. Bishop, S. A. Gbabrial, A. W. Jarvis, G. P. Martelli, M. A. Mayo, and M. D. Summers (eds.). 1995. *Virus Taxonomy: Classification and Nomenclature of Viruses. Sixth Report of the International Committee on Taxonomy of Viruses. Virology Division International Union of Microbiological Societies.* Vienna: Springer Verlag.

Musen, Mark. 1992. "Dimensions of Knowledge Sharing and Reuse." In *Computers and Biomedical Research* 25: 435–467.

Neisser, Ulric. 1982. "John Dean's Memory: A Case Study." In Ulric Neisser (ed.). *Memory Observed: Remembering in Natural Contexts.* San Francisco, CA: W. H. Freeman and Company, 139–159.

Neumann, Laura. 1995. "What about Leech Treatment? Nursing Classification and Professionalization." Working paper, Illinois Research Group on Classification.

*Newsweek.* 27 February 1967. "South Africa: A Shade of Difference." 69: 40 and 42.

*Newsweek.* 3 July 1970. "South Africa: White at Heart." 80: 30–31.

Nordenfelt, Lennart. 1983. *Causes of Death : A Philosophical Essay.* Stockholm: Forskningsrådsnämnden, Delegationen för långsiktsmotiverad forsking. Distributed by Editorial Service, FRN.

Norman, Donald. 1988. *The Design of Everyday Things.* New York: Doubleday.

O'Connell, Joseph. 1993. "Metrology: The Creation of Universality by the Circulation of Particulars." In *Social Studies of Science* 23: 129–173.

Olesen, V. L., and E. W. Whittaker. 1968. *The Silent Dialogue: A Study in the Social Psychology of Professional Socialization.* San Francisco: Jossey-Bass.

Ormond, Roger. 1986. *The Apartheid Handbook, 2nd edition.* Harmondsworth: Penguin.

Orr, Julian E. 1990. "Sharing Knowledge, Celebrating Identity: War Stories and Community Memory in a Service Culture." In D. S. Middleton and D. Edwards (eds.). *Collective Remembering: Memory in Society.* London: Sage, 169–189.

Orr, Julian E. 1996. *Talking about Machines: An Ethnography of a Modern Job.* Ithaca, NY: ILR Press.

Ortner, Sherry. 1974. "Is Female to Male as Nature is to Culture?" In Michelle Zimbalist Rosaldo and Louise Lamphere (eds.). *Woman, Culture, and Society.* Stanford, CA.: Stanford University Press, 67–87.

Park, Robert Ezra. 1952. *Human Communities; The City and Human Ecology.* Glencoe, IL: Free Press.

Piaget, Jean. 1969. *The Child's Conception of Time.* New York: Basic Books.

Pickering, Andrew (ed.). 1992. *Science as Practice and Culture.* Chicago : University of Chicago Press.

Pinsonneault, A., and K. L. Kraemer. 1989. "The Impact of Technological Support on Groups: An Assessment of the Empirical Research." In *Decision Support Systems* 5:197–216.

Poincaré, Henri. 1905. *Science and Hypothesis,* New York: The Science Press.

Porter, Roy. 1994. "The Ruin of the Constitution: The Early Interpretation of Gout." In *Transactions of the Medical Society of London* 110:90–103.

Porter, Theodore M. 1986. *The Rise of Statistical Thinking, 1820–1900.* Princeton, NJ: Princeton University Press.

Porter, Theodore M. 1994. "Information, Power, and the View from Nowhere." In Lisa Bud-Frierman (ed.). *Information Acumen: The Understanding and Use of Knowledge in Modern Business.* London: Routledge, 214–246.

Prins, Gwyn. 1981. "What Is to be Done? Burning Questions of Our Movement." In *Social Science and Medicine* 15: 175–183

Proust, Adrien. 1892. *La Défense de l'Europe contre le Choléra.* Paris: G. Masson.

Proust, Marcel. 1989. *A la Recherche du Temps Perdu, tome iv.* Paris: Pleiade.

Pullum, Geoffrey K. 1991. *The Great Eskimo Vocabulary Hoax and Other Irreverent Essays on the Study of Language.* Chicago: University of Chicago Press.

Pyne, Stephen J. 1986. *The Ice: A Journey to Antarctica.* Iowa City: University of Iowa Press.

Qian, Sima [Ssu-ma, Ch"ien, ca. 145-ca. 86 B. C.]. 1994. *Historical Records/Sima Qian.* Translated with an introduction and notes by Raymond Dawson. Oxford: Oxford University Press.

*Rand Daily Mail.* 1966. "Race Laws Are Condemned." December 14: 17.

*Rand Daily Mail.* 1983. "Lize? The Bureaucratic Headache of a Bundle of Joy." July 23: 10.

Ravetz, Jeremy. 1971. *Scientific Knowledge and Its Social Problems.* Oxford: Oxford University Press.

Rayward, W. Boyd. 1975. *The Universe of Information: The Work of Paul Otlet for Documentation and International Organisation.* Moscow: Published for International Federation for Documentation (FID) by All-Union Institute for Scientific and Technical Information (VINITI).

Reagan, Leslie J. 1997. *When Abortion Was a Crime: Women, Medicine, and Law in the United States, 1867–1973.* Berkeley: University of California Press.

Rich, Adrienne. 1978. "Cartographies of Silence." In Adrienne Rich, *The Dream of a Common Language: Poems, 1974–1977.* New York: Norton.

Ridley, M. 1986. *Evolution and Classification; The Reformation of Cladism.* London: Longman.

Ritvo, Harriet. 1997. *The Platypus and the Mermaid, and Other Figments of the Classifying Imagination.* Cambridge, MA: MIT Press.

Robbin, Alice. 1998. *The Politics of Racial and Ethnic Group Classification.* Unpublished manuscript, School of Information Studies, Florida State University, Tallahassee. At: arobbin@mailer.fsu.edu.

Rosch, Eleanor, and Barbara Lloyd (eds.). 1978. *Cognition and Categorization.* Hillsdale, N. J.: L. Erlbaum Associates.

Rose, Nikolas S. 1990. *Governing the Soul: The Shaping of the Private Self.* London: Routledge.

Roth, Julius A. 1963. *Timetables: Structuring the Passage of Time in Hospital Treatment and Other Careers.* Indianapolis: Bobbs-Merrill.

Roth, Julius A. 1966. "Hired Hand Research." In *American Sociologist* 1: 190–196.

Rothwell, David J. 1985. "Requirements of a National Health Information System." In R. A. Côté, D. J. Protti and J. R. Scherrer (eds.). *Role of Informatics in Health Data Coding and Classification Systems.* Amsterdam: Elsevier, 169–178

Ruhleder, Karen. 1995. "Reconstructing Artifacts, Reconstructing Work: From Textual Edition to On-line Databank." In *Science, Technology and Human Values* 20: 39–64.

Rymer, Russ. 1993. *Genie: A Scientific Tragedy.* New York: Harper Collins.

Sacks, Harvey. 1975. "Everyone Has to Lie." In M. Sanches and B. G. Bount (eds). *Sociocultural Dimensions of Language Use.* New York: Academic Press, 57–80.

Sacks, Harvey. 1992. *Lectures on Conservation, vol. 1 and 2.* Oxford: Blackwell.

Schachter, Daniel L. 1996. *Searching for Memory: the Brain, the Mind, and the Past.* New York: Basic Books.

Schaffer, Simon. 1989. "Glass Works: Newton's Prisms and the Users of Experiment." In David Gooding, Trevor Pinch and Simon Schaffer (eds.). *The Uses of Experiment: Studies in the Natural Sciences.* Cambridge: Cambridge University Press, 67–104.

Scherrer, J. R., R. A. Côté, and S. H. Mandil. 1989. *Computerized Natural Medical Language Processing for Knowledge Representation.* North Holland: Elsevier Science Publishers B. V.

Schivelbusch, Wolfgang. 1986. *The Railway Journey : The Industrialization of Time and Space in the 19th Century.* Berkeley, CA: University of California Press.

Schmidt, Kjeld. 1997. "Of Maps and Scripts: The Status of Formal Constructs in Cooperative Work." In *Proceedings of GROUP "97, Phoenix, Arizona, 16–19 November.* NY: ACM Press. Preprint.

Schmidt, Kjeld, and Liam Bannon. 1992. "Taking CSCW Seriously: Supporting Articulation Work." In *Computer Supported Cooperative Work (CSCW): An International Journal.* 1: 7–41.

Schmidt, Kjeld, and Carla Simone. 1996. "Coordination Mechanisms: Towards a Conceptual Foundation of CSCW Systems Design." In *Computer Supported Cooperative Work: The Journal of Collaborative Computing* 5: 155–200.

Schön, Donald. 1983. *The Reflective Practitioner: How Professionals Think in Action.* New York: Basic Books.

Schütz, Alfred. 1944. "The Stranger: An Essay in Social Psychology." In *American Journal of Sociology* 69: 499–507.

Scott, Karla Danette. 1995. "'When I'm with My Girls': Identity and Ideology in Black Women's Talk about Language and Cultural Borders." Ph. D. thesis, Department of Speech Communication, University of Illinois, Urbana-Champaign.

Scott-Ram, N. 1990. *Transformed Cladistics, Taxonomy and Evolution.* Cambridge: Cambridge University Press.

Serres, Michel. 1980. *Le Passage du Nord-Ouest.* Paris: Editions de Minuit.

Serres, Michel. 1990. *Le Contrat Naturel.* Paris: F. Bourin.

Serres, Michel. 1993. *Les Origines de la Géométrie.* Paris: Flammarion.

Shapin, Steve. 1989. "The Invisible Technician." In *American Scientist* 77: 553–563.

Shapiro, Alan E. 1996. "The Gradual Acceptance of Newton's Theory of Light and Cold, 1672–1727." In *Perspectives in Science: Historical, Philosophical, Social* 4: 59–140.

Shibutani, Tamotsu. 1978. *The Derelicts of Company K: A Sociological Study of Demoralization.* Berkeley, CA: University of California Press.

Shilts, Randy. 1987. *And the Band Played On: Politics, People, and the AIDS Epidemic.* New York: St. Martin's Press.

Simmel, Georg. 1950 [1908]. "The Stranger." In Kurt Wolff (ed.). *The Sociology of George Simmel.* Glencoe, IL: Free Press, 402–408.

Slaughter, Mary M. 1982. *Universal Languages and Scientific Taxonomy in the Seventeenth Century.* Cambridge: Cambridge University Press.

Sontag, Susan. 1977. *Illness as Metaphor.* New York: Farrar, Straus and Giroux.

Sorlie, Paul D., and Ellen B. Gold. 1987. "The Effect of Physician Terminology Preference on Coronary Heart Disease Mortality: An Artifact Uncovered by the Ninth Revision ICD." In *American Journal of Public Health* 77: 148–152.

South African Institute of Race Relations, Inc. 1970. "Statement on the Population Registration Amendment Bill." In *doc. RRV. 44/69. EH. 26/3/69.* Johannesburg.

Sowden, Lewis. 1968. *The Land of Afternoon: The Story of a White South Africa.* New York: McGraw-Hill.

Star, Susan Leigh. 1983. "Simplification in Scientific Work: An Example from Neuroscience Research. In *Social Studies of Science* 13: 205–228.

Star, Susan Leigh. 1985. "Scientific Work and Uncertainty." In *Social Studies of Science* 15: 391–427.

Star, Susan Leigh. 1989a. *Regions of the Mind: Brain Research and the Quest for Scientific Certainty.* Stanford, CA: Stanford University Press.

Star, Susan Leigh. 1989b. "The Structure of Ill-Structured Solutions: Heterogeneous Problem-Solving, Boundary Objects and Distributed Artificial Intelligence." In M. Huhns and L. Gasser (eds.). *Distributed Artificial Intelligence 2.* Menlo Park, CA: Morgan Kauffmann, 37–54.

Star, Susan Leigh. 1991a. "The Sociology of the Invisible: The Primacy of Work in the Writings of Anselm Strauss." In David Maines (ed.). *Social Organization and Social Process: Essays in Honor of Anselm Strauss.* Hawthorne, NY: Aldine de Gruyter, 265–283.

Star, Susan Leigh. 1991b. "Power, Technologies, and the Phenomenology of Standards: On Being Allergic to Onions." In John Law (ed.). *A Sociology of Monsters? Power, Technology and the Modern World: Sociological Review Monograph 38.* Oxford: Basil Blackwell, 27–57.

Star, Susan Leigh. 1991c. "Invisible Work and Silenced Dialogues in Representing Knowledge." In I. V. Eriksson, B. A. Kitchenham, and K. G. Tijdens (eds.). *Women, Work and Computerization: Understanding and Overcoming Bias in Work and Education.* Amsterdam: North Holland, 81–92.

Star, Susan Leigh. 1992. "The Trojan Door: Organizations, Work, and the 'Open Black Box'." In *Systems/Practice* 5: 395–410.

Star, Susan Leigh (ed.). 1995a. *Ecologies of Knowledge: Work and Politics in Science and Technology.* Albany, NY: SUNY Press.

Star, Susan Leigh (ed.). 1995b. *The Cultures of Computing. Sociological Review Monograph Series.* Oxford: Basil Blackwell.

Star, Susan Leigh. 1995c. "The Politics of Formal Representations: Wizards, Gurus, and Organizational Complexity." In Susan Leigh Star (ed.). *Ecologies of Knowledge: Work and Politics in Science and Technology.* Albany, NY: SUNY Press, 88–118.

Star, Susan Leigh. 1995d. "Work and Practice in Social Studies of Science, Medicine and Technology." In *Science, Technology and Human Values* 20: 501–507.

Star, Susan Leigh. 1998. "Grounded Classifications: Grounded Theory and Faceted Classifications." In *Library Trends* 47: 218–252.

Star, Susan Leigh, and James R. Griesemer. 1989. "Institutional Ecology, "Translations" and Boundary Objects: Amateurs and Professionals in Berkeley's Museum of Vertebrate Zoology,1907–39." In *Social Studies of Science* 19: 387–420.

Star, Susan Leigh, and Karen Ruhleder. 1996. "Steps Toward an Ecology of Infrastructure: Design and Access for Large Information Spaces." In *Information Systems Research* 7(1): 111–134.

Star, Susan Leigh, and Anselm Strauss. 1999. "Layers of Silence, Arenas of Voice: The Ecology of Visible and Invisible Work." *Computer Supported Cooperative Work: The Journal of Collaborative Computing* 8: 9–30.

Star, Susan Leigh. 1996. "From Hestia to Home Page: Feminism and the Concept of Home in Cyberspace." In Nina Lykke and Rosi Braidotti (eds.). *Between Monsters, Goddesses and Cyborgs: Feminist Confrontations with Science, Medicine and Cyberspace.* London: ZED-Books, 30–46.

Star, Susan Leigh, Geoffrey C. Bowker, and Laura J. Neumann. In press. "Transparency beyond the Individual Level of Scale: Convergence between Information Artifacts and Communities of Practice." To appear in A. P. Bishop (ed.). *Digital Library Use: Social Practice in Design and Evaluation.* Cambridge, MA: MIT Press.

Stone, Sandy. 1991. "The *Empire* Strikes Back: A Posttransexual Manifesto." In Julia Epstein and Kristina Straub (eds). *Body Guards: The Cultural Politics of Gender Ambiguity.* NY: Routledge, 280–304.

Stonequist, Everett V. 1937. *The Marginal Man: A Study in Personality and Culture Conflict.* New York: C. Scribner's Sons.

Stouman, K., and I. S. Falk. 1936. "Health Indices: A Study of Objective Indices of Health in Relation to Environment and Sanitation." In League of Nations, *Quarterly Bulletin of the Health Organization* V: 901–1081.

Strauss, Anselm. 1959. *Mirrors and Masks: The Search for Identity.* Glencoe, IL: Free Press.

Strauss, Anselm. 1978. "A Social World Perspective." In *Studies in Symbolic Interaction* 1: 119–28.

Strauss, Anselm. 1993. *Continual Permutations of Action.* New York: Aldine de Gruyter.

Strauss, Anselm, S. Fagerhaugh, B. Suczek, and C. Wiener. 1985. *Social Organization of Medical Work.* Chicago: University of Chicago Press.

Strong, P., and J. Robinson. 1990. *The NHS—Under New Management.* Bristol, PA: Open University Press.

Suchman, Lucy. 1987. *Plans and Situated Actions: The Problem of Human-Machine Communication.* Cambridge: Cambridge University Press.

Suchman, Lucy. 1988. In "Representing Practice in Cognitive Science." *Human Studies* 11:305–325.

Suchman, Lucy. 1994. "Do Categories Have Politics? The Language/Action Perspective Reconsidered." In *Computer Supported Cooperative Work* (CSCW) 2:177–190.

Suchman, Lucy, and Brigitte Jordan. 1990. "Interactional Troubles in Face-to-Face Survey Interviews." *Journal of the American Statistical Association* 85 (review paper no. 409): 232–253.

Suchman, Lucy, and Randall Trigg. 1993. "Artificial Intelligence as Craftwork." In Seth Chaiklin and Jean Lave (eds.). *Understanding Practice: Perspectives on Activity and Context.* New York: Cambridge University Press, 144–172.

Sudnow, David. 1967. *Passing On: The Social Organization of Dying.* Englewood Cliffs, NJ: Prentice-Hall.

*Sunday Times* (Johannesburg). 1955. "Concern at Methods of Classifying Coloureds." August 21: 4.

*Sunday Times* (Johannesburg). 1984. "Oh No, Not Again! Vic Is Reclassified for FIFTH Time!" November 4: 21.

Suzman, A. 1960. "Race Classification and Definition in the Legislation of the Union of South Africa, 1910–1960." In *Acta Juridica* 339–367.

Sweetser, Eve. 1987. "The Definition of Lie: An Examination of the Folk Models Underlying a Semantic Prototype." In Dorothy Holland and Naomi Quinn (eds.). *Cultural Models in Language and Thought.* Cambridge, UK: Cambridge University Press.

Szreter, Simon. 1988. "The Importance of Social Intervention in Britain's Mortality Decline c. 1850–1914: An Interpretation of the Role of Public Health." *Social History of Medicine* 1–37.

Taylor, John R. 1995. *Linguistic Categorization: Prototypes in Linguistic Theory, second edition*. Oxford: The Clarendon Press.

Taylor, James R., and Elizabeth J. Van Every. 1993. *The Vulnerable Fortress: Bureaucratic Organization and Management in the Information Age* (with contributions from Helene Akzam, Margot Hovey, and Gavin Taylor). Toronto: University of Toronto Press.

Thomas, W. I., and Dorothy Swaine Thomas. 1970 [1917]. "Situations Defined as Real Are Real in their Consequences." In Gregory P. Stone and Harvey A. Farberman (eds.). *Social Psychology Through Symbolic Interaction*. Waltham, MA: Xerox Publishers, 54–155.

Thompson, E. P. 1967. "Time, Work Discipline and Industrial Capitalism." In *Past and Present* 38: 56–97.

*Thunder Press*. 1969. "Is It Still a Harley?" July 1969: 1 and 69.

Timmermans, Stefan. 1996. "Saving Lives or Saving Identities? The Double Dynamic of Technoscientific Scripts." *Social Studies of Science* 26: 767–797.

Timmermans, Stefan. In press, a. *Saving Life or Saving Death? The Myth of CPR*. Philadelphia, PA: Temple University Press.

Timmermans, Stefan. 1999a. "When Death Isn't Dead: Implicit Social Rationing during Resuscitative Efforts." In *Sociological Inquiry* 69 (1): 78–101.

Timmermans, Stefan. 1999b. "Closed-Chest Cardiac Massage: The Emergence of a Discovery Trajectory." In *Science, Technology and Human Values* 24 (2): 213–240.

Timmermans, Stefan. 1998. "Mutual Tuning of Multiple Trajectories," In *Symbolic Interaction* 21 (4): 425–440.

Timmermans, Stefan, and Marc Berg. 1997. "Standardization in Action: Achieving Local Universality Through Medical Protocols." In *Social Studies of Science* 27: 273–305.

Timmermans, Stefan, Geoffrey Bowker, and Susan Leigh Star. 1998. "The Architecture of Difference: Visibility, Controllability, and Comparability in Building a Nursing Intervention Classification." In Berg, Marc, and Annemarie Mol (eds.). *Differences in Medicine: Unraveling Practices, Techniques and Bodies*. Raleigh, NC: Duke University Press.

Tomalin, N., and Hall, R. 1970. *The Strange Last Voyage of Donald Crowhurst*. New York: Stein and Day.

Tort, Patrick. 1989. *La Raison Classificatoire: les Complexes Discursifs - Quinze Etudes* Paris: Aubier.

Trigg, Randall, and Susanne Bødker. 1994. "From Implementation to Design: Tailoring and the Emergence of Systematization in CSCW." In *Proceedings of ACM 1994 Conference on Computer-Supported Cooperative Work*. New York: ACM Press, 45–54.

Tripp-Reimer, Toni, G. Woodworth, JoAnne C. McCloskey, and Gloria Bulechek. 1996. "The Dimensional Structure of Nursing Interventions." In *Nursing Research* 45: 10–17.

Turkle, Sherry. 1995. *Life on the Screen: Identity in the Age of the Internet*. NY: Simon and Schuster.

Turnbull, David. 1993. "The Ad Hoc Collective Work of Building Gothic Cathedrals with Templates, String, and Geometry." In *Science, Technology & Human Values* 18(3): 315–343.

United Nations Commission on Human Rights. 1968. *Study of Apartheid and Racial Discrimination in Southern Africa: Report of the Special Rapporteur of the Commission on Human Rights*. 17 December 1968 (United Nations documents E/CN.4/979 and Add. 1 and 6).

United Nations Office of Public Information. 1969. *Segregation in South Africa: Questions and Answers on the Policy of Apartheid*. New York: United Nations, Office of Public Information, May 1969.

Unterhalter, Beryl. 1975. "Changing Attitudes to "Passing for White" in an Urban Coloured Community." In *Social Dynamics* 1: 53–62.

Van Regenmortel, Marc H. V. 1990. "Virus Species, A Much Overlooked but Essential Concept in Virus Classification." In *Intervirology* 31: 241–254.

Vansina, Jan 1961. *Oral Tradition; A Study in Historical Methodology*. Chicago, IL: Aldine.

Verghese, Abraham. 1994. *My Own Country*. New York: Simon and Schuster.

Wackers, G., 1995. *Standardization: Reliability of Performance and Safety*. Typescript from author.

Wagner, Ina. 1993. "Women's Voice: The Case of Nursing Information Systems." In *AI and Society* 7: 1–19.

Walsh, J. P., and G. R. Ungson. 1991. "Organizational Memory." In *Academy of Management Review* 16 (1): 57–91.

Wannenburgh, A. J. 1969. "Parents Fight to Have Daughter Accepted as White." In *Sunday Times* (Johannesburg), March 9: 15.

Ward, C. M. 1993. "Progress Toward a Higher Taxonomy of Viruses." In *Research in Virology* 144: 419–453.

Waterson, A. P., and Lise Wilkinson. 1978. *An Introduction to the History of Virology.* Camridge: Cambridge University Press.

Watson, Graham. 1970. *Passing for White: A Study of Racial Assimilation in a South African School.* London: Tavistock.

Weick, K. E., and K. H. Roberts. 1993. "Collective Mind in Organizations: Heedful Interrelating on Flight Decks." In *Administrative Science Quarterly* 38: 357–381.

Werley, H. H., and N. M. Lang. 1988. *Identification of the Nursing Minimum Data Set.* New York: Springer.

Werley, H. H., N. M. Lang, and S. K. Westlake. 1986. "The Nursing Minimum Data Set Conference: Executive Summary." In *Journal of Professional Nursing* 2 (4):217–224.

Wertsch, James V. 1991. *Voices of the Mind : A Sociocultural Approach to Mediated Action.* Cambridge, MA: Harvard University Press.

Wertsch, James V. 1998. *Mind as Action.* New York: Oxford University Press.

White, Kerr L. 1985. "Restructuring the International Classification of Diseases: Need for a New Paradigm." Guest editorial. In *The Journal of Family Practice* 21(1): 17–20.

Winner, Langdon. 1986. "Do Artifacts Have Politics?" In Judy Wacjman and Donald MacKenzie (eds.). *The Social Shaping of Technology: How the Refreigerator Got its Hum.* Milton Keynes: Open University Press.

Wolfe, Susan J., and Julia Penelope Stanley (eds.). 1980. *The Coming Out Stories.* Watertown, MA: Persephone Press.

Woolgar, Steve. 1995. "Representation, Cognition, and Self: What Hope for an Integration of Psychology and Sociology?" In Susan Leigh Star (ed.). *Ecologies of Knowledge: Work and Politics in Science and Technology.* Albany, NY: SUNY, 154–182.

World Health Organization. 1991. *A Proposed Standard International Acupuncture Nomenclature: Report of a WHO Scientific Group.* Geneva: WHO.

Yates, Frances. 1966. *The Art of Memory.* Chicago: Chicago University Press.

Yates, JoAnne. 1989. *Control through Communication: The Rise of System in American Management.* Baltimore: Johns Hopkins University Press.

Yates, JoAnne, and Wanda J. Orlikowski. 1992. "Genres of Organizational Communication: A Structurational Approach to Studying Communication and Media." In *Academy of Management Review* 17: 299–326.

Yates, JoAnne, Wanda Orlikowski, and Julie Rennecker. 1997. "Collaborative Genres for Collaboration: Genre Systems in Digital Media." In *Proceedings of the Thirtieth Annual Hawaiian International Conference on System Sciences.* Washington, DC: IEEE Computer Society Press, 1997.

Young, Allan. 1995. *The Harmony of Illusions: Inventing Post-Traumatic Stress Disorder.* Princeton, NJ: Princeton University Press.

Zielstorff, R. D., C. I. Hudgings, S. J. Grobe, and the National Commission on Nursing Implementation Project (NCNIP) Task Force on Nursing Information Systems. 1993. *Next-Generation Nursing Information Systems: Essential Characteristics for Professional Practice.* Washington, DC: American Nurses Publishing.

Ziporyn, Terra. 1992. *Nameless Diseases.* New Brunswick, NJ: Rutgers University Press.

Zorbaugh, Harvey Warren, 1929. *The Gold Coast and the Slum: A Sociological Study of Chicago's Near North Side.* Chicago, IL: University of Chicago Press.

# Name Index

# Subject Index